THE MAGIC OF A NAME
THE ROLLS-ROYCE STORY
PART TWO

Our future lies on the turbine engine side. There are no new aircraft contemplated or being built which will take the Merlin or Griffon engines, or any other piston engines.

Ernest Hives to the Rolls-Royce Board on 30 May 1946

The scale of advance was enormous. The RB 211 was the largest turbofan built by Rolls-Royce, being roughly double the thrust of the Conway.

Philip Ruffles, Director of Engineering
and Technology, Rolls-Royce 2000

I'm not blaming anyone else . . . We had promised a bit more than we could perform. Things were not deplorable. Flight engines were four months late. Four months is four months but it is not deplorable. The deplorable thing was the cost . . . The accountants never got cash-flow into our heads.

Sir David Huddie in 1999 on the
Rolls-Royce receivership in 1971

We knew that if we were to stay competitive in the long-term we must be able to meet our airline customers' requirements over the whole of their range.

Sir Ralph Robins, Chairman,
Rolls-Royce plc 2000

THE MAGIC OF A NAME
THE ROLLS-ROYCE STORY
PART TWO
THE POWER BEHIND THE JETS
1945–1987

Peter Pugh

ICON BOOKS UK
TOTEM BOOKS USA

Published in the UK in 2001 by Icon Books Ltd.,
Grange Road, Duxford, Cambridge CB2 4QF
e-mail: info@iconbooks.co.uk
www.iconbooks.co.uk

Distributed in the UK, Europe, South Africa and Asia
by Airlife Publishing Ltd., 101 Longden Road,
Shrewsbury SY3 9EB

Published in Australia in 2001 by
Allen & Unwin Pty. Ltd., PO Box 8500,
83 Alexander Street, Crows Nest, NSW 2065

Published in the USA in 2001 by Totem Books
Inquiries to: Icon Books Ltd., Grange Road,
Duxford, Cambridge CB2 4QF, UK

Distributed to the trade in the USA by
National Book Network Inc.,
4720 Boston Way, Lanham,
Maryland 20706

Distributed in Canada by Penguin Books Canada,
10 Alcorn Avenue, Suite 300,
Toronto, Ontario M4V 3B2

ISBN 1 84046 284 1

Typesetting by Hands Fotoset, Woodthorpe, Nottingham

Design and layout by Christos Kondeatis

Printed and bound in the UK by
Biddles Ltd., Guildford and King's Lynn

CONTENTS

PHOTO CREDITS

The majority of the photographs have been provided by the Rolls-Royce Heritage Trust, most from the Derby Branch, but a significant number, particularly those of Bristol engines or aeroplanes, from the Bristol branch.

The photograph of Bill Gwinn and his Pratt & Whitney colleagues in the group between pages 22 and 23 is reproduced by kind permission of United Technologies Corporation (Archive), Hartford, Connecticut 06101.

The photographs of the Mustang and of HMS *Ocean,* also in the group between pages 22 and 23, and of HMS *Invincible* between pages 278 and 279 are reproduced by kind permission of the Imperial War Museum.

Most of the photographs of Rolls-Royce and Bentley motor cars in the group between pages 182 and 183 were provided by the Sir Henry Royce Memorial Foundation.

Most of the photographs of Rolls-Royce diesel engines and products with those engines powering them, in the group between pages 214 and 215, were provided by Brian Leverton and Peter Vinson.

The photographs of Admiral Rickover, the USS *Nautilus* and HMS *Dreadnought,* also in the group between pages 214 and 215, were provided by Andrew Rice, Communications Manager, Rolls-Royce Naval Marine.

THE VALUE OF MONEY

In a book about a business, we cannot ignore the changing value of money, and – with the exception of the inter-war years – the twentieth century has been inflationary. There is no magic formula for translating 1900 prices into those of 2000. Some items have exploded in price, others have declined. We have to choose some criterion of measurement, and I have chosen the average working wage.

The Victorian age was one of stable prices, but prices started to rise just before the First World War, and rose sharply at the end of it. (Wars are always inflationary, because they distort supply and demand.) Immediately after the war, prices were more than twice as high as in 1914, and although they declined somewhat in the depressed economic conditions of the 1920s and 30s, they remained about twice as high as those before the war.

Price controls and rationing were imposed in the Second World War, but as these were withdrawn, prices again doubled. Inflation continued at about 3 per cent a year through the 1950s and 60s, but then rose sharply, almost catastrophically, in the 1970s. Although it was brought under control by the end of that decade, there were two more nasty upward blips in the early and late 1980s before the more stable 1990s.

I have used the following formula:
Late nineteenth century – multiply by 110 to equate with today's prices
Early twentieth century – multiply by 100 to equate with today's prices
1918–45 – multiply by 50 to equate with today's prices
1945–50 – multiply by 30 to equate with today's prices
1950–60 – multiply by 25 to equate with today's prices
1960–70 – multiply by 18 to equate with today's prices
1970–74 – multiply by 15 to equate with today's prices
1975–77 – multiply by 10 to equate with today's prices
1978–80 – multiply by 6 to equate with today's prices
1980–87 – multiply by 3 to equate with today's prices

INTRODUCTION

As I said in the Introduction to the first part of this history, *The Magic of a Name, The Rolls-Royce Story, The First 40 Years,* many books have been written about Rolls-Royce over the years. The Bibliography on pages 328 to 332 shows nearly fifty specifically on Rolls-Royce cars or aero engines or both, and there are hundreds more which devote many words to Rolls-Royce's products and their importance. However, most of these books are either biographies of the leading personalities, or works that tackle specific aspects of the cars or aero engines.

The last publication to attempt a truly comprehensive history of Rolls-Royce was Harold Nockolds's *The Magic of a Name,* published by Foulis in 1938 and reprinted several times until a third edition was finally printed in 1972. This book, with the one published last year, and the third part to be published to coincide with the hundredth anniversary of the meeting of Royce and Rolls, is the first attempt to cover comprehensively the story of Rolls-Royce from its earliest days until the present. The Rolls-Royce name *is* magic, made so by the calibre of the people and its products, and the title *The Magic of a Name* is singularly appropriate. Foulis, now owned by Haynes Publishing, has kindly allowed us to use the title again.

The first part of this trilogy took the story from 1904 to 1945 with all the excitement of the meeting of the great engineer, Henry Royce, and the aristocratic motor car enthusiast and salesman, the Hon. C.S. Rolls, through the early pioneering days with the best car in the world, the Silver Ghost, on to the design by Royce and production by Rolls-Royce of the leading aero engine of the First World War, the Eagle. We moved on to Royce's design of an aero engine which enabled Britain to retain the Schneider Trophy. This association with the aircraft designer, R.J. Mitchell, led on to the Spitfire and Rolls-Royce's Merlin engines. As we saw, without the Spitfire and the Hurricane, and the Merlins that powered them, the Battle of Britain would have been lost. Finally, we saw how Ernest, later Lord, Hives realised the significance of the jet engine and how he backed Sir Frank Whittle to the hilt.

And that is where we pick up the story in this part, with Whittle and the jet engine. We see how Rolls-Royce adapted itself quickly from mass-producing piston aero engines to designing and producing gas turbine engines, initially for aero engines but later to power ships and power stations also. At the same time, Rolls-Royce returned to the motor car business and for the first time produced complete cars, not just engines and chassis. It also began to manufacture diesel engines.

We shall see how the leaders of the company realised that Rolls-Royce must win orders in the USA if the company was to survive as a major force in the aero engine business, and how this led to the design of an innovative and world-beating three-shaft engine, the RB 211. Unfortunately, we shall have to go through the pain of how the development of this engine so overstretched the company that it was forced into receivership.

However, we shall end on the happy note of Rolls-Royce emerging from sixteen years under Government ownership, facing a promising future with its reputation for designing and producing engines of the highest quality untarnished.

Peter Pugh
January 2001

AUTHOR'S ACKNOWLEDGEMENTS

In researching this book on Rolls-Royce I have read all of the books mentioned in the bibliography, and quoted from several of them. I am grateful to the authors and to the publishers. I have interviewed many people, some of them several times, and I am grateful to them too. Most of the interviews took place in Britain, but I travelled to the USA several times, both to research the archives held at Gilbert, Segall and Young, Rolls-Royce's legal advisers in New York, and those at Rolls-Royce Allison in Indianapolis, and also to interview a number of people who had been in close contact with Rolls-Royce. I also travelled to Australia to interview a number of people there.

I would like to mention especially: Roy Anderson, Steven Arris, Peter Baines, Sir Basil Blackwell, John Blatchley, Geoffrey Bone, Martin Bourne, Tom Bowling, Walter Boyne, Peter Byrom, Bill Castle, Gerry Collin, Charles Coltman, John Coplin, Roger Cra'ster, Ken Crooks, David Davies, Gordon Dawson, Michael Donne, Sir George Edwards, Cyril Elliot, Ernest Eltis, Chris Farara, Mac Fisher, Sir Ian Fraser, Lloyd Frisbee, Dick Garner, Phil Gilbert, Colin Green, Sir Arnold Hall, Fred Hardy, Ronnie Harker, Bill Harris, Sir Terence Harrison, Rhoddie Harvey-Bailey, Dr. Bob Hawley, Lionel Haworth, Denis Head, Paul Heiden, Alex Henshaw, Jock Heron, Dr. Mike Hicks, Sam Higginbottom, John Hodson, Ronnie Hooker, Ralph Hooper, Sir David Huddie, Mike Hudson, Gerry James, Stuart John, John Jones, Jim

Keir, Lord Keith, Lord Kings Norton, Kevin Kirby, Jock Knight, Carl Kotchian, Eberhart von Kuenheim, Brian Leverton, Gordon Lewis, Sir Douglas Lowe, Sue Lyons, Jackson McGowen, Sir Arthur Marshall, Sir Peter Masefield, Sir Robin Maxwell-Hislop, Tom Metcalfe, Hamish Middleton, Stewart Miller, David Mitchell, Dr. Gordon Mitchell, Fred Morley, Sir Ian Morrow, Michael Neale, Tom Neville, Alan Newton, Rex Nicholson, Sir Robin Nicholson, John Nutter, Gordon Page, Peter Pavey, Don Pepper, David Pickerill, Sir David Plastow, Ashley Raeburn, Melvin Reynolds, Andrew Rice, Tim Rigby, Jim Rigg, Ian Rimmer, Cliff Rogers, John Rose, Philip Ruffles, Trevor Salt, Brian Slatter, Sir Alec Smith, Guy Smith, Sir John Smith, Tom Smith, Reg Spencer, Gordon Strangeways, Bob Sunerton, Joe Sutter, Alan Swinden, Bill Thomas, Lord Tombs, Richard Turner, Peter Vinson, John White, Ian Whittle, Geoff Wilde, T. Wilson, Andy Wood, Jimmy Wood, Paul Wood, John Woodward, John Wragg, Ken Wright.

Unfortunately, since I interviewed them, a number have died – Sir David Huddie, Lord Kings Norton, T. Wilson, Ronnie Harker, Stewart Miller, Roger Cra'ster, Sir Arnold Hall, Lionel Haworth, Gordon Strangeways and Ken Wright. All made a great contribution to both Rolls-Royce and the aircraft industry.

I received constant help and encouragement from Sir Ralph Robins, the Chairman of Rolls-Royce plc. John Warren, Assistant Company Secretary at Rolls-Royce, was extremely helpful in providing the relevant documentation pertaining to board decisions.

Philip Hall, curator of the Sir Henry Royce Memorial Foundation at The Hunt House, Paulerspury, gave me full access to the archives held there and provided some of the photographs. Michael Donne, former air correspondent with the *Financial Times*, kindly sent me tapes of work he had done in preparing a book on Rolls-Royce which was never published. Peter Baines, Andrew Rice, Andrew Siddons, Peter Pavey, Tom Smith and Brian Leverton were all very helpful in providing information and photographs.

As with the first part of this history of Rolls-Royce, my undying gratitude goes to Mike Evans, formerly Chairman of the Rolls-Royce Heritage Trust. His help and advice have been unstinting and his meticulous attention to detail is an example to every author of a non-fiction book. As with the first book he has also helped enormously with the photographs, not only with advice on which to publish but also with sourcing them and helping with the captions. He has now retired and is beginning to do some of the things he has wanted to do for years. I hope he will find time to help me with Part Three.

As with the first book, I also received help from Richard Haigh and Julie Wood of the Rolls-Royce Heritage Trust.

Peter Pugh
January 2001

CHAPTER ONE

THE GAS TURBINE IS OUR FUTURE

'FRANK, IT FLIES!'
'THE TURBINE ENGINES HAVE ARRIVED'
'LET US TALK ABOUT THE ENGINEERING'
EXPERIENCED PERSONNEL
PRATT & WHITNEY
WESTINGHOUSE

'FRANK, IT FLIES!'

THE WAR WAS OVER. For six years in Europe, slightly less in the Far East, the leading nations of the world had indulged in an orgy of destruction. In 1943, Joseph Goebbels, the German Nazi politician, had asked the German people:

Do you want total war?

Now there was total defeat. Could the world recover? World War I, with its eight million victims, had seemed bad enough, but this war had brought over thirty million. France lost 620,000, Britain 260,000 – less than in the 1914–1918 war – but central Europe had suffered grievously. Poland lost more than 20 per cent of its population, including millions of Jews who had been murdered. Yugoslavia lost 10 per cent. It was difficult to know how many had died in the Soviet Union, but estimates ranged between twelve and twenty million. Germany lost five million. In the Far East, the losses and dislocation were also almost too great to grasp.

And material losses were greater than in World War I. Both the Germans and the Russians had pursued scorched earth policies, and hardly a major city in Europe or in Britain went unscathed. Housing for the survivors was of paramount importance, but in many countries almost a quarter of the houses were uninhabitable. Everyone was hungry, many faced starvation, the word 'famine' was on people's lips. To make matters worse, in 1945 Europe suffered a drought. Cereal crops fell from 59 to 31 million tons (excluding the Soviet Union). Even the normally rich agricultural France produced only half its crop of food grains. Everywhere there was food rationing, and the rations had to be cut and cut again. Anyone losing his (or her) ration book would probably lose his life.

As winter approached, fuel also took on life-and-death significance. Coal production had fallen drastically. From the Ruhr came only 25,000 tons per day compared with the 400,000 tons before the war. However, even if the coal had been produced, the means of transporting it had been destroyed. Some 740 of the 958 bridges in American and British zones in Germany – Occupied Germany had been split into four sectors run by the Americans, British, French and Russians respectively – were out of action. France had been left with only 35 per cent of both its railway locomotives and its merchant fleet. Production everywhere was way below pre-war levels and, of course, in such a situation of supply being well beneath demand, inflation was rampant. As Walter Laqueur said in his book, *Europe since Hitler*:

Whole countries were now living on charity and to say that economic prospects were uncertain would have been a gross understatement in the summer of 1945. There seemed to be no prospects.

In Britain, which at least was on the winning side and which had not suffered invasion, the situation was not quite so bleak. Nevertheless, the country was in a very weak state. To survive, it had been forced to sell a third of its overseas assets, reducing its annual income from overseas investments by more than half. The country's merchant fleet was less than three-quarters of its pre-war size and only 2 per cent of Britain's industry was producing for export. Most of its food and raw materials came from overseas and it was struggling to pay for them.

What were Rolls-Royce's prospects? The company had been totally geared to war production and, as we saw in the first volume of this history (*The Magic of a Name, The Rolls-Royce Story, The First 40 Years*), it had played a vital part in providing the engines for the fighters and bombers of the Allies. Aero engines for military purposes were not likely to be in such great demand in the coming years, though governments were aware that the rapid running-down of their armed forces, that had taken place after World War I, had not been in their best interests.

Rolls-Royce realised that, as well as supplying the air forces of the world, its best prospects lay in being a leading player in the field of civil aviation. We saw in the first volume that Ernest (later Lord) Hives appreciated that the gas turbine engine would be the future power source for aircraft.

The story of the modern jet engine had begun in the 1920s when two men, acting independently, had put forward their ideas for propulsion by means different from the reciprocating engine. In 1926, Dr. A.A. Griffith, of the Royal Aircraft Establishment's Engines Experimental Department, proposed the use of a single-shaft turbine engine with a multi-stage axial compressor as a means of driving a propeller through a reduction gear. In 1928, a young Royal Air Force officer, Frank Whittle, wrote a paper entitled 'Future Developments in Aircraft Design'. He envisaged aircraft flying at speeds of 500 mph, at a time when the fastest RAF fighter could not reach 200 mph, but felt it would be necessary for the aircraft to fly at great heights where the air was rare. At this stage, he was not sure about the means of propulsion, although he was already considering rockets and gas turbines driving propellers. However, a year later, in October 1929, he realised that it was not necessary to have a propeller because the exhaust from the gas turbine could be used to propel the aircraft.

The 1930s were a decade of frustration for Whittle as he tried to win backing for his idea. He was introduced to Griffith, but Griffith was somewhat dismissive, saying that his 'assumptions were over-optimistic'. Griffith himself was also frustrated by his superiors. Whittle's original work had been reviewed by the Engines sub-committee of the Aeronautical Research Committee in 1930, and the committee had concluded that:

At the present state of knowledge the superiority of the gas turbine over the reciprocating engine cannot be predicted.

Nevertheless, Whittle persisted and, in 1935, formed a company called Power Jets in conjunction with two former RAF colleagues – Rolf Dudley Williams and J.C.B. Tinling (later known affectionately by the Whittle children as Uncle Willie and Uncle Col). Financial backing was forthcoming from O.T. Falk & Partners. [Since the publication of *The Magic of a Name, The Rolls-Royce Story, The First 40 Years,* in which the progress of Whittle's jet engine is told more fully, I have read the report of M.L. Bramson, the consulting engineer used by O.T. Falk before they made their investment, and Bramson's comments thirty years later. I am grateful to the Bristol Branch of the Rolls-Royce Heritage Trust for reproducing these in their *Sleeve Notes Issue 7.*]

Bramson told of how Whittle appeared in his office one day in 1935 telling him that he needed finance for the development of a system of jet propulsion for aircraft, which he had invented. His material consisted solely of thermodynamic and aerodynamic calculations and diagrams, and there were no engineering designs. Bramson was initially sceptical of 'the eyebrow-raising improbability of his basic thesis that aeroplanes could be made to fly without propellers'. Nevertheless, he felt Whittle seemed to know what he was talking about and decided to study his theories. Bramson said later:

At the end of the period I got quite excited. First, because of the insight, clarity and accuracy of his presentation and calculations; second because my scepticism of any project based on internal combustion turbines (which had hitherto resisted all practical development efforts) disappeared when I realised that here, for the first time, was an application where maximum energy was needed in the turbine exhaust, instead of in the shaft. This was, of course, the reverse of all past objectives for such turbines. And thirdly, because of the dramatic advance in aviation technology implicit in Whittle's theories.

Bramson felt that Whittle's ideas must be financed and introduced him to the bankers O.T. Falk & Partners. Falk commissioned Bramson to investigate further and produce a detailed report. On the strength of his report, Falk financed Whittle through a newly formed company, Power Jets Limited. Work on a prototype engine began at British Thomson-Houston in Rugby. On 12 April 1937 an engine was run and Whittle wrote in his diary:

Pilot jet successfully ignited at 2,000 rpm by motor. I requested a further raising of speed to 2,500 rpm and during this process I opened valve 'B' and the unit suddenly ran away. Probably started at about 2,300 and using only about 5hp starting power . . . noted that return pipe from jet was overheating badly. Flame tube red hot at inner radius; combustion very bad . . .

The following two years were ones of financial hardship, research setbacks and component shortages, but by June 1939 test runs of the engine were beginning to look promising. Whittle said later:

On 23 June we reached a speed of 14,700 rpm; the next day we went to 15,700 and then on the 26th we ran up to 16,000. We did several runs up to this speed on succeeding days and on one of these occasions – 30 June – DSR [Director of Scientific Research] was present.

It was a critical day because the DSR, Dr. Pye, became completely converted to the project. He agreed that the Air Ministry should buy the experimental engine, but still leave it with Power Jets for continuing experimentation. For Power Jets there was the added bonus that the Ministry would pay for spares and modifications.

Considerable progress was made during 1940, including the production of an aircraft to carry Whittle's engine by the Gloster Aircraft Company. By April 1941 the Gloster/Whittle E.28/39 was ready, and on 7 April at Gloster's airfield at Brockworth, Gloster's chief test pilot, Gerry Sayer, taxied the aeroplane and took it off the ground three times and flew for 200 to 300 yards.

The first proper flight took place at Cranwell in Lincolnshire at 7.35 p.m. on 15 May 1941. It lasted seventeen minutes and Whittle recalled later:

I was very tense not so much because of any fears about the engine but because this was a machine making its first flight. I think I would have felt the same if it had been an aeroplane with a conventional power plant . . . I do not remember, but I am told that, shortly after take-off, someone slapped me on the back and said, 'Frank, it flies!' and that my curt response in the tension of the moment was 'Well, that was what it was bloody well designed to do, wasn't it?'

There was only one person from the Ministry of Aircraft Production present at this flight but, when news reached London, a large delegation led by the Secretary of State for Air, Sir Archibald Sinclair, went to Cranwell on 21 May for a further demonstration. John Golley, who collaborated with Whittle to write a book, *Whittle, the true story*, described the scene very well:

Gerry Sayer brought gasps from the uninitiated onlookers with a high-speed run downwind, when the strange whistling roar of the propellerless engine riveted their attention as they watched the E.28 pulling up into a steep climbing turn and shoot skywards. The absence of a propeller was a source of amazement, and few of those privileged to see Sayer could have had any doubts that they had witnessed the beginning of a new chapter in aviation history...

One of two officers watching the E.28 take off was heard to ask, 'How the hell does that thing work?' His companion replied, 'Oh, it's easy, old boy, it just sucks itself along like a Hoover.' [Whittle himself said that it 'sucks itself along like a bloody great vacuum cleaner!'] Dan Walker of Power Jets was amused to hear one officer – not knowing that Walker was one of the engineers intimately concerned – assure everybody in his immediate vicinity that the power plant was a Rolls-Royce Merlin engine driving a small four-bladed propeller inside the fuselage. He stated positively that he had seen it!

From this moment the future of the jet engine and, in particular, Whittle's engine, was assured. However, a prototype engine was one thing; production of reliable engines in quantity was quite another. And this was where Rolls-Royce came into the picture.

'THE TURBINE ENGINES HAVE ARRIVED'

Rolls-Royce would have been aware of developments in the Internal Combustion Turbine (ICT) field, and on 1 June 1939 it recruited A.A. Griffith, giving him the facilities to continue the development of his axial compressor units. Hives instructed him to 'go on thinking', and Griffith proposed the most advanced of his concepts from the RAE – a contra-rotating engine, the CR1, with a fourteen-stage high-pressure system and six-stage low-pressure ducted fans.

While encouraging Griffith, Hives also made contact with Whittle, whom he had met at Power Jets' factory in 1940 at the instigation of Stanley Hooker. As we have seen, Hooker had been recruited to Rolls-Royce just before the war and made an enormous contribution on the Merlin supercharger. Hooker had already been to see Whittle, and said later:

I first met Frank Whittle in January 1940. At that time he was located with a small team of engineers at an old disused foundry at Lutterworth, near Rugby. His firm was called Power Jets, and the work he was doing was Top Secret. I was taken to see his first jet engine by Hayne Constant who, at that time, was the Director of the Engine Research Department of the Royal Aircraft Establishment at Farnborough. Constant had specialized in both centrifugal and axial compressors, and had frequently visited Derby to discuss with me the development of the Merlin supercharger. There was snow on the ground when he took me from Derby to Lutterworth, and I saw for the first time the strange jet engine roaring in its test bed. Compared to the sophisticated design and manufacture of Rolls-Royce, it looked a very crude and outlandish piece of apparatus. Yet, standing near to it while it was running, I felt conscious that I was in the presence of great power. Whether it was useful power or not, I had no idea.

I cannot claim that I was an immediate convert to the jet engine. That took some months, while I did my own analysis of the gas-turbine engine and, more importantly, came under the spell of Frank Whittle's genius and super technical knowledge.

By August, Hooker was convinced that Whittle's engine was so revolutionary Rolls-Royce should become involved. Through the spring and summer as Hooker regularly visited Whittle's operation at Lutterworth, Whittle's engine improved in reliability and performance to the point where it could make quite long runs at 800 lb thrust.

This did not sound very much and when Hooker suggested to Hives that he go and meet Whittle, Hives said of the 800 lb thrust:

That doesn't sound very much. It would not pull the skin off a rice pudding, would it?

However, when Hooker pointed out that the Merlin in a Spitfire at 300 mph gave about 840 lb thrust Hives agreed to go to the Power Jets factory in Lutterworth, where he was shown round by Whittle. Whittle remembered Hives saying: 'I don't see many engines. What's holding you up?' Whittle explained the problems in having certain components made, whereupon Hives said: 'Send us the drawings to Derby, and we will make them for you.'

In 1994, Whittle recounted to Sir Ralph Robins, the Chairman of Rolls-Royce plc, that he also emphasised to Hives the simplicity of his engine. Hives thought about this and replied:

We'll soon design the bloody simplicity out of it!

Within a short time, Rolls-Royce was making turbine blades, gearcases and other components for Power Jets. For some reason, perhaps sensing their integrity, Whittle felt comfortable with Rolls-Royce, while he felt threatened by the likes of Rover and Armstrong Siddeley. Rover had been given a contract by the Air Ministry to work on the production of the Whittle engine following a meeting between Tinling, the Power Jets director, and Maurice Wilks, chief engineer of the Rover Car Company. Whittle recalled:

This was based on our respect for their outstanding engineering ability, particularly in the field in which we were engaged. I believe that this respect was mutual.

By the spring of 1941 Rolls-Royce was deeply involved, and on 28 May 1941, shortly after the first flights of the Gloster/Whittle E.28/39, Hives wrote to Hennessey at the Ministry of Aircraft Production:

We had a very successful meeting on Sunday afternoon at the Rover factory where we met the Wilks brothers and Thomas. I took with me our superintendent who looks after the production of all experimental pieces. It was agreed provisionally which parts we should undertake, and another meeting has been arranged for Friday when we hope to bring back the drawings, and get on with the job.

I also arranged with Wilks that I would send our Mr Rowledge, Dr Griffith and Dr Hooker to the factory at Clitheroe where it was hoped they would see one of the engines stripped . . . Any suggestions which we have to offer will in no way interfere with the production of the parts for Rovers, and we have agreed that in order to cover the finance and contract position, we will act as sub-contractors to Rovers on the production of pieces. We are expecting in six to eight weeks' time to have our own machine running.

During the summer of 1941, a Gas Turbine Collaboration Committee was formed. The idea almost certainly came from Hives. Whittle thought so too, writing in his book: 'I understand that the original suggestion for this committee was made by Hives.'

Rolls-Royce helped Whittle with his 'surging' problems, building a test rig with a 2,000 hp Vulture piston engine with a step-up gear made by putting two Merlin-propeller reduction gears in series and driving them backwards to achieve the 16,500 rpm required by Whittle's W.2 compressor.

On 12 January 1942, Hives wrote to Whittle inviting him to Derby to discuss a proposal that Rolls-Royce should build a version of his engine. As we have seen, Hives had already recruited A.A. Griffith (who had moved from the Air Ministry research laboratory to the RAE in Farnborough). However, Griffith's axial flow ideas were proving difficult to convert into a practical engine, and Hives liked the simplicity of Whittle's engine.

It was agreed that Power Jets would be the main contractor on the Rolls-Royce version of Whittle's engine, and that Rolls-Royce would be the sub-contractor. On 30 January 1942, Hives visited Power Jets and, having made the point that he wanted Rolls-Royce to be at the forefront of jet engine manu-facture, said that more than technical collaboration would be necessary. In the short term, however, not much happened except that the Ministry of Aircraft Production gave Power Jets a contract for the design and develop-ment of six engines, and Power Jets immediately placed subcontracts with Rolls-Royce.

Much of 1942 seems to have been spent struggling with attempted solu-tions to mechanical problems. According to Dr. Bob Fielden, while the W.1 engine was satisfactory the design of the W.2 was not. It was clear early in the year that Rover was wondering how it could extricate itself. Spencer Wilks was finding it very difficult to work with Whittle, and early in February 1942 he put forward a proposal to Hives whereby Rover and Rolls-Royce could collaborate.

Hives wrote to him on 11 February:

I have discussed with Mr Sidgreaves [Arthur Sidgreaves, appointed Managing Director of Rolls-Royce in 1929] the proposition you put forward last Saturday.

The decision we have arrived at is that it would be impossible to take advantage of your offer. As I pointed out to you, in agreement with MAP and Power Jets Ltd., we are producing a Whittle turbine to Rolls-Royce designs, and we have undertaken the development work. In connection with this project, we have agreed to act as sub-contractors to Power Jets Ltd. I am sure you will appreciate the impossible position, which would arise if we have any link-up with the Rover Company.

For his part, Whittle was becoming frustrated by Rover's lack of progress. Only the day before Hives wrote to Wilks, Whittle had also written to Rover:

We can summarise our views by saying that experimenting cannot go on indefinitely before a decision is made on the first production model, and in the meantime, if we have the correct picture of the situation, the mechanical side is being seriously neglected.

Rover's brief was to manufacture, but it was only allowed to change any design feature to ease production. However, it found what Whittle had given it could not be made and the company was therefore carrying out illicit development work without keeping Whittle informed.

Despite the many difficulties during 1942, Hives became more and more convinced that gas turbines were the future for the aircraft industry. He realised that Whittle's W.2B engine worked, because Rolls-Royce was test-flying it at Hucknall, as is clear from this letter from Ray Dorey to Flight Lieutenant W.E.P. Johnson at Power Jets, written on 14 April 1942:

Wellington Test Bed (W.2B into Wellington)
We had a small discussion with DRD after you left the other day, and provisionally fixed for a Wellington II to be sent to us for this job.

As Whittle made further progress on his W.2/500 engine, he suggested to Hives that Rolls-Royce should take over production, as Power Jets was clearly not in a position to do so. Hives and Sidgreaves visited Power Jets on 8 October, and Hives did so again on 4 December. At this latter meeting, Hives told Whittle that he and Sidgreaves had discussed Whittle's suggestion with Sir Wilfrid Freeman, by this time Chief Executive of the Ministry of Aircraft Production, and Air Marshal Linnell himself. Hives summarised Rolls-Royce's position, which was that the company was definitely entering the aircraft gas turbine field, it was interested in Power Jets' W.2/500 and would like to undertake the production, and because of its own commitment to Merlin production, Rolls-Royce would require extra facilities to produce the W.2/500.

Power Jets' willingness for Rolls-Royce to take over production is set out clearly in a letter from Tinling to Hives on 16 November 1942.

The W.2/500 is nearing the stage of readiness for production . . . We are of the opinion, against the background of some knowledge of other firms, that your Company is the best able to produce this engine as it should be, and among the few with whom we would feel entirely happy to collaborate.

As we have seen, Hives had been concerned that effort which should have been directed exclusively towards winning the war, was being wasted on jets. However, he had now become what he called 'a turbine enthusiast'.

Apart from our own contra-flow turbine, we have designed, and shall have running by the end of this month, a Whittle type turbine. Our approach to the Whittle turbine was that we set out to design a turbine with a modest output, but one which, we hoped, would run and continue to run. Our turbine will be the heaviest and biggest, and relatively give less thrust, than any of the others, but we shall be disappointed if it does not run reliably, and for sufficiently long periods for us to learn something about it, and we can then proceed to open up the throttles, in short, to follow the usual Rolls-Royce practice on development.

We are fortunate in as much as we work in the most friendly way with Whittle. We have a great admiration for his ability. We are also equally friendly with the Rover Company. We have shown our friendship in a practical way by producing in our Experimental Department quantities of difficult pieces, both for Power Jets and Rover.

We have been approached by Whittle to undertake the production of the W-500. To do this we should require extra facilities. We are confident of one thing, however, that jets are never going to make any real progress until some well-established firm [and, of course, he meant Rolls-Royce] becomes the parent or big brother, to get a move on.

A comparison of the rate of the U.S. development and ours shows us up badly. [In 1941, Whittle's drawings had been sent to the US Government, which had passed them on to General Electric, enabling it to develop the GE-1A engine, which powered the first US jet aircraft, the Bell XP-59A Airacomet. This aircraft made its maiden flight at Muroc Dry Lake, California, later to be named Edwards Air Force Base, on 2 October 1942.] It is already time that some of the various jet projects were brought together, and the researches pooled.

Some decisions should be taken as regards the Power Jets factory at Whetstone. I was astonished at the size of it, and the emptiness of it, when I visited it a short time ago.

Hives knew that Rolls-Royce must gain control of the development and production of Whittle's engine. In the first week of January 1943, Hives and Hooker met Spencer Wilks for dinner at the Swan and Royal in Clitheroe, and Hives asked Wilks: 'Why are you playing around with the jet engine? It's not your business, you grub about on the ground, and I hear from Hooker that things are going from bad to worse with Whittle.'

Wilks replied: 'We can't get on with the fellow at all, and I would like to be shot of the whole business.'

Then Hives said: 'I'll tell you what I'll do. You give us this jet job, and I'll give you our tank engine factory in Nottingham.'

And the deal was done.

Some of the jet engine work by Rover had been carried out at Waterloo Mill in Clitheroe as well as at Barnoldswick and one of the first decisions Hives made following his agreement with Wilks was to transfer everything to Barnoldswick. As Hooker said later:

10

This decision [the take-over of the 'jet job'] – which surely ought to have been taken at national level much earlier – changed the whole tempo of the development of the jet engine. Instead of small teams working in holes in the corner, in one stroke nearly 2,000 men and women, and massive manufacturing facilities, were focused on the task of getting the W.2 engine mechanically reliable and ready for RAF service . . . Armed with a letter written in red ink (blood we called it) by Sir Stafford Cripps, Minister of Aircraft Production, which stated that 'nothing, repeat nothing, is to stand in the way of the development of the jet engine' we were able to indent on the local factories for any expertise we required.

Prime Minister Winston Churchill had written to the Minister of Aircraft Production as long ago as 30 July 1941:

I shall look forward with interest to hearing of the success or otherwise of the trials of the Whittle engine in the fortnight's time. I hope they will be favourable, but I gathered from you that the present turbine blades were not working. We must not allow the designer's desire for fresh improvements to cause loss of time. Every nerve should be strained to get these aircraft into squadrons next summer, when the enemy will very likely start high-altitude bombing.

Rolls-Royce moved some key personnel (including Stanley Hooker) up to Barnoldswick and, in May 1944, the first Meteor Is, powered by Rolls-Royce Wellands (the production version of Rover's Whittle-type, the W.2B23) were delivered to the RAF. (It was decided to name the Rolls-Royce jet engines after rivers. The idea was to denote the continuous airflow/combustion process in a gas turbine.) Wing Commander H.J. Wilson had put together a unit known as CRD Flight under the auspices of the RAE at Farnborough, and by June 1944 this Flight had been equipped with six Meteors. Within a few weeks these aircraft were transferred to 616 Squadron at Manston, and they began operations against Germany's latest weapon, the V.I flying bomb, on 27 July 1944.

However, gas-turbine-powered aircraft saw little action in the war. As the *News Chronicle* wrote in July 1945:

[The Meteors] operated over the Channel outside the gun and balloon belts. Against such difficult and fast targets as the flying bombs the Meteors had considerable success.

In the meantime the German jet-propelled fighters and fighter-bombers were appearing in increasing numbers over the battlefields in Holland and Germany. These were the single-jet Messerschmitt 163 [in fact, it was rocket powered] and the twinjet Messerschmitt 262 and the Arado 234. The German pilots relied mainly on their high speed for their protection against interception by our fighters. Only the Tempests and one of the later marks of Spitfire had a hope of dealing with the

German jets. [The News Chronicle was being very patriotic. The American P-51 Mustang, admittedly powered by the Rolls-Royce Merlin, was also capable of, and indeed succeeded in, shooting down Messerschmitt 262s.]

Air Marshal Sir Arthur Coningham, commanding the RAF, Second Tactical Air Force, became more and more anxious to match the British jet against the German jet. It became increasingly obvious that only a jet could be set to match a jet; but no one had practical experience of air fighting at speeds in excess of 500 mph.

During the closing months of the war in Europe the Meteors were sent across the Channel. For a while they were stationed near Brussels until the pilots familiarised themselves with the conditions. Later the Meteor unit was moved forward into Holland and finally into Germany. During this period the operation[s] of the Luftwaffe declined until they reached vanishing point. The Meteor pilots waited in vain for the chance to show what they could do.

Nevertheless, Hives knew that the gas turbine aeroengine was the future and threw the weight of Rolls-Royce behind its development. However, he knew that production of jet engines would be pointless unless the airframe manufacturers designed aircraft to accommodate them.

On 15 November 1944 he wrote to his old friend, Sir Wilfrid Freeman.

Our own view is that it is certain we shall finish the war an entirely obsolete Air Force. I know this is the last thing you would like to see, having been responsible for maintaining the technical superiority of the RAF for so many years.

The turbine engines have arrived! Our recent success with the B-41 emphasises that on the engine side the efficiency and performance has been well demonstrated. The Service experience with jet engines in the F.9/40 [Gloster Meteor] shows that they can be made as reliable as present conventional engines.

There is no mystery about turbine engines: they allow one to crowd much more equivalent horsepower into a very much lighter power plant. For instance, the B-41 in a modern fighter is equivalent to a piston engine of 7,000 HP.

All we can see as regards aircraft for these new type engines is a small amount of planning on fighters, but there is no sense of urgency at the back of it. The urgency all appears to be concentrated on the Brabazon type of civil machines.

It is unfortunate that the timing of the introduction of turbines should coincide with the demand for civil aircraft, but we consider the RAF should have preference.

On jet engines there is very little invention: the fact that we could design, produce, manufacture and demonstrate the full performance on the B-41 in 6 months confirms this point.

You know better than I do how long it will take to replace the various RAF machines with modern types – the first thing to face up to is that the present machines are already obsolete, and that the Air Ministry, and the Cabinet, and the Nation should face up to it.

The Germans are already ahead of us on the practical use of jet machines. As regards the technical details of their engines, we should say that they are not as good as we can provide; but on the other hand, their development is handicapped because of the best materials not being available to them.

Rolls-Royce continued to devote resources to development of the jet engine. The Welland was quickly followed by the Nene and Derwent, which, by 1946, were demonstrating high performance and reliability. By the middle of 1946 over 20,000 hours of engine testing had been completed as well as 1,500 hours of experimental flight testing. At the same time 150 Meteor aircraft had completed 20,000 hours of service flying.

There were, of course, some problems. In his book *Rolls-Royce from the Wings*, Ronnie Harker wrote:

With the Meteor beginning to reach the squadrons, there were understandably teething troubles to be overcome with the engines. This meant frequent visits by the top engineers to obtain first-hand information. It was during some of these visits, when I was the pilot, that we had some rather hairy incidents, which had fate been unkind might have set back engine development considerably! Perhaps Stanley Hooker or Adrian Lombard [Rolls-Royce's chief engineer] had a jinx or maybe they just wanted to press on regardless!

On one occasion, I flew a party of engineers from Hucknall down to Manston in our Oxford to attend a dinner to celebrate 616 Squadron becoming operational. We had a pleasant evening with the pilots and their wives and on the morrow prepared to leave for the North. The weather was really foul with low cloud and rain and bad visibility and I thought about delaying the departure. However, the weather was obviously not going to improve and as Lombard as usual was keen to press on, I let him persuade me to take off.

We became airborne and were in cloud at 100 feet. We turned north to cross the Thames Estuary and were flying between the sea and the cloud base, when about halfway across, the port engine lost power. I throttled it right back, but having non-feathering propellers we were only just able to maintain height on one engine. We managed to turn round on the reciprocal course and we tried to find Manston again. [Manston was used principally as an airfield for damaged aircraft returning from missions to Europe. It had wide runways and was well equipped with fire tenders. This could have been why Harker chose it.] Everybody on board was now rather agitated, and all eyes were searching through the murk to try to find the coast line and the aerodrome. We skated over the hills with very little to spare and found the runway lights. I flew in, in a position to do a 40° turn, lower the under-carriage and flaps and go straight in but just as I was in the middle of doing this a Mosquito loomed right up in front of me, having turned inside us from the left. It was in full pitch with a lot of throttle on, motoring in to land in front of us. We flew straight into his

13

slipstream which tilted us up vertically on to our port wing tip with only about 100 feet in which to pick the wing up, there was no hope of going around again. By putting on full top rudder and full aileron and full power on the starboard engine we levelled out as we hit the runway but with a lot of drift on. Fortunately nothing broke and we were on the ground in one piece. It was [a] near thing!

In spite of such adventures, the flights eventually proved trouble-free and maintenance was carried out by one third of the personnel required for Spitfire squadrons, even though the Meteor was fitted with two engines as opposed to the Spitfire's single piston engine. So successful was the Meteor that 3,800 were built between 1944 and 1955. On 7 September 1946, a Derwent-powered Gloster Meteor F Mk 4 – EE549, flown by Group Captain E.M. Donaldson, set a new world air speed record of 615.81 mph. The present thrust of the Derwent and Nene was 3,500 lb and 5,000 lb respectively, but advanced engines were under development, which by 1947 would deliver 4,000 lb and 5,750 lb thrust. As well as these two centrifugal compressor engines, an axial-flow engine, the AJ-65, was also under development and was expected to be ready for sale in 1948.

'LET US TALK ABOUT THE ENGINEERING'

Hives was convinced that Rolls-Royce's future lay with the jet engine. On 30 May 1946 he told his board:

Our future lies on the turbine engine side. There are no new aircraft contemplated or being built which will take the Merlin or Griffon engines, or any other piston engines.

Hives was perfectly correct that the future lay with the jet engine, but Rolls-Royce's first entry into the civil aviation market was with the Merlin engine. However, civil aviation was not going to be easy. The Americans, who would provide the competition both in airframes and aero engines, were way ahead of potential British competitors as they had continued to develop their transport aircraft during the war. Nevertheless, Trans-Canada Airlines (TCA, now Air Canada), which had operated Merlin-powered Lancastrians during the war, chose the Merlin to power its Canadair DC-4M North Star airliner, effectively a Douglas DC-4 with a pressurised cabin. The maiden flight took place in Montreal in July 1946. Similar aircraft were operated by several airlines – including the newly formed British Overseas Aircraft Corporation (BOAC), which called them *Argonauts*. Merlins also powered the York, which had first been used in Transport Command of the RAF. It was based on the Lancaster wing and power plants, and was used by a number of airlines.

The Trans-Canada Airlines business was considered important enough for

Hives to send J.D. (Jim, later Sir Denning) Pearson to Canada to negotiate with TCA and Canadair, which were to build the aircraft. He set up a Rolls-Royce subsidiary to support the operation. When Pearson was recalled, Jimmy Wood, who had served as a graduate apprentice during the war and who would later be awarded the OBE for 'services to British commercial interests in Japan', went out to replace him. There were plenty of problems as Alec Harvey-Bailey, son of R.W.H. Bailey, made clear in his book *Rolls-Royce, the Sons of Martha*:

The Merlin in TCA was taking time to settle down and Jimmy Wood was sent to the TCA base in Winnipeg and EW-D [Eric Warlow-Davies] was virtually full-time in Montreal, with me handling the Derby end . . . Hs [Hives] soon replaced Harry Cantrill, who was then running Merlin and Griffon development because time was not of the essence to him and TCA were demanding rapid cures. Hs's choice was David [later Sir David] Huddie, who was then one of the rising stars in engineering. Based on service priorities he reshaped the development programme on the engine, but it was not an easy task. During the war Merlin had been finessed to the point where he had little room to manoeuvre, but steady improvements were made in the operation. With hindsight our lack of experience in civil aviation, and particularly in engine management, were a factor. TCA were a technically oriented airline, with a powerful engineering department with somewhat dogmatic views that did not help. As an example, they were wedded to a particular brand of oil, which added to our problem.

In the end, it necessitated intervention by Hives to placate TCA. Gordon McGregor, the airline's president, demanded that Hives fly out to discuss the problems with him. Before McGregor could launch into a tirade about the unreliability and therefore extra cost of the Merlin, Hives took the wind out of his sails by giving him a cheque and saying:

That is to recognise some of the trouble we caused you. Now let us talk about the engineering.

According to Harvey-Bailey:

[In] those days the cheque was like an international telephone number and McGregor, a man of integrity, responded and set the tone of the meeting.

However, while TCA were satisfied with the engineering programme, they were still unhappy about potential costs. After letting the arguments run for some time, Hs said: 'You are the experts, you tell us what the engine should cost.' TCA came up with a parts cost of $4 an engine flying hour and Hs replied: 'Right you pay us $4 an hour and we will pick up the balance.' This did two things, it quietened TCA and gave Rolls-Royce some very clear if tough targets. It was the first flying hour agreement and established a policy that exists to this day in various forms. Equally, Derby responded to the challenge and finally achieved profitability at the figure.

Back in Europe, the Merlins were still giving service in helping to thwart the Soviet blockade of Berlin in 1948. (As had been predicted by some including, dare it be said, Adolf Hitler, the real threat to post-war peace came not from fascist Germany but from the communist Soviet Union and it was not long before their expansionist aims in Western Europe were threatening World War III.)

Most of the aircraft involved in the Berlin Airlift were American, but the British made a significant contribution with Yorks, Hastings, Sunderlands and Tudors. Harvey-Bailey wrote:

The effect of the short stage length and the high percentage of take-off power gave all engines a hard time. On the Merlin the main problem was exhaust valve deterioration. This was tackled by changing cylinder blocks at shorter lives and embodying the latest modifications, as well as trying new coatings on both valves and valve seats. Facilities had to be increased to meet demand for repaired blocks and engines, but a good standard of reliability was sustained.

The year 1949 brought more problems with the Merlin as BOAC began to operate their Argonauts. The main problem was a spate of intercooler pump gland failures. The intercooler system differed from earlier civil engines in that it gave full intercooling at take-off, but mixed main engine coolant in the intercooler flow at cruise, to raise the charge temperature and stop plug leading. The danger was that an intercooler pump gland failure could drain the main cooling system. In the air, or rather on the ground, were six BOAC Argonauts stuck between London and Colombo with pump failures and the rest of the fleet grounded. (Before the war and for a short time after it, Colombo was an important centre for British companies trading in the Far East. In 2000 it is insignificant, thanks to the liberal capitalist regime in Singapore as opposed to the communist one set up in Sri Lanka, formerly Ceylon.)

The Test Department in Derby said no leakage had ever occurred on test and no one seemed able to suggest a cure. Hives became more and more irate at the failure to find a solution and said:

The Transport Department is full of Rootes vehicles with packless gland water pumps. They don't leak. Crewe is full of motor cars with pumps that don't leak and here we are unable to fix a simple job like this. This has got to stop, someone is going to fix the coolant pump job in twenty-four hours or be fired.

The man he chose was Alec Harvey-Bailey, who later said that Hives's famous answer when asked by someone how many people worked at Rolls-Royce, 'About half of 'em', was made at this time when Hives was under great stress. Harvey-Bailey himself was now under stress, but came up with the solution. As he put it:

Forced to reappraise the mixing scheme . . . it was shown that at cruise powers the charge temperature was sufficient to avoid plug leading without charge heating. By deleting the mixing scheme and using full intercooling for take-off and zero intercooling at cruise by putting a stop valve in the circuit, the requirements of the operation were met. Intercooler pump seals did wear in service, but it was a problem that could be lived with and finally improved materials produced satisfactory seals.

This saved the fleet from being 'grounded'. Once that happens, getting a fleet flying again can prove very difficult as British Airways and Air France will surely find with Concorde. If the Merlin was proving not entirely satisfactory for the relentless slog of civil operations, the costs of entering the jet age were not going to be light either, as these minutes from the board meeting of 5 September 1946 testify:

The Managing Director supplemented his report on Aero Division matters by drawing the attention of the Board to the enormous cost of production of turbine engines. The production cost of the Nene was twice that of the Merlin, and the Dart was proving much too expensive. The reason for this appeared to be that we were at present much too short of knowledge. The amount of scrap was terrific. Whittle's theory always was that turbines could be produced cheaply, and a main problem for us was how to produce them at a reasonable price.

A controversial decision was made by the recently elected Labour Government in September 1946 when it gave permission for the sale of twenty-five Rolls-Royce Nene and thirty Rolls-Royce Derwent engines to its ally, the Soviet Union. The engines were copied and powered the MiG-15, MiG-17, IC-28 and Tu-14 combat aircraft used by both the Soviet Union and its allies. During the Korean War (1950–53), the Royal Navy and American air forces found themselves fighting against them.

Sir Stafford Cripps is usually given credit – if that is the right word – for this generous act, although there is some evidence to suggest that Rolls-Royce was also enthusiastic to make the sale. And indeed, why not? The Soviet Union was Britain's ally. 'Uncle Joe' Stalin was a popular hero and, more important, Rolls-Royce employed thousands of people and needed to keep them busy. Perhaps, if Hives and Cripps had visited Moscow and suffered the privations of the British Embassy there, they would have looked on Soviets less kindly. Lord Gladwyn wrote in his *Memoirs*:

Bevin [Foreign Secretary in the post-war Labour Government] liked to tell the story of Brindle, the Petersons' elderly Scotch terrier and how he, Bevin, suffered in the pursuit of His Majesty's Foreign Policy. Towards the end of our little meal, Ernie would recall, 'they gave me, with the cheese, what I took to be a dog biscuit, so I said, by the way of a joke like, and quite unsuspecting: 'Lady Peterson, these look like dog biscuits.' 'As a matter of fact they are, Secretary of State,' she said, 'but we like them.' Well, after a bit of a pause, I said, 'Lady Peterson, I 'ope I'm not depriving Brindle of his dinner?' And do you know what she replied? 'That's quite all right, Secretary of State, Brindle won't eat them.'

By 1948, as the Cold War intensified when the Soviets blockaded Berlin, even Cripps was sufficiently disillusioned with Britain's former ally to prevent the sale of hardware or technology.

EXPERIENCED PERSONNEL

Whatever promise the Soviet Union held for new business for Rolls-Royce, the United States of America clearly promised much more and Hives was quick to respond to a cable he received from Philip Taylor in New Jersey on 30 January 1946. The cable read:

MAY I DISCUSS WITH YOU IN ENGLAND A LICENSE FOR THE MANU-FACTURE AND SALE OF YOUR TURBINE ENGINES IN THE UNITED STATES. PLANS INCLUDE CREATION OF A NEW COMPANY WITH ADEQUATE UNITED STATES CAPITAL AND EXPERIENCED PERSONNEL.

Hives immediately sent a cable to Jim Pearson in Montreal to check out Taylor's background and qualifications. (As we have seen, Pearson, one of Rolls-Royce's most promising young engineers of the 1930s, had been sent to Montreal to liaise with Trans-Canada Airlines and the Royal Canadian Air Force, both of which had ordered the Canadair-built DC-4 powered by the Merlin.)

Taylor had been chief engineer of Curtiss-Wright which had been producing air-cooled piston engines throughout the war. He had fallen out with the Curtiss-Wright board and now wanted to promote the jet engine in the USA.

Hives also sent a holding cable to Taylor, which said:

We are free to discuss licence but would need satisfying on the quality and experience on the engineering side. Have booked telephone call to you ten o' clock our time Saturday morning [5 a.m. for Taylor and, note, transatlantic telephone calls had to be booked days in advance in 1946] in order to avoid unnecessary journey.

After talking to Taylor, Hives was reluctant to grant him a licence or even to write a letter to Taylor saying that Rolls-Royce was willing to negotiate. Conversations had been taking place with Continental Motors, which had made Merlins under licence and Hives wrote to Pearson on 2 February 1946:

We are much more interested in the Continental set-up. There we have a recognised aero engine firm, which is supported by the US Army Air Corps, which has complete technical staff and manufacturing facilities.

I think it might help you in your dealings with Continental if you can say that there are other offers being made.

Nevertheless, Rolls-Royce sent a letter to Taylor agreeing not to conclude any agreement with any other organisation in the USA for thirty days.

Taylor had done his homework thoroughly when commissioned by his former employer Pan American Airways, which he had joined when he left Curtiss-Wright, to review the gas turbine field, a project sponsored by the US Army Air Corps. He had noted the effort that Rolls-Royce had put into the development of the turbojet engine and reported:

The technical excellence of these engines and the rate at which development has been carried out is well known in the aircraft industry and the Services in this country.

Competition in the USA in the immediate future was expected to come from two sources: the Allison Engineering Division of General Motors and the Westinghouse Electric Company. Allison was building the General Electric I-40, a derivative of Whittle's engine, and TG-180 turbojet engines. Westinghouse, under Navy sponsorship, had developed an engine designated as the 24-C but, as in mid-1946, had no manufacturing facilities to produce it.

The General Electric engines had their origin in the procurement by the US Army Air Force (USAAF) in the summer of 1941 of the designs of the W.2B turbojet engine developed by Power Jets Ltd. The USAAF gave the designs to the General Electric company (GE), requesting the company to develop an American version of the engine. The first engine, the I-16, was used to power the Army Bell P-59 and the Navy Ryan FR-1. However, it was soon recognised that the engine's power was too low for aircraft in the pipeline. GE therefore developed the I-40, an engine with 4,000 lb thrust, somewhere between the Derwent and the Nene. At the same time as the I-40 with its centrifugal compressor, GE worked on an axial-flow compressor engine, the TG-180.

Westinghouse had also entered the aircraft engine business in 1941, at the request of the Navy Department. They produced the Model 19 with approximately 1,500 lb thrust and this engine made its maiden flight in the first Navy turbojet fighter, the McDonnell Phantom. It was also quickly seen to be too low in power and Westinghouse designed the Model 24 of 3,000 lb thrust. This was still too low in power for projected military aircraft.

Until the British Meteor, powered by a Rolls-Royce Derwent engine, showed that aircraft could be flown at speeds over 600 mph, the armed services in the USA had concentrated on demanding turbine propeller engines giving greater *range* rather than greater speed. A fighter, if it were to attain speeds of 600 mph, would *have* to be powered by a turbojet engine. In the middle of 1946, no USA manufacturer could produce such an engine. Rolls-Royce had a distinct competitive advantage. The Derwent and the Nene were superior to any engine domestically manufactured in the USA.

The production position in the USA, even of the low-powered turbojet engines, was no better. In 1943, the USAAF had placed two engines in production – the General Electric I-40 and TG-180. The I-40 was built by GE itself in Syracuse and also by Allison in Indianapolis, while the TG-180 was built by Chevrolet in Tonawanda, a suburb of Buffalo, New York. The Navy placed orders for the Westinghouse Model 19 with Pratt & Whitney, where they were to be built at its plant in Hartford, Connecticut.

On VJ-Day in August 1945, all these production orders were cancelled. Allison underbid GE for the production of the I-40 and GE closed its Syracuse plant. GE, asked to take over production of the TG-180 from Chevrolet, refused to do so. A reduced production order for the TG-180 was transferred to Allison. The Navy cancelled nearly all Pratt & Whitney's order for the Model 19.

Looking at the US companies involved in jet engine development and production, some were in a position to become competitive with Rolls-Royce in the future, others not.

GE, of course, was a huge company, and had submitted a proposal to the USAAF for the design and development of a 6,000 lb turbojet engine to be interchangeable with the TG-180. However, the development period was likely to be at least two years – and even then it was calculated that Rolls-Royce's AJ-65 would produce 500 lb more thrust, weigh 800 lb less and have 10 per cent better specific fuel consumption. GE was also redesigning the I-40 to increase its thrust to 5,000 lb, which would make it competitive with Rolls-Royce's Nene, but by the time it was ready, the advanced Nene would be available.

Westinghouse was also a huge company but had not, up to early 1946, made a large commitment to the development of turbojet engines. The Navy, using Westinghouse as its contractor, could force the company to increase its commitment.

Allis-Chalmers had also entered the turbojet engine business, at the insistence of the Navy, by taking a licence to build the British de Havilland H-1 turbojet engine. Not much progress had been made before VJ-Day when the Navy's order was cancelled.

Packard, which had manufactured some 50,000 Rolls-Royce Merlin engines, was left with a plant at Toledo largely idle after VJ-Day. Its answer was to make an expendable turbojet engine with characteristics similar to the I-40 but with a total life endurance of ten hours. This was clearly not going to be a serious competitor.

Allison was a different matter. An aircraft engine manufacturer since the 1920s, the company was very close to the USAAF and was likely to stay in the forefront of turbo engine development.

Three others – Menasco Manufacturing Company (now a leading manufacturer of aircraft undercarriages), Wright Aeronautical Corporation and Frederic Flader Inc. – were all engaged on turbo-propeller engine development, but none was deemed likely to provide competition to Rolls-Royce in the immediate future.

The potential in the USA, in military sales alone, was enormous. The Commander of the USAAF, General Carl A. Spaatz, had made clear his intentions with regard to jet-propelled aircraft.

Not only will the Army Air Force replace its P-51 Mustang and P-47 Thunderbolt fighters with P-80 Shooting Stars and other new type jet-propelled fighters as soon as they are available, but bomber squadrons will be equipped with jet-propelled aircraft in the near future. Demand over the next few years could be 16,000 aircraft.

Nevertheless, securing some of that business for Rolls-Royce would not be easy as was made clear when Captain Baird of the US Navy called on Colonel Darby in Rolls-Royce's Conduit Street offices in London on 1 August 1946. Darby reported to Hives:

Captain Baird called on me to-day in connection with the position as between ourselves and the Taylor Turbine Corporation, and he was very anxious to find out the actual situation. I pointed out that while we were still negotiating, the final agreement had not yet been signed, and I tried to find out the American Government's reaction to the Taylor Turbine Corporation.

I rather gather the American Government are a little jealous that we should be ahead of their development on Turbine engines, but they admitted this and stated they might be inclined to consider buying engines from us for a maximum period of, say, 12 months. This would be at the rate of 10 engines per month, and they anticipate that by this time they will be able to look after their own requirements.

As to the Taylor Corporation, he made it quite clear that they would get no support from the Government at this stage, and that if they wanted to sell engines to the

United States authorities they would have to satisfy them that there was a very strong technical organisation behind the Taylor Corporation, and that we would have to agree to give the Taylor people the benefit of our future research and development. I pointed out that this was a general condition of licence contracts, subject to any security regulations.

Captain Baird appeared to be very anxious to get hold of a couple of engines to put them through an American form of type test as, unless they passed this, they would certainly not be accepted either for Civil or Military purposes. I rather felt he was hoping that we would send a couple of engines, that they would fail on their test and that would solve the problem. If we send engines to the U.S. for test, I think it is quite clear we shall have to send one or two very reliable men to live with them until they have been through their test. Even then, the manufacture and sale of a British engine under licence in America is not going to be very easy, but I suppose we must expect that.

On 19 September 1946, Rolls-Royce signed an agreement with Philip Taylor giving him an exclusive licence to sell 'Rolls-Royce Turbine Engines known as the Nene and Derwent for use in aircraft, and all spare parts and accessories manufactured by the Company for such engines'. The licence was to last for two years but could be terminated after one year by the company if it had not received from Taylor, or by his introduction, orders for ten engines. There was also provision in the agreement for Taylor to manufacture Rolls-Royce engines under licence.

PRATT & WHITNEY

By this time Taylor was in consultation with the Grumman Aircraft Corporation over the supply of Nene engines, but by the end of 1946 he was also in negotiation with Pratt & Whitney, part of the United Aircraft Corporation. On 12 December, Taylor sent a telegram to Hives saying:

Would very much appreciate your assurance of support to Pratt & Whitney following our initial contact with them.

And, in early January 1947, Taylor again cabled Hives:

Pratt & Whitney, after conference with me, contact with the Navy and examination of the Nene endurance test engine, has expressed a definite interest in manufacturing the Nene engine to meet proposed Navy schedule and other US requirements... This is equivalent to Taylor Turbine giving a sublicence to Pratt & Whitney for the manufacture of the Nene engine only, which is the extent of their present interest. This arrangement seems to be a practical means of providing an immediate source of

Frank, later Sir Frank, Whittle. Rolls-Royce quickly appreciated
the value of his jet engine development.

ABOVE AND OPPOSITE: As Rolls-Royce tested a
Whittle-type W.2B engine in a Wellington bomber,
Hives became convinced that gas turbines
were the future engines of the aircraft industry.

Lord Hives – one of the greatest British businessmen
of the twentieth century.

Spencer Wilks, the Managing Director of Rover. Hives
and Hooker met him for dinner at the Swan and Royal
in Clitheroe in January 1943, and Hives said to Wilks:
'Why are you playing around with the jet engine?
It's not your business, you grub about on the ground.'

This Gloster Meteor performed the first flight of
a turboprop engine in the world with its Rolls-Royce
Trent engines on 25 September 1945.

Trans-Canada Airlines, which had operated Merlin-powered
Lancasters during the war, chose the Merlin to power
its DC-4M North Star Airliner. This photograph shows
the aircraft flying over Ottawa with Canada's
parliament buildings clearly visible in the background.

An early Meteor powered by a Rolls-Royce Derwent.
Note the anti-Daunts in the engines. These were
fitted after Michael Daunt was sucked into an engine
in 1943 while checking if any oil was leaking from
the motor. Fortunately he was not killed but was heavily
bruised and severely shaken. After the accident,
steel grilles were fitted over the air intakes.

OPPOSITE: On 7 September 1946, a Derwent-powered
Gloster Meteor, flown by Group Captain E.M. Donaldson,
set a new world air speed record of 615.81 mph.

The Rolls-Royce Nene, the first clean sheet turbojet to come from
Stanley Hooker and his team at Barnoldswick. It was scaled
down to create the Derwent V which became the standard engine
in the Gloster Meteors.

THE POWER PLANT

Published by PRATT & WHITNEY AIRCRAFT   DIVISION, UNITED AIRCRAFT CORPORATION

VOL. IV No. 3   EAST HARTFORD   CONNECTICUT   MAY 15, 1947

Committee Votes $899.72 Awards For Suggestions

Eagleson High Man With $122 Award

Two suggestions he made to improve machining and tooling for parts on the R-2000 won awards of $122.50 for Alex Eagleson of Department 432 — the top prize voted at the first meeting of the Employe Suggestion Plan Committee. Awards totalling $777.22 were allowed other persons making early suggestions.

Richmond Awarded $120.27

Second highest winner was Ralph A. Richmond of department 820, green assembly, who also had two suggestions accepted, for a total of $120.27. He suggested improving the assembly tool used to drift R-4360 outer roller bearings into place. He also recommended an adapter to be used together with the prop shaft device to hold an R-4360 support and gear assembly as it is being made into a complete support and pinion assembly.

Wilbur C. Fletcher, department 815, testing, was awarded $105.60 for suggesting a method of adapting liquid flowmeters on hand in place of purchasing new equipment.

For a suggestion on improving gear tooth grinders, Rudolph Fregin was voted $100.

Other Winners

Other winners under the new plan include: Gordon W. Lepper, department 412; Oscar J. Reeves, department 820; Zenas Mikelis, department 632; Charles Koenig, department 825; Carl W. Skold, department 142; Morris Weinstein, department 312; Edward E. Griffin, department 241; Ivan E. Brown, department 421; John Dube, department 825; Clarence Aronson, department 821; Leslie Fleming, department 810; Leo A. Slopak, department 55; Frederick Bense, Jr., department 55; Harold

ATC Has Clean Slate

A perfect safety record — that's the score the Army's Air Transport Command — Pacific Division chalked up for 1946.

In setting this record 460,-192,553 passenger miles were flown by the division's Douglas C-54 Skymasters. They carried nearly 110,000 passengers on about 20,000 separate trans-Pacific flights.

The airplanes are powered by P&WA R-2000 Twin Wasp engines.

Smoking To Be Allowed In Factory From July 15

W. P. Gwinn, general manager, announces that smoking will be permitted on the factory floor on or before July 15. All employes will be allowed to smoke except those who work on hazardous operations or within areas where the practice would be dangerous.

The decision to remove the "no smoking" ban came after a plant-wide survey which showed that smoking would not create safety hazards except in certain areas. These will be designated by signs. The present ban will remain in force until the go-ahead signal is posted. Since it is expected that complete safety controls cannot be set up in the dangerous areas until about the middle of July, the new regulations probably will not become effective until that time.

The new policy is based on a premise that everyone will cooperate in observing the safety rules, Gwinn says, adding that its continuation will depend on that cooperation.

Cancer Drive Successful

The P&WA Cancer Fund Drive ended with a total of $3,256 contributed by Aircrafters to the Greater Hartford quota of $47,500, campaign officials announce.

Of the total P&WA donations, $80 represents a special contribution by department 32, master crib.

Pratt & Whitney Aircraft Acquiring Powerful "Nene" Jet Engine; To Begin Tooling At Once For Its Manufacture

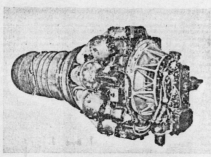

The Rolls-Royce "Nene" turbojet engine.

Pratt & Whitney Aircraft has acquired the option for the American manufacturing and sales rights for the British Rolls-Royce Nene jet engine, and will begin tooling up immediately for its manufacture.

That announcement came early this week from William P. Gwinn, general manager. Negotiations for the Nene, the most advanced and most powerful jet power plant flying today, are being conducted through the Taylor Turbine Corporation of New York. Rolls-Royce has long been one of Great Britain's outstanding aircraft power plant designers and manufacturers and has risen to pre-eminence in the jet field in that country.

(Nene is pronounced "Neen". The engine is named after a river which flows from the center of England to The Wash, on the North Sea coast. The name Nene follows Rolls-Royce's practice of naming all their engines after English rivers.)

Proved Production Engine

"Getting the Rolls-Royce Nene engine," Gwinn said, "puts Pratt & Whitney Aircraft in the enviable position of leadership in both the jet and piston engine fields.

"The Nene is a proved production engine," he pointed out. "It is a power plant which incorporates a decade of engineering development. It has demonstrated its ability, both on test stand and in flight.

"The manufacture of the Nene is also expected to strengthen Pratt & Whitney Aircraft's position in the industry with respect to its piston engines. We sincerely believe that there is need for both piston and jet-type power plants. Some aircraft require the high speed delivered by turbo-jets; others demand the economical fuel consumption provided by piston-type engines. With our company in a position to supply either type of

Wasp Majors Power Super-Bombers; Kenney: "Engines Exceptionally Good"

The airplanes of the nation's two very heavy bombing groups will consist exclusively of Consolidated Vultee B-36's and Boeing Superfortresses, all powered with Pratt & Whitney Aircraft Wasp Majors, General George C. Kenney, chief of the Strategic Air Command, has announced.

He said he expected B-50 bombers to begin reaching the SAC's two groups "shortly", with deliveries of B-36's commencing soon after the satisfactory completion of the test flights now being conducted.

Increase in engine power to 3500 horsepower in the B-50 has greatly improved all operating characteristics of the Superfortress — take-off, rate of climb and speed.

Plane Is A-Bomb Weapon

The B-36 heavy bomber could carry an atomic bomb to any inhabited region in the world and return home without refueling in the event of an enemy attack, the Army Air Force has disclosed.

The giant craft is powered by six Pratt & Whitney Aircraft pusher type Wasp Majors, developing a total of 18,000 horsepower, enough to power 200 average passenger cars. Use of two superchargers on each engine, together with pressurized cabins, enables the B-36 to attain a ceiling of 40,000 feet.

Kenney Writes Gwinn

General Kenney, writing to W. P. Gwinn, general manager, about the Wasp Major

The US engine manufacturer Pratt & Whitney acquired a licence to build the Nene. They called it the J42 Turbo Wasp and eventually manufactured 1,137 of them.

The first Pratt & Whitney manufactured Nene,
the J42 Turbo Wasp, on 5 January 1949.
LEFT TO RIGHT: J.L. Bunce (factory manager),
Wright A. Perkins (engineering manager), William P. Gwinn
(general manager), and Leonard C. Mallett
(assistant general manager).

After the war, the US Navy wanted to buy engines from its traditional supplier, Pratt & Whitney, who supplied J42s (a licence of the Nene) and J48s (a licence of the Tay). This photograph shows a Grumman Cougar powered by a J48.

Ernest Hives, Managing Director of Rolls-Royce,
discusses the supply of jet engines to our allies with
Sir Stafford Cripps, Minister of Economic Affairs.
Harold Wilson, President of the Board of Trade
and future Prime Minister, looks on.

Both the USAAF and, in this case, the South African Air Force
used Mustangs, powered by Rolls-Royce Merlin engines, during
the Korean War. They were superior to the Yak piston-engined
fighters of the North Korean air force, though not to the
Chinese jet-powered MiGs.

HMS *Ocean* off the coast of Korea during the
Korean war, with Fairey Fireflies (the nearest four in each
line) and Hawker Sea Furies. The Fireflies were
powered by Rolls-Royce Griffon engines, the Sea Furies
by Bristol Centaurus engines.

Nene engines at the same time leaving unaltered the position of Rolls-Royce and Taylor Turbine with respect to the sale and manufacture of other Rolls-Royce engines in the US . . . I assume from our phone conversation that you agree in principle to our making such arrangements. We would appreciate your comments by cable so that we may continue our negotiations.

Hives was certainly in favour of a deal whereby Pratt & Whitney would manufacture the Nene under licence. With Curtiss-Wright, Pratt & Whitney had been the major piston aero engine supplier to the US Government both before and during the war. Indeed, so important was the company that the US Government had precluded it from research and development on the jet engine while it concentrated on continuous refinement of the Wasp series of air-cooled piston engines. Nevertheless, it clearly possessed the facilities, manpower and contacts to become a force in jet engine production and sales and, as we have seen, it was already involved with producing Westinghouse's Model 19 for the US Navy. Hives will also have taken on board Captain Baird's warnings about breaking into the US Military either from abroad or through the unknown and untested Taylor Turbine Corporation.

Hives had given open house to C.J. McCarthy, a vice-president of United Aircraft Corporation (the owners of Pratt & Whitney) when McCarthy and his wife had visited Britain the previous August; on 24 December 1946, when he knew of Taylor's contact with Pratt & Whitney, he cabled the UAC's H. Mansfield ('Jack') Horner, saying:

We are informed by Philip B Taylor that you have expressed interest in project for manufacture in America of gas turbine engines to our design.

We would welcome your association with this project and confirm that both Taylor and your corporation would be assured of our wholehearted support.

Pratt & Whitney had already laid down plans whereby the company would be either abreast or ahead of the other gas turbine manufacturers by 1950. In the meantime, involvement with Rolls-Royce's Nene would be a useful stopgap. As stated in Pratt & Whitney's own history, published in 1950:

The other unexpected event that altered Pratt & Whitney's Aircraft's post-war course was the Navy's decision to bring the Rolls-Royce Nene to this country. When the Navy made this decision, late in 1946, the Nene offered more power and better reliability than any turbojet likely to appear soon on the American scene. Its rating of 5,000 pounds static thrust was attained with a total weight of 1,700 pounds. It was this low weight-powered ratio [should read 'high power to weight ratio'] plus its durability that attracted the Navy to an Americanized version of the Nene for a new jet fighter (F9F) then under development by the Grumman Aircraft Engineering

23

Corporation at Bethpage, Long Island. The Navy was also interested in stimulating competition among American jet engine manufacturers. So successful was the Nene in attaining this objective that it quickly became known as the 'Needle engine'.

To Pratt & Whitney this appeared an excellent opportunity to plunge into the gas-turbine competition long before its own jet developments were ready for production. The American rights permitted Pratt & Whitney to do its own production engineering on the Rolls-Royce design and make any changes it felt necessary without prior permission from the British firm. Pratt & Whitney had a high regard for Rolls-Royce engineering ability in all forms of aircraft power, particularly in the gas-turbine field where they had been active since the early development of the Whittle design.

By February, Hives felt negotiations with Pratt & Whitney were so important that he must go to the USA himself and on 15 March he sailed for New York on the *Queen Elizabeth*. By early May, Pratt & Whitney's negotiations with the Navy were completed and on 26 May the United States Navy issued this press release:

The Navy Department today ordered from Pratt & Whitney Aircraft Division of United Aircraft Corporation a number of Rolls-Royce Nene turbojet engines. These will power a new carrier fighter designed by the Grumman Aircraft Engineering Corporation. The first of these planes are expected to be flown in test late this summer. Two of them will be equipped with British-made Nenes brought to this country for experimental purposes. Tests are expected to lead to further development and modification of the British engine by Pratt & Whitney.

The US Navy did not burn its boats elsewhere. L.S. (Luke) Hobbs, Vice-President for Engineering at United Aircraft Corporation, wrote to Hives on 27 June 1947:

As Gwinn [William ('Bill') Gwinn, manager of Pratt & Whitney] has no doubt told you, the Navy is providing competition in the I-40, and General Motors (Allison) are out to get this business if possible . . . they should not be underrated.

At the end of 1946, as negotiations with first Phil Taylor and then Pratt & Whitney got under way, Rolls-Royce had appreciated that it would need the services of a legal firm in New York, and William Gill (the Rolls-Royce Financial Director) and David Huddie (of whom much more later), were sent in search of someone suitable.

Huddie was later to tell the story of visits to several firms that all professed intimate knowledge of gas turbine engines and how perfectly suited they were to represent Rolls-Royce. Finally, he was advised to interview Thayer and Gilbert, a two-man operation at 52 Wall Street. Shown into Phil

Gilbert's office with its linoleum-covered floor, Huddie was told by Gilbert that he and Thayer knew nothing about gas turbine engines. Huddie was impressed by this honesty and appointed the firm as Rolls-Royce's legal advisers in the USA, a position the firm – now Gilbert, Segall and Young – still holds today.

In early 1947, Thayer and Gilbert were soon in the thick of advising Rolls-Royce – first on how to ensure a smooth transition from representation by the Taylor Turbine Corporation to a licensing deal with Pratt & Whitney, and then by tying up a trouble-free contract with Pratt & Whitney.

William Gill wrote to Thayer on 25 February 1947 giving him the background of the company's unsuccessful attempts to interest US aero engine manufacturers in its jet engine technology in 1945, its subsequent granting of selling rights plus an option on a manufacturing licence to Taylor and the present possibility of a licensing deal with Pratt & Whitney. He continued:

In the course of the negotiations between Taylor and the US Navy authorities, it became apparent that the latter were not anxious to see a new engine-building organization enter the field, but that they would be very much in favour of seeing this work given to one of the existing manufacturers. Taylor was therefore encouraged by the promise of a substantial contract from the US Navy to approach Pratt & Whitney. Conversations between Taylor and Pratt & Whitney have now been proceeding for some two months, and we believe the time has now arrived when we must make direct contact with Pratt & Whitney in an endeavour to bring these negotiations to a head. It is for this purpose that Mr Hives proposes to visit America.

I should explain that Mr Taylor formed the Taylor Turbine Corporation for the purpose of exploiting the selling rights, which we had given to him. It is difficult to see what future there is for this Corporation in connection with our work in the event of the actual manufacture being undertaken by Pratt & Whitney. It is to be hoped that this position will be clarified by Mr Hives during his visit.

In the event, the position was clarified and to the satisfaction of Phil Taylor who wrote to Hives on 1 December 1947.

Dear Hs,
Now that the work of Taylor Turbine Corporation has been accomplished I wish to express my appreciation for the many actions on the part of Rolls-Royce which made the project possible. Your own long-range views had their full measure of effect in the final accomplishment.

While I regret that Taylor Turbine Corporation could not carry forward as originally planned I believe that the tie between Rolls-Royce and United Aircraft Corporation is of greater advantage to Rolls-Royce than anything which Taylor Turbine Corporation could have provided.

Negotiations with Pratt & Whitney to conclude a final agreement proved more difficult, primarily because Pratt & Whitney saw themselves as being already a jet engine manufacturer and did not want to bind themselves into an agreement whereby they would be paying Rolls-Royce a royalty on *all* jet engines that they produced.

Luke Hobbs had begun studies on gas turbines in 1939 and, although discouraged from further development during the war by the US Government, he and his board had decided in 1945 that the future was the gas turbine and set themselves a five-year plan to catch up with their competitors.

Pratt & Whitney were also concerned that Rolls-Royce was going to turn its development efforts on to other engines and discontinue further development of the Nene. Hives reassured them on this point and the wrinkles in the agreement were gradually ironed out.

On 28 November 1947, Gwinn cabled Derby to say:

Grumman Nene was flown yesterday for one hour, 15 minutes. Everything OK. Pilot very pleased and snap rolled machine.

The next day Gilbert cabled:

United closed with Taylor today after I approved contracts and Taylor released all claims against Rolls.

On 9 December, Bill Gwinn wrote to Hives:

I have just returned to the office with a group of us who attended a Navy and Press preview of the new Grumman Nene fighter. About fifty newspaper and picture service people were in attendance, plus a lot of Navy people, including Admiral Oster. The show went off splendidly and all are very pleased with the airplane engine combination. Today marked the ninth flight in approximately ten days' time and the engine hasn't given one bit of trouble. All commented on the absence of any smoke trails, especially in high-speed flight and aerobatics. In the show today he did plenty of the latter, such as inverted flight and right and left hand snaprolls in a climb.

As Pratt & Whitney's production of the Nene progressed, Rolls-Royce continued to develop new engines, notably the axial-flow compressor engine, the AJ-65 (by this time called the Avon) and the turbo-propeller engine, the Dart. Hives wanted to know Pratt & Whitney's views on these two engines. He had been visited in early 1949 by Guy Vaughan of the Wright Company, who had made it clear that Wright wanted to be involved in the gas turbine engine business. On 14 February, Hives wrote to Gwinn:

We have just been given permission by our Ministry to sell the Avon engine in the USA only. The Avon engine and its variants is undoubtedly going to be very good. Although there was a time when it was thought that the Metro-Vick Sapphire would be a serious competitor, this is no longer a major worry to us. On the other hand, we would hate to see the Sapphire sold in the USA, which might happen if we neglect our opportunities with the Avon.

I have no wish to press Pratt & Whitney for a decision under the terms of our agreement. What I put forward for your consideration is that it might be worthwhile your placing an order for, say, 6 Avon engines, delivery 12 to 15 months, 4 with a view to installing into an aircraft, and 2 for model testing.

And in May 1949 Hives visited the USA with William Gill. As a preliminary to this visit he wrote to Hobbs saying *inter alia*:

I am anxious to have a frank discussion because our association up to date has been on the most friendly and helpful basis. It will not affect our attitude at all if Pratt & Whitney decide that they have no interest in our future engines; it would not make the least difference to our continued enthusiastic support for the Nene and Tay project.

We have been frequently asked by visitors from the USA whether Pratt & Whitney are taking up the Avon, and we would obviously like to know the answer to give them a direct answer.

Hives went on to say that he fully understood that it was only a matter of time before Pratt & Whitney produced gas turbine engines to their own design to build on their reputation in the reciprocating engine field. He felt that as Rolls-Royce had been early in the gas turbine field the company could speed Pratt & Whitney's move into self-sufficiency in this new area.

Following meetings with Pratt & Whitney and its parent, United Aircraft Corporation, Hives concluded that the American company did not want any further engines from Rolls-Royce. Before he sailed back to Britain, he wrote by hand from The Gotham Hotel to Horner, President of UAC:

Before sailing I wanted you to know that while I am sorry that Pratt & Whitney do not wish to take up any future Rolls-Royce engines I am grateful to you for making your position so clear.

He felt that a slight modification to their agreement was necessary so that he could continue discussions with Curtiss-Wright. He concluded:

If this is satisfactory to Pratt & Whitney please let me know.

Horner replied on 21 June saying that Hives had misunderstood their position. Maybe Hives had deliberately misunderstood their position so that the position could be made absolutely clear. Did Pratt & Whitney want other engines from Rolls-Royce or not?

Horner said that, although Pratt & Whitney was not interested in the Avon, it could well be interested in future engines. Horner visited Britain in August, and was royally entertained by Rolls-Royce including the mandatory dinner at Duffield Bank House followed by darts and snooker. However, he resisted the offer of a special consulting basis contract and made no commitments on the Dart turboprop engine beyond requesting the receipt of 'any information, which you might wish to pass on relative to your visitation from the American Airlines people'.

By the middle of 1950, production of the Nene was finally running smoothly in East Hartford (there had been considerable delay and more problems than originally anticipated in reaching this happy position), but the Pratt & Whitney/Rolls-Royce co-operation was clearly winding down. Gwinn wrote to Hives on 13 June and, after congratulating him on his elevation to the peerage, said:

Now that the Nene engine is rolling along smoothly in production at our plant, and the Tay development is rapidly drawing to a close, particularly at your plant, and we have turned out our first two production engines, we feel that the time has arrived when we should discontinue our engineering representation at your plant.

It was now clear that Pratt & Whitney wanted to develop its own engine and on 28 January 1952 Hives wrote to Horner:

I feel that the Pratt & Whitney engine policy is now settled, and you have decided that, in future, you will only develop and market your own design. I think you can say right away that you have no interest in the Rolls-Royce engines such as the Avon and the Dart.

We have no desire or intention of having another Licensee in the USA. As you are aware, we have started a facility in Canada and, although it is outside the Licence territory, in resolving future policy we should like to avoid any misunderstandings.

As the 1950s progressed and Rolls-Royce added the bypass engine, the Conway, to its range, it continued to keep Pratt & Whitney informed of all developments and any negotiations with other companies. As far as the bypass engine was concerned Rolls-Royce felt that it could not be separated from the axial-flow engine and offered both to Pratt & Whitney.

Those on offer were the Avon RA-3 giving 6,500 lb thrust, the Avon RA-7 giving 7,500 lb thrust and the Avon RA-14 giving 9,500 lb thrust. These were all axial-flow engines. And then there was the Conway giving 11,500 lb thrust and the RB 93 giving 2,700 lb thrust, which were bypass engines.

The terms of any agreement would be £1 million for a ten-year licence with royalty payments of 3 per cent of the US selling price, with a minimum royalty payment of £100,000 per annum. Horner felt that the two types of engine should be kept separate and wrote to Hives on 2 March 1953:

While we agree that it is entirely logical to group your axial flow powerplants in one arrangement, we do not feel it is logical to also include in such a package another basic concept such as the by-pass design.

However, Hives was adamant on this score, replying:

The broad specification of the Bypass engine consists of a twin-spool axial jet engine incorporating a bypass bleed. The design and technical difficulties we know from experience follow the problems connected with axial engines. From the practical side I am sure you will be the first to realise the difficulty of trying to operate two separate licences. The information from the bypass engine spills over into the straight axial type. We cannot see how it would be possible to operate two separate licences with different firms.

After a month of negotiations, United Aircraft Corporation turned down the possibility of a licensing arrangement. Horner wrote to Hives on 23 April. He stated the terms, which had been changed somewhat from the original proposal, and said:

We are not interested in entering into a license agreement upon these terms. It is our thought that by rejecting the offer at this time, Rolls-Royce is left free to grant a license to any other potential licensee.

However, Horner did insist that if *better* terms were offered to another potential licensee, he should have first refusal.

Hives replied on 30 April thanking Horner for releasing Rolls-Royce from all obligations on future engines and saying that although this meant a break in direct association, Rolls-Royce would always remember 'the fair treatment we have always received from you'. He also said that, although the two companies would inevitably find themselves in competition this did not 'prevent us extending a sincere invitation to you or any of your people to visit us at any time'. He concluded by mentioning that Horner probably knew that discussions with Westinghouse had been taking place and these would now continue.

WESTINGHOUSE

Indeed, four months earlier in December 1952, Hives had written to the lawyer Walter Thayer alerting him to the possibilities with Westinghouse.

It will interest you to know that we are being ardently courted by some of our friends in the USA . . . Although the approach came from an individual – one of the permanent officials of the Bureau of Aeronautics – we found the real interest had emanated from the US Navy. They are rapidly realising that in spite of the millions of dollars they have poured into Westinghouse they are getting no return, and neither do the future prospects hold out any hope . . . In addition . . . I had a letter from Westinghouse saying they would very much like to have a discussion with me.

In February 1953, Phil Gilbert, after visiting Washington and talking confidentially to Government officials, confirmed to Hives the possibilities now open to Rolls-Royce because of Westinghouse's mixed performance as a supplier to the US military. On the one hand, the Government was very pleased with Westinghouse's 'remarkably good job in the field of developing atomic power for aircraft' where the company was way ahead of other companies in the field such as General Electric. On the other hand, the Government was not so happy with them over their failure to supply the gas turbine engines ordered by the Navy.

Gilbert believed that, on balance, the authorities wanted to keep Westinghouse as a supplier but would probably welcome the engineering skills of Rolls-Royce. He assured Hives that it would not be necessary for Rolls-Royce to send over large numbers of personnel, 'but the hope is that a reasonable number of qualified Rolls-Royce engineers whose reputation in the Navy is, of course, unassailable, would make it possible for Westinghouse to organise the engineers it has and get them on the right track'.

At the same time Gilbert made Hives aware of the mounting competition:

I was also informed by the same informant that Pratt & Whitney's J-57 engine is going to be used more generally than any other engine manufactured in this country. He also told me that Pratt & Whitney have a new engine, called the J-75, which is supposed to be coming along in good shape. It is hard for me to believe, but I was informed that the J-57 engines now being turned out by Pratt & Whitney have a thrust of 20,000 lbs. I was also informed that Ford is hiring 30,000 people to make the J-57 engine and that Allison has the greatest production of any manufacturer in this country, most of which devoted to the J-22. I was also informed that Chrysler is going into production of a sizeable number of parts for the J-46 engine.

In August 1953, Rolls-Royce signed the following agreement with Westinghouse.

1. 10 year agreement.
2. Full exchange of information during first 8 years.
3. Westinghouse have right to manufacture R.R. engines in the United States and right to sub-licence when requested by United States Government.
4. R.R. has reciprocal rights in England.
5. Westinghouse pay R.R. £1,000,000 [£25 million in today's terms] by January 1954.
6. Westinghouse to pay royalties, at following rates, on :–
   (a) All J.34 engines manufactured and sold after March 1st 1955.
   (b) All other engines manufactured and sold after March 1st 1954.
       2½ % until $5,000,000 has been paid.
       2 % thereafter until an additional $5,000,000 has been paid.
       1½ % thereafter.
       Royalties are payable whether or not they incorporate R.R. patents and information.
7. Minimum royalties of $560,000 per year.
8. All licences to R.R. are royalty free.

Hives sent a copy to Jack Horner of United Aircraft Corporation to show him that the terms were no better than those offered to Pratt & Whitney.

Although Hives was pleased to have signed the agreement with Westinghouse, he knew, especially after a visit to the USA in the autumn of 1953, that much needed to be done – on the selling as well as the engineering front – if the maximum was to be extracted from the agreement. While he had been in Washington, apart from a courtesy visit to meet President Eisenhower, Hives also met officials in the US Navy Department and in the US Air Force. He was concerned by the conversations and wrote to G.A. Price at Westinghouse on 21 November 1953.

My summing up at the end of the Washington visit was that the conditions in dealing with the Government are rather similar to the conditions over here. Our policy is that we sell our engines to the aircraft constructor but the Government places the orders for the engines. This is one of the chief weaknesses I believe exist at Westinghouse; they are not in sufficiently close contact with the constructors . . .

On the West Coast we had long discussions with Mr Donald Douglas senior and all his technical staff, and they were very impressed and interested in the Rolls-Royce range of products. We also had similar discussions with Mr Kindelberger of North American and his staff, Mr Gross of Lockheed, General McCartney and his staff at Convair and Mr Ryan and his staff at Ryan Air.

There's tremendous goodwill for Rolls-Royce and its products. There is plenty of potential business there, but there is a terrific lot of work to be done to restore their confidence in Westinghouse.

Hives went on to recommend his old friends Phil Taylor and Robbie Robinson who, as we saw, helped to set up Rolls-Royce with Pratt & Whitney in 1946. He had noticed when talking to the aircraft manufacturers that they still enjoyed a high reputation. He wrote to Price:

I would recommend that when you come to a satisfactory arrangement as regards using Taylor and Robinson you should not keep it a secret; it should be publicly announced, because I am sure it would be taken as an indication of the determination of Westinghouse to come to the front on aero engines.

Unfortunately, Rolls-Royce's association with Westinghouse did not prove to be a very happy one. Hives made another visit to the USA in early 1955 and, on his return, wrote to Price. He began:

One cannot help but be disappointed that the association of Westinghouse and Rolls-Royce has not so far produced any positive results.

He went on to say that he had examined in detail whether Rolls-Royce had either not given Westinghouse full support or had let it down in any way, and had concluded that it had not. He was very depressed by his visit to Washington:

After a discussion with Mr Roger Lewis, Assistant Secretary, it was apparent that the USAF had no interest in or any desire to support Westinghouse. I then saw Mr Smith, Assistant Secretary of the US Navy Air Force, and I was disappointed at their attitude towards Westinghouse. I came away with the impression that it would be necessary for you, and nobody less than yourself, to get a statement of the future policy with regard to Westinghouse.

Hives had felt that criticism of Westinghouse was indirectly criticism of Rolls-Royce and certainly of Rolls-Royce's decision to tie themselves to Westinghouse. He was anxious to show Price that Rolls-Royce's reputation was, and always had been, very high in the USA. Ever since the highly successful P.51, fitted with Rolls-Royce Merlin engines and the first fighter to be put back into operation on the outbreak of war in Korea, Rolls-Royce engines had proved successful in the USA, and now surely if only the combined Westinghouse/Rolls-Royce resources could be properly harnessed together the result would be a great success.

Hives discussed his concern with Phil Gilbert and on 22 March Gilbert wrote an internal memorandum which said:

This morning I had a long discussion with Hives about the Westinghouse situation. Lord Hives had in his pocket and showed to me the latest payment by Westinghouse which came to something in excess of $2,000,000. [In 1949 the pound had been devalued from $4 to $2.80. The sum of $2 million was probably worth £20 million in today's terms.] Lord Hives said, however, that the Westinghouse situation was terrible, and that he thought the time had come to decide what were the rights of the parties in the event the US Government withdrew any support of Westinghouse. He wants to know (1) what can Westinghouse do, and (2) what can Rolls-Royce do. I told him it was my impression that all Westinghouse could do was continue making the annual payments as required until such time as Rolls-Royce released Westinghouse. Under the agreement the parties can assign the agreement but only with the consent of the other party which consent is not to be unreasonably withheld. Rolls-Royce, of course, might want to act if they can because so long as the Westinghouse agreement exists, if Westinghouse is wholly inactive, Rolls-Royce will be stuck with the impossibility of doing anything with their products in this country.

Gilbert went on to say that the real problem was one of lack of management in the gas turbine division at Westinghouse. Hives had some ideas on who might be suitable, but was reluctant to say anything as he felt it was Westinghouse's business.

There was no improvement by the autumn of 1956. By this time David Huddie was convinced that the arrangement with Westinghouse was holding Rolls-Royce back. He told Phil Gilbert that he was very bearish about Westinghouse and that Rolls-Royce's only way forward was to set up an operating company in the US. Airlines in the USA would all insist that Rolls-Royce could supply spares from depots within the country.

Hives made another visit to the USA in November 1956 and Gilbert talked to him about his reactions just before he flew back to Britain on 15 November. The president of American Airlines, C.R. Smith, had spoken to Hives about his concern over his airline's purchase of the Lockheed Electra with Allison engines, in spite of the assurance of General Motors that it would pour in as much money as necessary for Allison to get the engine up to scratch. Smith told Curtis that he should talk to Hives while he was in the US because 'Rolls-Royce was the only organisation that really knew anything about the business'. Curtis made contact with Hives and the possibility of a tie-up between General Motors and Rolls-Royce was explored, but Hives said he was not in a position to discuss any arrangement with General Motors because of Rolls-Royce's agreement with Westinghouse.

Hives was again depressed by his conversations with the Bureau of Aeronautics. Pratt & Whitney was in a class of its own. General Electric was also highly regarded but this was largely on the reputation of the J-47, which the company had not made for some time. Also in the frame was Allison. Then there was Curtiss-Wright and trailing them all was Westinghouse, Rolls-Royce's partner.

Gilbert tried to push Hives towards clearing up the Westinghouse situation and opening negotiations with General Motors. From what Hives told him of the current engine situation it seemed an ideal situation. Hives told him that the standard engine in the US was Pratt & Whitney's J57 – a large, heavy engine with a rated 10,000 lb of thrust. This engine was suitable for transcontinental service on the Boeing or the DC-8 but was not suitable for transoceanic flight. On the drawing board Pratt & Whitney had a new engine, the J-75, with a claimed thrust of 13,000–14,000 lb. If this were developed successfully it would be capable of transoceanic service. The only current engine capable of transoceanic flight was Rolls-Royce's Conway, which although some years away from full production, had already tested successfully. It was already rated at 13,500 lb thrust and Hives claimed it could be stepped up to 14,500 or even 15,000 lb. General Motors's Allison had nothing except some old J-40s and therefore looked the ideal partner for Rolls-Royce.

In spite of Rolls-Royce's dissatisfaction, the agreement stayed in place for another four and a half years, and it was only in September 1959 that David Huddie, by this time Director and general manager of Sales and Service, Aero Engine Division (Hives had retired in 1956), was able to write to Gilbert telling him that Huggins of Westinghouse had told him that 'they are out of the aero engine business'. Huddie went on to say:

They have verbally freed us to negotiate with anyone in the USA on all our engines, but before we start, could you please give me as soon as you can your legal opinion on the anti-trust aspects of our associating with either Pratt & Whitney or General Electric. All our evidence says that all future Navy or Air Force engine contracts will be awarded to one or other of these two.

The Westinghouse agreement had clearly not lived up to expectation. Indeed, the conclusion that it had held Rolls-Royce back is inescapable. Geoff Wilde, who was called on to help solve Westinghouse's problems with its J-40 turbojet, said to Jim (later Sir Denning) Pearson that he felt the whole exercise had been a colossal waste of time and effort when Rolls-Royce should surely have been concentrating on developing and selling four major new engines – the Dart turboprop, Avon turbojets, the Conway bypass engine and the Tyne turboprop. Pearson could only reply that Westinghouse's royalty payments were a great help in providing finance at a difficult time for the company.

Meanwhile, how did Rolls-Royce fare with the design, development, manufacture and sale of its turboprops, turbojets and new bypass engines? But before we look at those engines, we must remind ourselves that Hives was nearing the end of his career and this is an appropriate moment to summarise his achievements. In the first part of this history, we saw how he reorganised the company when he was appointed general manager and a Director in the mid-1930s, and how he galvanised the company into producing Merlin engines to help win the war while at the same time positioning it to take advantage of the new jet age.

While he was preparing the company to provide the jet engines for the new civil aircraft he also made sure that Rolls-Royce remained a key supplier to the military, selling Merlins to the air forces of Britain's allies after the war. He also negotiated licences around the world. We have just seen the success in the USA where Pratt & Whitney manufactured 1,137 Nenes and 4,021 Tays under licence. Fabrique Nationale (FN) in Belgium built 1,000 Derwents; Hispano-Suiza in France, 1,070 Nenes; the Commonwealth Aircraft Corporation of Australia, 130 Nenes; Hispano-Suiza, 581 Tays; Bristol, 240 Avons; Standard Motor, 410 Avons; Napier, 200 Avons; Svenska Flygmotor AG, 543 Avons; Commonwealth Aircraft Corporation, 178 Avons; and Svenska Flygmotor in Sweden, 560 Avons.

Hives had been offered a knighthood during the war but refused it. He said later:

I do not think I am giving any secrets away now when I say that I was pressed to accept a title in the early days of the war, and later, but I was a 'conscientious objector' to accepting any Honour which was available to the men in the Forces who were making great personal sacrifices. It was because of my objections that I was awarded the Companion of Honour which came within my specification, because it was not awarded to the services. It is generally given to the Church, to Poets and Men of Letters, with of course, Churchill and Attlee. I used to boast that I was the only plumber in the Union but this was disputed by Mr Essington Lewis, the Chairman of the Broken Hill Company in Australia who was awarded it in the same year as myself and was a brother plumber!

On 16 June 1950 it was announced that Hives had been made a Peer. As he felt his job was almost complete, he was prepared to accept the honour and never was one more richly deserved. Much has been written about Hives, especially in the publications of the Rolls-Royce Heritage Trust. Nevertheless, this leading businessman of the twentieth century deserves a full-scale biography.

Chapter Two

Effortless Take-off, Vibrationless Cruising

The Dart
The Avon
De Havilland
The Tyne
The Conway
The Spey
Rationalisation in the Aerospace Industry

THE DART

ONCE ERNEST HIVES HAD DECIDED that Rolls-Royce's future lay with gas turbine aero engines, development was given top priority and, furthermore, the centre for that development was going to be Derby, not Barnoldswick. To gain some experience with the sort of problems that would be met when driving airscrews with turbines a Derwent engine was equipped with a propeller drive as early as 1943 and in this form was known as the Trent. Later, two of these engines were installed in a Meteor which, in September 1945, became the first turboprop aircraft in the world to fly. The Trent ran for 633 hours on test and 298 hours in flight and the information gained on control and gearing difficulties proved very valuable on later types of turboprop.

An important engine along that development road, the Clyde (a two-shaft propeller turbine engine), was a joint effort by a team at Barnoldswick under Stanley Hooker and Adrian Lombard, and a Derby team under Lionel Haworth, one of Rolls-Royce's leading engineers in the post-war period. Roy Heathcote, who enjoyed a long and distinguished career in Rolls-Royce culminating in his appointment as Director of Engineering (Derby), wrote in his book for the Rolls-Royce Heritage Trust, *The Rolls-Royce Dart*:

In many ways the Clyde was ahead of its time, it was certainly a very ambitious engine . . . The first test as a complete engine was on 5 August 1945 and it became the first British propeller engine to complete satisfactorily the then new official combined Civil and Military Type Test. Approval certificates were issued by the Air Registration Board on 22 January 1948 and by the Ministry four days later. However, further development of the engine was abandoned in 1949 for lack of a suitable home.

The other engine on this development path was called the Trent (not to be confused with the supremely successful Trent series of engines manufactured by Rolls-Royce today). This engine was used to gain practical experience of the control problems of a propeller/turbine combination at reasonable forward speeds. Heathcote remembered that:

This was done by installing and then flying the engine in a Gloster Meteor aircraft. The installation work on the aircraft was carried out by the Hucknall Flight Test Establishment and it first flew on 20 September 1945. This flight was the world's first by an aircraft powered by a turboprop engine.

In Heathcote's opinion, Rolls-Royce gained some extremely valuable experience from its work on these two engines. The Clyde showed the effects of gear tooth meshing on the creation of destructive resonances within the reduction gear, while the Trent showed the benefits from changing to helical gearing for the high-speed trains. Heathcote also felt that work on the Trent control system led directly to the interconnected propeller/engine controls and safety locks in the propeller that were a feature of the Dart.

Although the proposal to build the Dart, or the RB 53 as it was known initially, came from Barnoldswick, Hives decided that the main design and development work should be carried out at Derby.

Other the years there has been considerable debate about the origin of Rolls-Royce prefixing its aero engine numbers with the letters RB. The custom began in the summer of 1944. Rolls-Royce had inherited Whittle's method of numbering engines when it took over the Rover gas turbine business but the Ministry of Aircraft Production (MAP) asked for a change to prevent confusion between engine types and bomber aircraft types. Before this the W.2 derived engines produced by Rover, and later by Rolls-Royce, were known by Whittle's type-number with suffixes denoting the sub-type – W.2B/23, W.2B/26, W.2B/37 and W.2B/41. Inevitably these were abbreviated to B.23, B.26, B.37 and B.41 which caused confusion with bomber aircraft numbers used by MAP. The engines were reclassified using the prefix, RB, to denote Rolls Barnoldswick.

Designed by a small team under Lionel Haworth the RB 53, at 1,000 hp, was originally aimed at Ministry of Aircraft Production trainers, the Boulton Paul Balliol and the Avro Althena, which required a single turboprop and three seats, with the intention that a second pupil could look over the shoulders of the instructor while the first pupil sat beside him. However, an engine in this class was also required for the Brabazon Committee's 11B specification for a short-range turboprop airliner, prototypes of which were being built by Armstrong Whitworth and Vickers-Armstrong. When the Ministry ordered two prototypes of the Vickers V.630 Viscount on 9 March 1946, the engine specified was not the Rolls-Royce RB 53 but the Armstrong Siddeley Mamba. Also in the field was Sir Roy Fedden with his axial engine, the Cotswold, and Napier with a 1,600 hp engine called the Naiad. Even at this stage, in spite of Rolls-Royce's undoubted lead in the field of gas turbine engines, competition was already severe.

The Viscount was conceived in response to the recommendations of the Brabazon Select Committee, which deliberated from 1943 until 1945 on the type of civil aircraft that would be required after the war. The Brabazon Committee recommended a '24-seat aircraft, powered with four gas turbine engines driving airscrews for European and other short-to-medium range services'.

In the choice of engine it was Rolls-Royce's experience that won the day. Air Commodore F.R. ('Rod') Banks, by this time Air Ministry Director of Engine Research and Development, felt that Rolls-Royce was the company likely to have the resources necessary for success and awarded it a development contract for the Dart. Hives's decision to move the development work to Derby had probably helped to build Banks's confidence that Rolls-Royce was the right team.

Banks may also have been influenced by the fact that the Dart was simple and used the proven centrifugal compressor technology of the Merlin, Griffon, Derwent and Nene. The competitors were using the still unproven axial compressors.

George Edwards (later Sir George Edwards, OM) wanted the Dart as opposed to an engine based on the Griffith axial compressor. In 1948, he told the Royal Aeronautical Society:

Centrifugal compressors [as superchargers] have been running in piston engines for many years and their rugged construction and freedom from icing will not be present in an axial compressor.

In other words, let's stick with what Rolls-Royce has made to work in practice rather than what it hopes might work, however good it might appear on paper.

Nevertheless there were initial problems. The first Dart to fly in a Lancaster in October 1947 was overweight and underpowered, and Captain Maurice Luby, who succeeded Banks, wanted to cancel the Ministry's commitment to the engine. However, he was constrained by Banks who assured him: 'Rolls will get it right.' In spite of this, there seemed to be no future for the Dart as the Royal Air Force trainer had switched to a piston engine, Rolls-Royce's Merlin 35, and the Viscount was about to be cancelled as British European Airways (BEA), the most obvious customer, said it preferred the piston-engined Ambassador.

Fortunately, at this point, Peter (now Sir Peter) Masefield, joined BEA as chief executive and convinced his chairman, Lord Douglas, that the gas turbine had to be the engine of the future.

Rolls-Royce pressed on, encouraged by Edwards, at that time the general manager and chief engineer of the Aircraft Division of Vickers-Armstrong. Edwards had kept the Viscount development funded.

By this time Edwards had emerged as a very powerful figure within the British aircraft industry. Born in 1908, he was educated at the South-West Essex Technical College and at London University where he gained his BSc (Eng). After seven years in general engineering he joined Vickers Aviation Ltd. in 1935 and worked on the development of the Wellington bomber. In 1939 he was appointed experimental works manager and when Vickers's chief designer, Rex Pierson, died in 1945, Edwards succeeded him. Alive to the possibilities of turbine-engined aircraft, he wanted the Dart as the power unit for the VC2, or Viceroy, ultimately the Viscount. (When India gained its independence in 1947, Viceroy seemed an anachronistic title.)

Once Vickers had decided that the Dart was the engine for its VC2, it was inclined to push Rolls-Royce into making commitments before the necessary development work had been complete. On May 8 1947, Sir Hew Kilner, the managing director of Vickers, wrote to Hives:

You will have received from us an invitation to quote for the supply of 'DART' engines for incorporation in the V.C.2., . . . and asking whether you are prepared to give a guarantee of performance.

We are asking this because we have been invited to make a firm tender to the Ministry of Supply for the V.C.2. in quantities . . . I shall, therefore, be interested to learn whether you are able to make a firm quotation of price, delivery and performance.

But Hives was too experienced to give guarantees before he could be sure of delivering them. He replied:

Much as we would like to, it is quite impossible at the present time for us to quote price and guarantee performance. The position of the Dart is that, although we have had our usual troubles, the engine is making steady progress but that means of course it is a long way off being an engine we should recommend for a civil airline. We should require at least another six months' development work before we could with any confidence give you the information you are asking for.

There is no doubt we have all been much too optimistic as regards the development time required for gas turbine engines.

We have already seen that Hives had hoped this would be the case so that few firms would have the necessary skills to compete.

Encouraged by Edwards's support, Rolls-Royce set to work to eradicate the problems. In reducing the weight from around 1,100 lb to 800 lb, a large redesign exercise was necessary and one of the biggest changes made was in the reduction gear. The original compound star arrangement with three layshafts, in which both high- and low-speed trains were spur gears, was replaced by one in which the high-speed train was helical. Rolls-Royce took the decision to work this gear at higher stress levels than the all-spur gear and was able, as a result, to use a much lighter gear.

Work on the air intake casing also produced a saving of weight. As Heathcote recalled:

In a particularly ingenious piece of design, the tank was made integral with the air intake casing by forming it between the outer wall of the intake passage and the cylindrical wall that comes rearward to the middle of the front vertical wall of the low pressure compressor casing. In effect, this redesign killed three birds with one stone because not only was there an overall weight saving, but the hot oil kept the intake passage free of ice and the outer wall acted as a form of prop which helped stabilise the compressor casing wall.

At the same time as the weight problem was tackled, the Rolls-Royce designers went to work to increase the power – an exercise made even more necessary because of the extra power demanded by the Viscount.

The prototype Viscount made its first flight on 16 July 1948 from the grass strip at Wisley, near Weybridge, and was flown by Captain J. Mutt Summers, Vickers's chief test pilot. (We have already seen in the first volume of this history, *The Magic of a Name, The Rolls-Royce Story, The First 40 Years*, that Summers had flown the first Spitfire on its maiden flight in March 1936.) After a continuous flight-testing programme, BEA introduced the Viscount on to its London–Paris and London–Edinburgh services in July 1950. It was the first aircraft powered by a propeller turbine – indeed, by a gas turbine of any sort – to be used on a scheduled passenger service. The original Brabazon idea of a twenty-four-seater aircraft had been abandoned for a thirty-two-seater, and Vickers soon developed a forty-seven-seater.

The Viscount was a breakthrough in passenger air transport, surpassing anything seen before in terms of comfort and reliability. Publicity was achieved not only by entering an England–New Zealand air race, but also, and more significantly, by the use of photographs showing how free the aircraft was from vibration in flight by placing both a threepenny bit and a pencil on their ends.

Nevertheless, it was not absolutely certain that BEA would buy it. Peter Masefield said later:

The Viscount would not have gone ahead if B.E.A. had not grasped it. They were going to turn it down and go instead on to the Ambassador on the grounds that it was unwise to gamble on what was then an unknown turbo-prop. Hives told me soon after I became chief executive of B.E.A. that they were going to give up the Dart unless we ordered it within the following fortnight. I recommended to Sholto Douglas that we did it, and I then sat down with George Edwards of Vickers and said the Viscount 630 then on the drawing board was too small with only twenty-four seats, while Hives was told that the engine would have to have a big increase in horsepower. From this emerged the Viscount 700 with an initial forty seats.

The minutes of a Rolls-Royce board meeting on 15 January 1948 noted:

The choice of British European Airways of the Ambassador in lieu of the Vickers Viscount had created strong feelings of resentment in the Vickers company . . . Vickers were anxious to ventilate their grievance in the press and had asked us to join them but the Managing Director stated that in his view it was inadvisable for the company to do so.

In the summer of 1950, BEA ordered twenty V.701 Viscounts (an order that was later increased) and by 1953, orders were flooding in for the Dart-powered aircraft. In its 'Review of Industry' in August 1953, *The Times* wrote:

A third of BEA's total network – about 5,000 miles of unduplicated routes – is now being flown by Viscounts . . . Since it was first introduced on BEA's services on April 18 last the Discovery class Viscount 701, with its smoothness, quietness and relatively high speed, has proved not only popular with passengers but profitable for the operator . . . Peter Masefield (Chief Executive of BEA) praised the reliability of the airline's Rolls-Royce Dart engines which, he said, had been 'remarkably free of troubles'.

The Viscount cut thirty minutes from the London–Stockholm journey and, as *The Times* put it:

The inaugural flight in both directions was characterised by the smoothness, effortless take-off and the quiet, almost vibrationless cruising which are already coming to be regarded as characteristic of the Viscount.

The biggest boost to the Viscount's sales prospects from round the world came from Vickers's success in selling it in North America. In 1952, Edwards went to Montreal and secured an order from Trans-Canada Airlines for fifteen Viscounts. The airline eventually bought fifty-one. In 1953 an upgraded Dart, the Mk 510, was introduced and this made the Viscount even more attractive to a US carrier, Capital Airlines, with its predominantly short-haul routes.

In 1954 the sales team at Vickers, led by Edwards, achieved a break-through when they sold three Viscounts, with an option on thirty-seven more, to Capital. The Viscount proved to be the perfect solution to enable the airline to compete with Capital's much larger rivals. *Aero Digest*, the US magazine, was full of praise both for the president of Capital, J.H. ('Slim') Carmichael, for his courage in buying outside the United States of America and for the Viscount itself, especially the Rolls-Royce Dart turboprop, which 'solves the [noise] problem at its source. The noise and vibration are practically nil in flight. This means a virtually irresistible passenger appeal, as well as crew comfort and increased efficiency.'

The US journal, *Aviation Week*, wrote:

The purchase could prove a significant blow to the US aircraft industry, particularly since it is the second time in two years a foreign manufacturer has cracked the market in the country with quantity orders of modern transports. Pan American Airways has ordered three de Havilland Comet 3s, with an option for an additional seven . . . The British were quick to press their new-found advantage by promising early delivery of the first three Viscounts. Vickers reports delivery had been hastened by British European Airways' consent to postponement of three of its Viscounts in favour of Capital. BEA notes it is doing this 'in the national interest'.

By the end of 1954, Capital turned its option on thirty-seven Viscounts into a firm order and took an option on twenty more, bringing the total order for Viscounts to 177 valued at £53 million (about £1.3 billion in today's terms).

Some controversy arose as reports circulated that Rolls-Royce had given a promise to George Edwards that the company would not sell the Dart to anyone else. On 17 December 1954, Phil Gilbert (Rolls-Royce's legal adviser in the USA) telephoned Hives to tell him that *Time* magazine had asked Gilbert to confirm a report in the *Wall Street Journal* that Rolls-Royce had agreed with Capital Airlines not to sell Dart engines to anybody else in the USA.

Hives denied this was the case, but pointed out that production was entirely taken up for the Viscount. As a result, all other aircraft manufacturers had been told that no Darts would be available until January 1958. US manufacturers wanted the improved Dart 2, whereas Vickers was currently buying the Dart 1. When the Dart 2 was available in January 1958, Vickers would have first crack and other manufacturers could then follow.

Hives told Gilbert he understood that Edwards had agreed with Capital that he would not sell the Viscount to any other US airline for six months after delivery of the order to Capital. This was an agreement between Vickers and Capital, and Rolls-Royce was not a party to it. Trans World Airlines (TWA) was interested in the Viscount and wanted one for testing. This would probably be supplied by TCA so that the agreement was not broken.

Jim Pearson elaborated on Hives's telephone comments in a letter to Gilbert on 29 December 1954 in which he wrote:

I do not know that I can add anything [to Hives's comments] except to say we consider this a perfectly normal commercial undertaking, which we have given to Vickers partly to induce them to spend large sums of money on expansion of Viscount manufacturing facilities, which of course in turn increases the market for our engines. There is no question of any sort of monopoly, because Westinghouse are perfectly free to make the same engine any time they like without any restrictions. Furthermore, there are competing engines of somewhat similar size available both in this country and in America.

The fact that we have offered our very latest engine, the RB.109 [which became the Tyne], to Douglas, illustrates that we have no restrictive thoughts about supplying our engines to America, in fact the very opposite. All we have done with the Dart is a perfectly normal bit of commercial business.

And the development continued. The Dart was uprated again, the Viscount was lengthened and the result was the Viscount 802, which could carry seventy passengers. This was followed by the Viscount 810 with the Dart RDa.7. In ten years the Viscount had moved from an aeroplane weighing 34,000 lb, carrying twenty-four passengers and cruising at just 300 mph, to one weighing 72,500 lb, carrying 70 passengers and cruising at 350 mph.

As Stewart Wilson wrote in his book *Viscount, Comet and Concorde*:

The world's first production turbine powered airliner, the Vickers Viscount, remains Britain's most numerically successful airliner.

Ordered by some 60 operators around the world, the Viscount was very much an aircraft of its time, bringing revolutionary standards of comfort to its passengers and outstanding operating economics to its owners. The resale of second hand examples added substantially to the number of operators who flew Viscounts at some stage, the tally finally reading over 200, including many military and corporate operators. At its peak, it was estimated that somewhere in the world, a Viscount was taking off or landing every 27 seconds.

As for the Dart engine, Wilson wrote:

Fundamental to the success of the Viscount and many other aircraft was its powerplant, the Rolls-Royce Dart turboprop.

The initial Dart engine, first bench-tested in 1946, produced 990 hp. Nine variants later it produced 3,400 hp. In its forty-year production cycle 7,100 engines were built, powering such aircraft as the Viscount, as well as the Fokker F27 Friendship, Hawker Siddeley HS.748, Hawker Siddeley Argosy, Handley Page Herald, Grumman Gulfstream 1, NAMC YS-11, Fairchild Hiller FH227, AW Argosy, Breguet Alizé and Convair 600.

THE AVON

We have already seen how Dr. A.A. Griffith had advocated the axial as opposed to the centrifugal compressor, and in 1945 the AJ-65 ('A' for axial, 'J' for jet and '65' for 6,500 lb thrust) began to evolve. The first engines were known as RA.1s. Stanley Hooker was asked by Hives to design and develop the engine at Barnoldswick.

In his excellent book *Rolls-Royce, Aero Engines*, Bill Gunston wrote:

By 1945 Hives was champing at the bit to get a world-beating axial into production. Throughout the war Griffith had openly scorned anything with a centrifugal compressor and inevitably this made Hives regard even the Nene and Tay as mere temporary stop-gaps. Tens of thousands of these centrifugal jets were made outside Britain, but the company that designed them hardly pushed them at all, and in any case from 1945 until 1950 suffered a near-desert in procurement of British combat aircraft. Hayne Constant at the National Gas Turbine Establishment and the technical experts at the Ministry all urged Hives to get cracking on axials.

And, as Gunston pointed out, there was no doubt that an axial turbojet would give more thrust per unit frontal area. But producing a practical and commercial axial jet engine was not going to be easy.

Hooker wrote later in his book, *Not much of an Engineer*, that he was not altogether happy with being given this assignment.

At that time Rolls-Royce had no real experience in the design of axial compressors, and I hated moving into the unknown on what was obviously such an important innovation in the new field of turbine engines. The main supporters of axial compressors were Constant [Hayne Constant, Director of the Engine Research Department at the Royal Aircraft Establishment during the war], and Griffith. The two companies, which had toyed with axials, were Metropolitan-Vickers at Manchester and Armstrong Siddeley at Coventry. Metrovick had followed the RAE lead, and with Dr D M Smith in control had made the F.2 turbojet, designed for 2000-lb thrust. Although this engine flew in 1943 in the F.9/40 Meteor, its program had been relatively slow. Apart from its smaller diameter it showed no advantages from the supposed higher efficiency of the complicated axial compressor.

Hooker remembered having many discussions with Pat Lindsey – whom he described as 'their brilliant young engineer' – after Armstrong Siddeley had taken over Metrovick's jet engine development under pressure from the Government, and becoming convinced that there were many aerodynamic and mechanical problems to be overcome before the axial compressor could be considered as a sound substitute for the reliable and proven centrifugal compressor. The problems were compounded by Griffith's proposal of a 7:1 pressure ratio compared with Whittle's 4:1.

Here is Hooker again on the problems they encountered:

We started the design at Barnoldswick, and it was not long before Lombard was in my office saying that there was no way we could meet the projected weight of the engine – in fact we would be 50 per cent heavier than Griffith's optimistic estimate. At the same time Harry Pearson was worried about the high pressure ratio, and so I took the decision to drop this to 6.3:1, and told how the weight must stay at his design estimate. Morley, who was doing the mechanical design of the compressor, was mindful of the fact that the blades all stuck out like tuning forks, and that resonant vibrations was a major hazard, causing the blades to break off at the roots.

In spite of the complexities, an AJ-65 engine was put on test at Barnoldswick in 1946 but proved difficult to start, would not accelerate, broke its first-stage blades and was only gradually coaxed up to 5,000 lb thrust.

Hives, unimpressed by the delays, decided to bring the project back to Derby. Apparently his impatience with Barnoldswick was accentuated when he learned that Marshal of the Royal Air Force, Sir Charles Portal, had called at Barnoldswick and not Derby. He said:

We can't have that. **Derby** is the centre of Rolls-Royce, not Hooker's bloody garage at Barnoldswick.

The Avon made its first public display in an Avro Lancastrian flown by

Captain R.T. Shepherd, Rolls-Royce's chief test pilot, at the Society of British Aircraft Constructors' Display at Farnborough in 1948 and in the following year it appeared again, this time in the Canberra bomber built by English Electric. However, what really stole the show was a Gloster Meteor with two Avon engines installed at Hucknall, which displayed a climbing power unmatched by anything else from the UK or from the USA or from the Soviet Union. The outbreak of the Korean War in 1950 brought a reversal of the running-down of military orders, and the Avon was soon being manufactured not only in Rolls-Royce's factories in Derby and Hillington but also, at the Government's insistence, under licence at the Bristol Aeroplane Company, at Napier and at the Standard Motor Company. At the same time an overseas licence was given to the Commonwealth Aircraft Corporation, which manufactured Avons for Australian-built Canberras and North-American F-86 Sabres. Avons were also sent to Canada to fly in the Avro Canada CF-100.

And from 1950 onwards, the Rolls-Royce Avon was chosen to power many of the country's most important military aircraft. As well as the Canberra, it powered the Vickers Valiant bomber and the Hunter and Swift fighters. Additionally, it was licensed in Sweden for production by Svenska Flygmotor.

At the Farnborough Show of 1951, the Avon was seen in the Hawker Hunter and the Vickers Valiant prompting the *New York Herald Tribune* to write:

The cold facts are that England demonstrated the finest jet fighter and jet bomber operating in the world today.

On the civil front, de Havilland ordered the Avon for its Comet series to be delivered in 1954 following the enormous success of its initial Comets.

Joe Sutter, who became chief design engineer at Boeing and who is best remembered for his work in developing the 747, said of the Comet:

Boeing came to look at the Comet at Farnborough. It made such a big impression they decided that the civil business was the place to be. We built the 367-80 prototype or Dash-80 where I was in charge of the aerodynamics. The Dash-80 became the 707.

When George Schairer, Boeing's supreme aerodynamicist and a man responsible as much as anyone for taking Boeing into the jet age, showed Lord Hives the 707, Hives said: 'This is the end.'

'The end of what?' asked a puzzled Schairer.

'The end of British aviation.'

DE HAVILLAND

The Comet story is an appropriate place to trace the history of the de Havilland company whose aircraft interests were destined to be absorbed by Hawker Siddeley and whose engine interests went to Bristol and hence to Rolls-Royce.

Geoffrey (later Sir Geoffrey) de Havilland was one of the pioneers of the aviation industry. He was working as a draughtsman at the Motor Omnibus Construction Company of Walthamstow in north London when, in 1909, he was bitten by the flying bug. Unfortunately he had no money and his father, a modestly paid rector, could not help him. More fortunately, his maternal grandfather, an Oxford businessman called Jason Saunders, was prepared to back him and invested £1,000 (about £100,000 in today's terms).

De Havilland took premises in Fulham in south-east London and with his friend (later his brother-in-law), Frank Hearle, a mechanic with the Vanguard Omnibus Company, he began to design not only his first aeroplane but also the four-cylinder, water-cooled 45 hp (35 kW) engine to power it. The Iris Motor Company built the engine for him for £250 (about £25,000 today).

By November 1909, de Havilland's Aircraft Number One, a canard biplane with a single engine driving two propellers through bevel gears, was ready but unfortunately the first flight ended in a crash in which de Havilland was hurt – though not seriously.

The second flight in September 1910 was more successful and de Havilland, running short of money, was relieved to sell the aircraft to the Government Balloon Factory at Farnborough. De Havilland went to Farnborough himself to supervise production under the control of what became known as the British Aircraft Factory. He remained there until just before the outbreak of World War I in August 1914. One of his designs was the BE.2 reconnaissance scout, which first flew in 1912 and saw action with the Royal Flying Corps in the early part of the war. Some 4,000 BE.2s were built.

De Havilland served for a short time as inspector of aircraft in the Aeronautical Inspection Directorate, but quickly returned to his first loves – designing and flying – with the Aircraft Manufacturing Company (Airco). His designs were acknowledged by having the 'DH' prefix attached to the model number of each type. And he enjoyed some great successes including the DH.4 two-seat, single-engined bomber, of which 1,449 were built in Britain and 4,846 under licence in the USA. Some 5,500 of a modified and more powerful version of the DH.4, the DH.9, were built, serving with several air forces and being used in commercial operations by a number of airlines. The world's first international air service – between London and Paris – was flown on 25 August 1919 in an Air Travel and Transport DH.4A.

The end of World War I brought the inevitable sharp downturn in orders for aircraft companies and although Airco was taken over by the Birmingham Small Arms Company (BSA) this was only for its plant, not for its flying activities. De Havilland took the aviation assets and set up the de Havilland Aircraft company on 25 September 1920 at the Stag Lane airfield in north-west London. The company would later move to Hatfield, north of London, where it would remain a production site until closed by British Aerospace in 1994.

In the 1920s de Havilland designed and built a number of aircraft and, to keep the cash flowing, also flew charters and dusted crops. Its most success-ful aircraft of that era was the two-seat light biplane, the DH.60 Moth, the aircraft that really made the company as it revolutionised private flying in Britain and elsewhere. In his book, *Boxkite to Jet*, Douglas Taylor explains how Frank Halford worked with Geoffrey de Havilland to buy old Renault V8 engines from the Air Disposal Company and use them to design a four-cylinder engine which they called Cirrus. It was this engine which powered the DH.60 Moth which first flew in February 1925. The DH.60 was followed by the DH.83 Fox Moth, the DH.85 Leopard Moth and the DH.87 Hornet Moth. The most successful was the DH.82 Tiger Moth, which was used as the standard *ab initio* trainer of the Royal Air Force and other Com-monwealth air forces during World War II, when it was manufactured in large numbers not only in Britain but also in Canada, Australia and New Zealand.

During the 1930s, de Havilland designed and produced no fewer than fifteen new aircraft types – including the twin-engined biplane DH.84 Dragon and the DH.89A Dragon Rapide. It was in a Dragon Rapide that General Franco returned from the Spanish Army in North Africa to take control of the Nationalist forces in the Spanish Civil War in 1936. [Here at Duxford, a Dragon Rapide is still used to give visitors a flight 'round the bay'.]

As well as the Dragons, de Havilland produced the DH.90 Dragonfly (one of the first aircraft intended for executive use), the four-engined DH.86 airliner, the twin-engined DH.88 Comet (long-range race-winner of the 1934 England to Australia air race), and the four-engined DH.91 Albatross (an airliner that could carry up to twenty-three passengers). All of these aircraft were of mainly timber construction, but in 1938 de Havilland produced its first all-metal airliner, the DH.95 Flamingo. During this period, de Havilland was not only building all these aircraft, but also powering them with its own Gipsy engines. Halford again helped with the design. Initially an alternative to the Cirrus, once uprated the Gipsy became de Havilland's prime power plant throughout the 1930s and into the Second World War. As Peter Stokes said in *From Gipsy to Gem*, 'The seal was set on the Gipsy's prime feature, reliability'. At the same time it was producing thousands of propellers both for its own and other manufacturers' aircraft.

During World War II, de Havilland concentrated on manufacturing the versatile Mosquito. Some 7,781 were built in Britain, Canada and Australia in its various forms as a bomber, night fighter, fighter-bomber/interdictor, photographic reconnaissance aircraft and trainer. As with Armstrong Siddeley, de Havilland also became involved with jet development. Its Vampire of twin-boom design first flew in September 1943 and was eventually developed into a fighter, night fighter, fighter-bomber and trainer, remaining in production in Britain and several other countries until the late 1950s.

The Vampire was powered by the de Havilland Goblin, a centrifugal flow turbojet. The designer of the Goblin was Major Frank Halford, who had begun work on the engine in 1941 following a visit to Cranwell with Geoffrey de Havilland to see Frank Whittle's work and also to see a flight by the Gloster E.28/39. Halford moved quickly. The first drawings of his jet engine were issued in August 1941, a prototype was run on the test bed eight months later and a month after that the British Government gave permission for the development of an aircraft that would be powered by the new engine, the Vampire.

By the beginning of 1943 the Halford H.1 was ready for flight-testing and, as the Vampire was not ready, two engines were installed in a Gloster F9/40 Rampage later to be renamed Gloster Meteor. At this stage the Whittle engine designed for the Meteor was not ready either. When the Meteor flew on 5 March 1943 at Cranwell it was powered by a Halford H.1/de Havilland Goblin jet engine. The first Vampire flew in September 1943, and in October 1943 a Goblin was sent to Lockheed in the USA and flew in the prototype XP-80 Shooting Star in January 1944. As a result, the initial flights of the first production jet fighters in both Britain and the USA were powered by Frank Halford's de Havilland engine.

After World War II, de Havilland's frenetic activity continued. As well as the DH.100 Vampire there were the DH.103 Hornet and Sea Hornet, the DH.104 Dove light transport, the DH.108 tail-less jet research aircraft, the DH.110 all-weather fighter (later developed into the Sea Vixen naval fighter), the DH.112 Venom fighter-bomber and Sea Venom all-weather naval fighter, the DH.113 Vampire night fighter, the DH.114 Heron feeder liner and the DH.115 Vampire trainer. At the same time de Havilland built the Chipmunk trainer designed by its Canadian subsidiary. It was also building its range of jet and piston engines, propellers, rocket engines and air-to-air missiles such as Blue Jay (later called Firestreak) and the Blue Streak long-range ballistic missile.

As Ronnie Harker pointed out in his book *Rolls-Royce from the Wings*:

On the question of rivalry, the main competition at the time [in the years immediately after the war] was between de Havillands and Rolls-Royce. Bristol came in rather later on the turbine scene. De Havilland under the leadership of Major Halford and

Doctor Moult had designed the Goblin engine, the dominant feature of which was the single-sided impellor on its centrifugal blower which drew air direct from the forward facing air intake so that there was little interference. Rolls-Royce favoured the double sided impellor, which required a plenum chamber to collect the air which entered the engine on both sides of the impellor but which enabled the engine diameter to be kept smaller. Not much was known at the time concerning aerodynamics of air intakes, thus there was much scope for debate and salesmanship as to which was the better method. In practice the installed thrust of the Goblin was in the order of 3,500 lb which it also gave on the test bed, whereas the Nene gave 5,000 lb thrust on the test bed and only 3,500–4,000 lb when installed in the aircraft. The discrepancy was due to intake losses. On the other hand the Nene was considerably smaller in diameter and so the nacelle drag was lower. When installed in the Vampire both engines gave similar top speeds, although the Nene climbed better. On faster aircraft the Nene gained an advantage as its smaller diameter and less drag was of greater importance.

This resultant competition between the two firms kept everybody engaged in sales promotion on their toes.

In the competition to decide which engine – the Rolls-Royce Nene or the de Havilland Goblin – should power the RAF Vampires, the Goblin was chosen, according to Harker, because of:

. . . the extra air intakes, known as 'elephant ears', that were required on the Nene which caused buffeting in flight, thereby reducing the Mach limitation.

The de Havilland Comet project had begun life in 1943. Indeed, it was considered by the Brabazon Committee, which reported to the Cabinet in that year. The Committee called this craft the Type 4, and assumed it would be merely a carrier of mail. However, the progress made with the jet engine meant that by the end of 1945, the mail-carrying concept had been abandoned in favour of an airliner, coded the DH.106, which would carry twenty-four passengers non-stop from London to Newfoundland against the strongest headwinds.

In late 1947 the DH.106 was formally given the name Comet. De Havilland realised that an axial-flow engine, probably the Rolls-Royce Avon with its greater growth potential and better specific fuel consumption, would be the right engine for the Comet in the long run, but in the short term, while such engineers were still being developed, the de Havilland Ghost was used as the power plant. The four engines were mounted in pairs close together and buried in the wing roots. This meant less drag but also structural complexity in the wing. Additionally, there was a tendency for the cabin to be noisier than with more conventional engine configurations.

John Cunningham, the de Havilland test pilot, took the aircraft on its first flight on 27 July 1949 and said later:

The flight was an absolute joy. It all went like clockwork. We climbed to 10,000 feet and landed after 31 minutes. I was entirely happy with all that had gone into it, and thought that it flew just like a great big fighter. My overall impression was that here was an aeroplane which was so far removed from the grinding piston-engine machines that, if somebody could make it work, here was the next stage of aviation.

On 27 August 1953, Cunningham flew the Comet 2 with four Rolls-Royce Avon engines. As we saw, the Comet was initially powered by four de Havilland Ghost engines but all subsequent developments, including the Comet 2, were powered by the more powerful and less thirsty Rolls-Royce Avons. It was the first of twelve ordered by BOAC to use on their South American routes in 1954. The chairman of BOAC, Sir Miles Thomas, said of the flight from Zurich to London:

On this record-breaking flight we were not even trying. This new Comet promises to surpass even the world-beating Comet 1. It is extremely smooth and very quiet in the air.

The previous autumn, Pan American Airways (Pan Am) had ordered three Comet 3s that would be half as big again as the Comet 1 and Pan Am's President and Chairman, Juan Trippe, said:

The Comet III will be capable of carrying a full load of passengers, mail and cargo for 2700 miles against the 50 mph headwind with adequate reserves and will be the first jet transport able to operate efficiently over the principal routes of the Pan-American system.

This order from Pan Am suggested that de Havilland would have broken into the world market with the Comet once it had developed the aircraft sufficiently to carry seventy-six passengers. The original Comet, powered by de Havilland's own Ghost engines, was limited to carrying only thirty-six passengers at 460 mph for 2,000 miles. Nevertheless, at the time it was more attractive than piston-engine-powered aircraft, and BOAC operated it profitably between 1952 and 1954. However, other airlines preferred to wait for the more powerful developments.

In the event, they had to wait a long time as tragedy overtook the aircraft, de Havilland and indeed the British aircraft industry when in January 1954 a Comet, taking off from Rome airport, unaccountably plunged into the Mediterranean, killing everyone on board. All existing Comets were grounded and minutely inspected, but nothing could be found to explain the crash, so on 23 March permission was given for the Comets to resume service. Within days, another Comet crashed, by coincidence also after taking off from Rome.

In his autobiography, *Sky Fever*, Sir Geoffrey de Havilland wrote:

Looking back, I think most of us at Hatfield were – perhaps fortunately – too numbed to realise fully the vast difficulties that lay ahead . . . The Air Ministry were most helpful in making available research facilities at Farnborough. The Royal Navy started the immensely difficult task of salving the wreckage of the first Comet . . . It is impossible to over-emphasise the achievement of the Royal Navy in overcoming the vast difficulties involved . . . Hard work and great patience resulted in the essential parts being recovered. These were sent to the Royal Aircraft Establishment at Farnborough where the almost impossible task of identifying hundreds of pieces of battered metal and placing them in their correct relative positions progressed for many months.

However, the solution came not from analysing the pieces of metal but from subjecting a Comet fuselage to continuous pressure in a specially constructed water tank. De Havilland continued:

One day Bishop and I flew over to Farnborough to see how the tests were progressing. Sir Arnold Hall was then Superintendent of the Royal Aircraft Establishment and had organised with his staff the vast job of testing out the wreckage, building the tank and doing much other work in a miraculously short time. We were talking in his office when he was rung up and told that the pressure cabin in the tank had failed after a period representing nine thousand hours' normal flying. We went at once to examine the failure, and found a rent in the side on the cabin which appeared to start from a rivet at the corner of a window-frame. The rent was repaired and the test resumed, and a similar failure occurred . . . we had reached the end of the trail. There was a definite evidence of weakness in the cabin structure.

De Havilland worked furiously to correct the fault, but it was four years before the Comet 4 could be launched, by which time the Boeing had launched the 707 and Douglas the DC-8. The Comet's lead had been lost.

The 707 first emerged as a sketch doodled by some Boeing engineers on a rainy Sunday afternoon in 1949. It went into cold storage for two years and was dusted off in 1951 when Boeing realised the US Air Force would probably need a jet-powered tanker able to refuel in mid-air its B52 bombers, which were due to enter service in 1955. In April 1952, the Boeing board sanctioned a prototype for an aircraft that could not only be used by the United States Air Force, but which could also carry 139 passengers at speeds up to 600 mph.

The gamble paid off. The USAAF bought the 707, to be called the KC135 in military service. On 13 October 1955, Pan Am Chairman Juan Trippe announced an order for twenty 707s. It was a cruel blow for de Havilland. Pan Am had been an early customer for the Comet. Although United ordered thirty of the rival Douglas DC-8, American Airlines ordered thirty 707s when Boeing committed itself to a long-range 707, the Intercontinental. Orders flooded in for the 707 from the major airlines of the world – Continental, Braniff and TWA in the USA, Sabena in Belgium, Air France, BOAC, Lufthansa, Qantas, and Air India. At the same time, Douglas captured Eastern, National and Delta in the USA, KLM in Holland, Scandinavian, Swissair and Japan Air Lines for its DC-8.

There can be little doubt that the Comet would have been bought by many of these airlines but for the crashes. Pan Am, definitely a leader that others tended to follow in the 1950s, had taken out purchase options. It would have put the Comet into the all-important US market and much would have followed from that.

The Comet did come back and a greatly improved Comet it was, with Rolls-Royce Avon engines as opposed to Ghost engines and a new fuselage with the latest fail-safe techniques up to stringent new standards laid down by the Air Registration Board. It could carry more people, faster, over longer distances. But it was too late. Only seventy orders were forthcoming – nineteen from BOAC, fourteen from BEA, twenty-six from the RAF and a few more from small airlines.

THE TYNE

In October 1955, when the Vickers factories were working flat out to produce the Viscounts ordered, the Viscount's successor, the Vickers Vanguard, was announced. The design was selected following close consultations between Vickers and BEA. Peter Masefield said that a high-wing configuration had been considered – passengers liked it because it gave them a good view – but abandoned because of the difficulties of arranging undercarriage retractions. George Edwards said that they looked at sixty variants before opting for the Vanguard powered by Rolls-Royce Tyne turboprop engine RB 109. The aircraft would be able to carry about 100 tourist passengers or forty first-class and forty-five tourist-class passengers. The Tyne was expected to give 2.8 times the cruise power of the Dart when it first went into airline service, with 40 per cent lower specific fuel consumption.

Edwards had looked at several engine possibilities, including both turbojet and bypass engines, and had chosen airscrew turbine power. The Rolls-Royce Tyne fitted the bill and Rolls-Royce's reputation with Vickers was secure after the continuing success of the Dart, which was expected eventually to give the Viscount a speed of 400 mph.

The Tyne, which made its initial test run in the spring of 1955, was a two-shaft high-pressure-ratio engine that developed 4,020 shp plus 1,175 lb of thrust at take-off. At a cruising altitude of 25,000 feet and a speed of 425 mph, the total power was 2,470 hp of which 2,270 was shp and 150 lb jet thrust. The Tyne's specific weight was 37 per cent less than the Dart, but its cruising power was 2.8 times greater.

Vickers claimed that the Vanguard, with a range of 2,500 miles, could operate 86 per cent of the world's international air services, including all the major US domestic services. And on 4 January 1957, Edwards was able to announce Britain's biggest single dollar-earning export order when Trans-Canada Airlines ordered twenty Vickers Vanguards. The decision was taken after one of the most exhaustive evaluations over a period of two years, including tests in both US and British airframes, and with both turbo-propeller and pure jet engines.

Not everyone at Rolls-Royce was in favour of developing the Tyne for the Vickers Vanguard. Lionel Haworth, for instance, was one of the dissenters. He remembered a meeting in the boardroom at Elton Road with Hives, Albert Elliott, Arthur Rubbra, David Huddie, Adrian Lombard and Lombard's engine performance and preliminary design teams present representing Rolls-Royce; while Sir George Edwards, Hugh Hemsley, Basil Stephenson, Eric Allwright and the performance and preliminary design teams were there from Vickers.

Lombard emphasised the low fuel consumption and the Vickers team spoke of the requirement of Bob Morgan, chief engineer of British European Airways, for greater capacity than the Viscount and a large freight hold. Lombard also emphasised the low cost per seat per mile flown. Haworth recalls, commenting:

I think we should make a bypass engine for the Viscount successor. We have just done the design of the RCo3 and, if we made another bypass engine, we could have it flying in three years or so. The engine we are discussing here today is a big job. At the proposed higher cruising flight speed the control of the propeller itself presents a major problem to keep it safe in all the failure modes, and I guess it will take eight years to get it right.

Hives reminded him that the Dart with propellers had done well and questioned why should they abandon this successful formula. Haworth replied:

Vickers and Rolls-Royce were in front with the Viscount; but the axial flow bypass engine is the engine of the future, and it is easier to do. If we adopt it now we will be out in front again.

A Vickers man reminded him about the 'seat mile cost'. Haworth countered

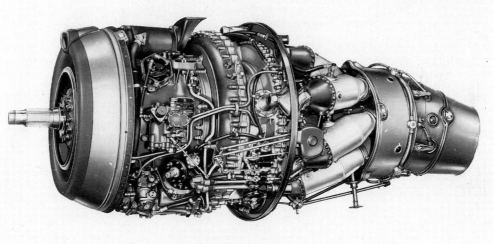

TOP: A British European Airways Viscount, powered by the Dart. *The Times* said of it: 'Characterised by the smoothness, effortless take-off and the quiet, almost vibrationless cruising which are already coming to be regarded as characteristic of the Viscount.'

BOTTOM: The Rolls-Royce Dart. Originally sponsored for a military trainer, its crucial application was in the Vickers Viscount. George Edwards of Vickers wanted the reliability of Rolls-Royce.

Following the success of the Viscount, Vickers saw
the opportunity for a big brother and launched
the Vanguard (OPPOSITE). Rolls-Royce designed the
Tyne (ABOVE) which offered unrivalled levels of
fuel efficiency and power. Unfortunately, by the time
the Vanguard was launched, the public had decided
it preferred jet-powered aircraft.

An early member of the Avon family of the type which
powered the English Electric Canberra and Hawker Hunter
(SEE OPPOSITE).

OPPOSITE TOP: An English Electric Canberra.

OPPOSITE BOTTOM: A Hawker Hunter.

The test pilot, Neville Duke, with his world air speed record
Hawker Hunter in 1953.

A Supermarine Swift, powered by the Avon. Although it
wrested the world air speed record from the Hawker Hunter later
in 1953, it never achieved the success enjoyed by the Hunter.

The Fairey Delta 2, with a reheat Avon, established a new
world air speed record of 1,132 mph in 1956, flown by
Peter Twiss. This was the first record over the 1,000 mph mark.

OPPOSITE: A later Avon (BOTTOM) powered the English Electric
Lightnings (TOP).

The de Havilland Comet 4 with Rolls-Royce engines.
It was a superb aircraft, establishing the first transatlantic
jet service, but the Comet's lead had been lost with the
tragic crashes in 1954. By 1958 when the Comet 4
was launched, the Boeing 707 and Douglas DC-8 had
established themselves.

The Conway was the world's first bypass fan jet
to enter service.

The Conway RCo12 (shown on the previous page) powered
versions of the Boeing 707 (TOP) and the Douglas DC-8
(BOTTOM). It was the first Rolls-Royce engine selected by a US
aircraft manufacturer.

The Handley Page Victor 'V' bomber was powered by the
Conway RCo11.

The VC10, powered by four Conways, suffered
as a political football and never achieved the sales
its performance deserved.

The RB 163 Spey (BOTTOM) succeeded the RB 141 Medway
when BEA decided the original DH.121 airliner was too big and
opted for the smaller Trident (TOP).

TOP: As well as the Trident, the Spey also powered
the successful BAC 111.

BOTTOM: An Italo–Brazilian AMX powered
by a licence-built Spey.

this by saying that passengers might prefer the speed and quietness of bypass engines, and under such circumstances seat mile costs were meaningless.

But Haworth did not convince the others and Rubbra chaired a number of meetings to organise the Tyne development. Gradually, the usual teething problems were overcome and the engine became a saleable reality. Eleven aircraft manufacturers considered using it – Vickers for the Vanguard, Lockheed for the Electra, Fairey for the Rotodyne, Canadair for the CL44, Breguet for the Atlantique, Shorts for the Belfast, Transall for the C160, Lockheed for the C130, Grumman for the W.2F, the Chinese for the YAK8, and Aeritalia for the G222L. Ultimately, only six manufacturers bought the engine: Vickers bought 238, Canadair 227, Breguet 275, Shorts 60, Transall 605 and Aeritalia 62.

Haworth had been right. Passengers had come to prefer pure jets. In March 1962, as Vickers handed over the last of the twenty Vanguards ordered by British European Airways, the company announced that it had no more orders but that it was 'taking the precaution of cocooning the jigs so that production could start up again if necessary'.

THE CONWAY

The Conway is the most dramatic engine development since the war. It shows such promise that it may well see everything else off on the North Atlantic run ten years from now.

This is what George Edwards said in an interview in April 1953. It had already been announced that the Conway would power the Vickers V1000 military transport aircraft as well as its civil derivative, the VC7 airliner, designed for a dual role over transcontinental and transatlantic routes, carrying between 100 and 150 passengers.

In a 'simple' jet engine, such as the axial compressor Avon, air passes through the intake and is compressed and passed to the combustion chamber. In this chamber fuel is pumped in and the air reaches a very high temperature. The heated air is expanded through the turbine and finally ejected through the jet pipe at high velocity.

In a bypass engine only part of the air is compressed and heated. The remainder bypasses the combustion chamber and the turbine and joins the heated gases in the jet pipe. This bypassed air lowers the temperature of the gases and the mixture is ejected at a lower velocity than that of the simple jet.

This lower velocity of jet discharge gives a better propulsive efficiency. In practice it gives lower fuel consumption for any given thrust, fuel consumption lower than for any simple jet engine produced up to that point. The bypass engine would be specially suitable for long-range subsonic transport and bomber aircraft.

Originally designed in 1948 for a bomber, which was cancelled, the Conway was then ordered by the Ministry of Supply for the Vickers V1000 military transport. When this project was also cancelled, which additionally meant that BOAC's interest in the VC7 airliner version was lost, official support was continued with the promise that the Conway would be chosen to power at least one of the V range of bombers. It was also chosen as an alternative power plant for the Boeing 707 and Douglas DC-8, and a number of airlines specified the Conway in their orders for 707s or DC-8s.

One of the early purchasers was Trans-Canada Airlines, which ordered four DC-8s and specified Rolls-Royce Conway engines. H.W. Seagrim, Trans-Canada's general manager of Operations, said that the Conway had been carefully evaluated against Pratt & Whitney's J-75 and TCA had decided that the Conway was better for a number of reasons.

It is the least expensive in terms of first cost. The application of the engine to the airframe gives lower operating costs.

It could be expected to make the least noise. (The engines would be fitted with noise suppression nozzles as well as reverse thrust mechanism to ease the braking load when landing.)

It is lighter than the J75.

The Conway will give 300 to 400 miles more range, or, alternatively, a bigger payload can be carried. The guaranteed economics of the Conway are rather better than those of the J75.

And it was on Canadian, European and Australian airlines that Rolls-Royce expended its sales efforts, feeling it had little chance of winning US airlines away from using the Pratt & Whitney J-75. BOAC ordered fifteen 707s to be fitted with Conways and all to be delivered in 1960.

Following its order in 1956 for fifteen Boeing 707s, BOAC ordered thirty-five Vickers VC10s, all to be powered by higher-powered Rolls-Royce Conways, in 1957 with delivery beginning in 1963. The new aircraft were to be used on BOAC's African, Far Eastern and Australian routes in conjunction with the nineteen Comet 4s and fifteen Bristol Britannia turboprop airlines that the Corporation had also ordered. The VC10 was designed as a medium- to long-range aircraft with a cruising speed competitive with other jet aircraft operating in the mid-1960s – that is, around 600 mph. It was also planned that the four Conway engines would be mounted at the tail of the aircraft similar to the principle adopted in the French Caravelle. Sir George Edwards, Managing Director of Vickers-Armstrong, said:

The decision to go ahead with the VC10 means that we will be offering a big propeller-turbine airline, the Vanguard from 1960, and the jet VC10 from 1963. There is a large market for both which are, in many ways complementary to each other . . . Once again the Vickers–Rolls-Royce partnership has brought about a side-by-side development of engine and airframe which will be as outstanding in the VC10 as it had been in previous aeroplanes.

The *Financial Times* said of Rolls-Royce's involvement:

The order will also bring good business to Rolls-Royce. The Conway is the world's first by-pass turbo-jet aero-engine. It is claimed to have a lower specific fuel consumption than a 'straight' turbo-jet engine, giving a greater range, as well as being lighter in weight, thus permitting a heavier payload and enabling extra fuel to be carried.

The Conway has already been chosen to power the 15 American 707 jet airliners for BOAC and a fleet of 707s for Lufthansa, the German airline. The Conway will also power a fleet of Douglas DC-8 jet airliners for Trans-Canada Airlines.

With the forthcoming new order – 140 engines for the 35 VC10 aircraft apart from any spare engines involved in the deal – the Conway will be well on the way to becoming one of the world's most important aero-engines, with a big export market in front of it.

But the big market was never achieved. According to Bill Gunston:

[The Conway] was designed to achieve 575 mph at 50,000ft, and the company's proposal offered a take-off thrust of 9,250lb for a weight of 2,984lb. This weight was appreciably less than that of the competing turbojet, the Bristol BE.10 Olympus. Throughout the 1950s the two engines were to compete, and Rolls-Royce considered they could show on paper that, had the reheat Conway been selected for such aircraft as the TSR.2 and Concorde, the result would have been either a lighter or a longer Rolls-Royce engined aircraft. From the outset it was appreciated that, because of the large flow of fresh air in the outer part of the jet, the by-pass engine would be particularly amenable to after-burning and thus to the propulsion of supersonic aircraft. Sadly, Ministry policy was to ignore the great possibilities, and the Conway had only one limited production application in a British military aircraft [i.e. the Handley Page Victor].

We have already seen how Vicker's V1000 was cancelled, a national tragedy in the view of many people. Rolls-Royce was concentrating on getting the Conway on to Boeing's 707 and the Douglas DC-8, two aircraft that did exist and were selling well. It succeeded on both but only to a limited extent. Of

1,519 707s and DC-8s built, only sixty-nine were powered by Conways in spite of the fact that it gave more thrust, had lower fuel consumption and was considerably lighter than the rival Pratt & Whitney JT-4. The final total production of Conway engines was 907. Following the unfortunate cancellation of Vicker's V1000, the Government at least gave support to Vicker's VC10 – though, as we shall see, the national airline, BOAC, was not so helpful.

The VC10 was conceived by Vickers as a medium-range airliner with four engines of around 14,000 lb thrust although, as the weight grew, the necessary thrust required was forced up to 20,000 lb. Vickers insisted on an order for thirty-five if it was to proceed with production and BOAC then gave them the order. The purchase was made against the advice of BOAC engineers who felt the VC10 would be less efficient than the 707.

By the time the VC10 was ready for service in 1964, a new Chairman, Sir Giles Guthrie, had been installed at BOAC with instructions to cut the Corporation's heavy losses. Guthrie was a merchant banker, but not without experience of the aircraft industry. He had won the Portsmouth–Johannesburg Air race in 1936 and spent the war in the Fleet Air Arm.

Guthrie became Chairman of BOAC on 1 January 1964. Before taking over he had received a letter from the Government, which said that the choice of aircraft was to be a matter for the Corporation to decide. It should act in accordance with its commercial judgement.

Bearing this in mind, Guthrie set out to find savings that would wipe out the large accumulated deficit of £77 million (about £1.4 billion in today's terms). He soon noticed that part of that deficit was attributed to development cost on the Comet and the Britannia. In other words, in his view BOAC was not only giving orders to British aircraft manufacturers but also supporting their development costs as well. Adding to this feeling of disenchantment came the news from his planners that the plan on which the VC10s had been ordered was unrealistic, and that only seven new large jets would be needed in the next five years. One of his first steps, after a row with the Government, was to cancel seventeen of the VC10s on order. And the BOAC engineers proved to be right; the running costs were significantly higher than those of the 707. Even the stretched Super VC10, introduced in 1965, proved to be about 10 per cent more expensive to operate than the 707. One of the reasons, perhaps the main one, was that the VC10 was designed for the old 'Empire' routes and therefore for short field performance. On the other hand, the 707 was designed for operation on long concrete runways. As a result, combined with BOAC's obvious reluctance to buy more aircraft, sales were poor and by the end of 1966 only forty had been sold to other airlines. By contrast, by 1979, 941 Boeing 707s and 556 McDonnell Douglas DC-8s had been built and were on order.

THE SPEY

In 1957, British European Airways (BEA) decided that the aircraft it needed for its European routes was one with a cruising speed of Mach 0.875. There were several British aircraft manufacturers competing for the contract with most proposing a four-engined aircraft. However, after a detailed evaluation in which Rolls-Royce was involved, de Havilland proposed a three-engined aircraft. De Havilland argued that such an aircraft would have lower operating costs.

The French Caravelle, with eighty-seat capacity and a 1,000-mile range, was already available, but BEA aimed for 121 seats and a range of 2,000 miles.

The engines offered by Rolls-Royce were the RB 140 with 8,000 lb thrust for the four-engined aircraft, and the RB 141 with 12,000 lb thrust for the three-engined aircraft. The contract was awarded to de Havilland with its three-engined DH.121 and the engine was to be Rolls-Royce's RB 141. Rolls-Royce moved very quickly to develop the RB 141. Its initial rating was 13,790 lb thrust but 20,000 or even 30,000 lb with reheat could be anticipated which meant its prospects, both civil and military, were good. However, with the development programme well under way, BEA decided that the DH.121 was too large, and requested a smaller aircraft with a seating capacity of ninety-seven passengers and range of 921 miles. De Havilland produced the Trident and Rolls-Royce offered the RB 163 (later known as the Spey), an engine with just under 10,000 lb thrust.

Like the successful Conway, the first bypass engine to go into service, the Spey was also a two-shaft bypass engine, but in this case suitable for two- or three-engined aircraft of shorter range from 200 to 1,500 miles and speeds up to Mach 0.875. The most important requirements were low operating costs and low take-off and landing noise, especially from short runways.

The engineers struggled to achieve a performance with which they were satisfied and the salesmen initially failed to secure the key orders for which they hoped. The most crucial was a contract to power Boeing's 727. Although launched a year later, orders for the 727 soon overtook those for Trident, as it became clear that BEA's forecast was wrong. In collaboration with Allison, Rolls-Royce submitted the AR 963-L engine with 12,170 lb thrust. In 1961, Boeing's project design office under Jack Steiner offered the 727 with this engine and the lead airline for the aircraft, United Airlines, accepted it. However, Eddie Rickenbacker, President of Eastern Airlines, rejected the AR 963 saying that the aircraft should have an American engine. He is reputed to have said: 'I won't have a limey engine at any price.'

There were suggestions that Rickenbacker's choice was not made entirely for technical or financial reasons. Those with long memories recalled that when, in 1917, he had been transferred to Britain as a fighter pilot in the US

Navy Air Force when the USA entered the war, his German-sounding name had led to his arrest as an enemy alien and imprisonment in Liverpool gaol. Bureaucratic bungling added to the insult by keeping him there for a week. Others pointed to Britain's supposedly ruthless holding of the Schneider Trophy (an international award presented to the nation with the fastest seaplane over a predetermined, measured course) in 1927 when other contestants' aircraft were not ready.

In his book *Legend and Legacy, the Story of Boeing and its People,* Robert Serling confirms that Boeing initially wanted the Spey on the 727, writing:

Boeing originally wanted to equip the 727 with a Rolls-Royce Spey derivative that would be built under licence in the United States by Allison, a proven power plant with a good reliability record. But Eddie Rickenbacker refused to accept the Spey. Boeing turned to Pratt & Whitney and convinced it to modify an existing military turbine for commercial use. It became the famed JT-8D, one of the most successful engines ever built.

Whatever the reasons, the door was open for Pratt & Whitney to develop its JT-8D engine, which powered the 727 – for a long time the most successful civil aircraft in terms of numbers in the history of passenger flying. Boeing eventually sold 1,832. It has since been overtaken by the 737, also initially powered by the JT-8D, which has received orders for 4,770 aircraft, with 3,762 delivered up to August 2000.

Meanwhile, Rolls-Royce was left with an order for supplying twenty-four Tridents. In 1961, the two-engined BAC 111 was announced. Initially, it was designed to carry seventy-nine passengers with a maximum take-off weight of 78,500 lb and a speed of up to Mach 0.78 – considerably slower than the Trident. Whereas the Trident could not match Boeing's 727, the BAC 111 could not match Douglas's DC-9.

However, the Spey built itself a very good reputation as the engine for both the Trident and the BAC 111, and also for the military aircraft the Buccaneer S-2. By 1966 it was designated for the maritime reconnaissance Comet, the Grumman Gulfstream II, the Lear Liner Model 40 transports, the Dutch F28 Fellowship airliner and, in reheated form, the McDonnell F4 Phantom jets for the Royal Navy and the RAF.

Throughout the first half of the 1960s, Rolls-Royce had been working hard to persuade the US military aircraft planners to accept the Spey. It succeeded in 1966 when the TF41 Spey, jointly developed with Allison, was adopted for the USAAF version of the A-7 Corsair fighter-bomber.

And the Spey was still winning large orders in the 1980s. In April 1981 it was announced that it had been chosen for the Italian–Brazilian light fighter aircraft, the AMX. A total of 300 were planned, 200 for Italy and 100 for

Brazil. Both the Italians and the Brazilians said that they felt the Spey was well proven, readily available and reasonably priced.

The Spey was also sold under licence to China. The initial breakthrough had been made in 1963 when the Chinese bought Dart-powered Viscounts. Stanley Hooker waxed lyrically about his experiences in helping to secure this deal in his book *Not much of an Engineer*. Originally he went to China in 1972 with Alan Newton and John Oliver from Rolls-Royce Derby and Trevor Powell from Bristol. Alan Newton talked to the Chinese about vertical take-off and landing (VTOL) and the Spey, Trevor Powell about the Olympus 593 and Viper, and John Oliver about the Dart. After discussions it became clear that the Chinese were interested in taking a licence to manufacture the Spey. Hooker wrote later:

They did not say which type of Spey, and I did not press them because we had no clearance from our Government to make any definite commitment. All four of us – and our Ambassador – felt that our visit had succeeded beyond our wildest dreams. It was now up to the Powers that Be in Whitehall to give us the go-ahead. In fact, nothing happened until Sir Kenneth Keith arrived as Chairman in November 1972. Then, in very short order, he took up the matter of co-operation with the Chinese with Edward Heath. Soon we were cleared to propose an agreement on the civil Spey engine, which the Chinese had been using for years in the de Havilland Tridents which they had purchased.

In November 1972, at a reception at the Chinese Embassy in London given by Ambassador Sung, I again met Mr Chen from Beijing. He took me on one side and said quite bluntly that the engine in which his government was interested was the military Spey 202, with afterburner, as used in British Phantoms. I could see all sorts of political difficulties with our Allies. When I told Kenneth, he hit the roof, 'It's taken me two months to get the civil Spey agreed. I'll never manage this one, so we must go back to the Chinese and tell them to take a more flexible approach – first the civil Spey, which will keep them busy for a few years and then perhaps the 202 can follow along. Or we could help them adapt the civil Spey for military use, which is how we did the 202 anyway? In the meantime, you produce a proposal for the civil Spey only.'

I consulted with Denis Jackson at Derby, who masterminded most of the company's foreign licences and is a man of great drive and energy. We produced a very broad-brush 'heads of agreement' document. Kenneth said it was a 'pretty thin effort' but told me to take it to the Chinese Commercial Office in London for transmission to Beijing. This thin effort happened to be the only piece of paper recording any discussion between our two countries; it was to be several years before the Chinese put a single word on paper, and everything was done by word of mouth – and with no tape recorders! In early 1973 Mr Peng, the Chinese Commercial Counsellor in London, invited me to return to Beijing for further discussions. Taking the same team as before I arrived in the Chinese capital in March 1973. I told Kenneth that, if things went well, he might be asked to join us. We were received with the usual hospitality, and on our arrival at Er Li Co I emphasized that Sir Kenneth was a most important man who should be invited to join the discussions.

The chairman of the meeting was Madame Wei, a fierce-looking lady of ample proportions. Suddenly it dawned on me that all was not well; there was a distinctly frosty atmosphere. Then I suddenly noticed that Madame Wei was clutching an envelope from Rolls-Royce with the address heavily underscored in red. She spoke harshly: 'This letter is addressed to the Republic of China. We are The People's Republic of China.' It really was an imperial brick, because the Republic is Taiwan. I apologised profusely, and ate all humble pie around. She went on 'People who address us incorrectly are not our friends, and we do not do business with them.' I crawled on my belly, as Hs used to say, and explained that it must have been a typist's error. She replied 'Correspondence should not be left to people who make mistakes. We are severely displeased, and we shall therefore accommodate you in the Friendship Hotel, which is reserved for common foreigners.' She arose, and the meeting was over.

[The author stayed in the Friendship Hotel in 1999 and found that its standard of comfort and service put most British hotels to shame.]

And the problems of selling the Chinese the civil, as opposed to the military, Spey were not over. When Kenneth Keith arrived back in Beijing it was made clear that the Chinese wanted the Spey Mk 202, not the civil Spey 512.

Hooker continues:

Kenneth said he fully understood the position, but pointed out that many nations (by implication, not Britain) were afraid of the might of China, and he said he doubted that he could swing the Government to agree to license the military engine.

In fact, that is just what he did do. As soon as we got back to England Kenneth began to lay about Whitehall with all his great energy and influence. He tackled the Minister of Aviation, Michael Heseltine, the Minister of Defence, Lord Carrington, the Foreign Minister, Sir Douglas Home, and the PM, Mr Edward Heath. In parallel, delicate negotiations were necessary with our NATO allies. Eventually we received full permission to proceed with negotiations on the Spey 202, an afterburning engine designed to fly at over Mach 2.

In the autumn of 1973 I was back in Er Li Co, this time with a different team specializing in the Spey 202. Day after day we toiled through every detail of the engine, though we still had no formal understanding of any kind, only the spoken words that they were interested in having a proposal. Suddenly I was summoned to the office of the Minister of Foreign Trade, Li Chang. He produced a slip of paper and read out that his government had decided to acquire the licence for the Spey 202, and that I could now take that as a firm decision.

However, it took a further two years until the licence agreement was signed on 13 December 1975. Denis Jackson, an engineer who had moved into the Commercial Department, worked day and night with a team of people to complete the negotiations. Finally the deal, worth £100 million, was signed and the Spey was produced at Xian, 600 miles south-west of Beijing.

RATIONALISATION IN THE AEROSPACE INDUSTRY

By the middle of the 1950s it was becoming clear to the more far-sighted that the centre of the world's aircraft industry was the USA. The huge cost of developing not only new aircraft but also new engines meant that long production runs were absolutely vital if those development costs were to be recouped.

The overwhelming need to break into the US civil airline market is made absolutely clear by the figures for aircraft built and on order between 1956 and 1979, which show that, out of a total of 8,558 aircraft, 6,621 (or 77 per cent) were, or would be, built in the USA.

In the mid-1950s, the British aircraft industry – especially the aero engine part – could still claim that it was at the forefront in a world context. On 1 September 1958, Denning Pearson, the Rolls-Royce Chief Executive, was able to write in the *Financial Times*:

The only gas turbine engined aircraft in commercial service in the Western World are British, and this background gives British engine manufacturers a unique experience . . . The prestige of the industry had never been higher. Using British gas turbine engines, the Comet, Viscount and Britannia were each first in their own particular sphere.

Pearson produced a chart showing that the percentage of world airliner orders specifying British engines had been 30 per cent in 1953/4, 34 per cent in 1954/5, 19 per cent in 1955/6, 58 per cent in 1956/7 and 54 per cent in 1957/8. But he sounded a warning note. The cost of necessary research and development was escalating rapidly and, he continued:

[T]he problem of finding these large sums is a major factor especially as the nature of the market demands that the selling price remains strictly competitive with the American product. Indeed the margin in our favour between British and US engine prices is already not as wide as might be imagined. The British aero engine industry cannot make the best use of its natural competitive advantages because of its large contributions to research and development costs compared with the US industry where today practically all the development of commercial aero engines stems from a continuous substantial Government defence programme for similar engines.

The Comet disasters and the results at least prompted an overdue rational-isation of the British aircraft industry. It had entered the post-war period with its head held high justifiably proud of its record. In his book *The Audit of War*, Correlli Barnett may have criticised the performance of some of the industry during the war but the fact remains that whereas in 1935 only 893 aircraft had been built for the RAF, in 1941 over 20,000 were built, with the engines to power them, 80 per cent of which were produced by Rolls-Royce and Bristol. Indeed, in 1940 Britain produced 71 per cent more aircraft than Germany and 63 per cent more aero engines. As we saw in the first volume of this history, Rolls-Royce performed particularly well, led by the incompar-able Hives. Nevertheless, there were too many airframe and engine manu-facturers for all to be competitive and prosperous during peacetime.

There were no fewer than twenty-seven airframe manufacturers: Airspeed, Sir W.G. Armstrong Whitworth Aircraft, Auster Aircraft, Avro (A.V. Roe & Co.), Blackburn Aircraft, Boulton Paul Aircraft, The Bristol Aeroplane Company, Cierva Autogiro Company, Cunliffe-Owen Aircraft, English Electric Company, Fairey Aviation Company, Folland Aircraft, General Aircraft, Gloster Aircraft Company, Handley Page, The de Havilland Aircraft Company, Hawker Aircraft, Martin-Baker Aircraft, Miles Aircraft, Percival Aircraft, Portsmouth Aviation, Saunders-Roe, Scottish Aviation, Short Brothers, Vickers-Armstrong (Weybridge), Vickers-Armstrong (Supermarine) and Westland Aircraft.

And there were eight aero engine manufacturers: Alvis, Armstrong Siddeley Motors, Blackburn Aircraft (Cirrus), The Bristol Aeroplane Company, The de Havilland Engine Company, Metropolitan-Vickers Electrical Company, D. Napier & Son and Rolls-Royce.

The Brabazon Committee under Lord Brabazon of Tara, a distinguished aviator who had been a close friend of C.S. Rolls and had worked for a time at C.S. Rolls & Co. before World War I, prepared a plan for post-war civil aircraft development. (More than one committee was involved. The first, called the Transport Aircraft Committee, was formed in 1942 and com-prised mainly senior civil servants. The second, formed the following year, included a range of aviation industry and airline representatives including Captain Geoffrey (later Sir Geoffrey) de Havilland. The Committee recom-mended that the British aircraft industry should compete in all sectors of the market. Specifically, it suggested that four wartime aircraft be converted into civil transport aircraft and proposed five new projects: a large piston-engined airliner for transatlantic routes, a Douglas DC-3 replacement for short-haul routes in Europe, a medium-range airliner for Empire routes, a mail-only jet aircraft capable of flying non-stop to the USA and a small feeder aircraft carrying eight to ten passengers. The incoming Labour Government in 1945 accepted these proposals and added some more so that the five projects grew to sixteen – four conversions, four 'interim' aircraft and eight new aircraft.

As Sir Geoffrey Owen pointed out in *From Empire to Europe, The Decline and Revival of British Industry Since the Second World War,* his excellent analysis of key British manufacturing industries, it was too many.

Doubts about the industry's ability to handle so many projects began to surface in the late 1940s, and grew more insistent during the 1950s as several of the new programmes ran into difficulty. Some technical misjudgements had been made on the civil side, and three of the aircraft recommended by the Brabazon Committee were abandoned as uneconomic before entering service [the Brabazon long-range, piston-engined airliner, the Apollo short-range turboprop airliner and the Avro 693 medium-range turboprop airliner].

The biggest disaster to befall the industry was the mysterious series of crashes of the new and potentially world-beating de Havilland Comet. Ahead of its US rivals at the time of its launch in 1952, by the time the revamped Comet 4, with Rolls-Royce Avon engines, was relaunched in 1958 both the Boeing 707 and the Douglas DC-8 were available. In their scholarly work *The Technical Development of Modern Aviation*, Donald Miller and David Sawers wrote:

The early start that Sir Frank Whittle had given the British engine industry kept it a year or two ahead of the American manufacturers in design for nearly a decade after the war, but British airplane manufacturers were mostly as much behind. When the Comet failed, no other companies were ready to exploit the lead in engine design; by the time jet airliners entered service generally in 1958, the British lead was no longer significant.

Looking back on the Brabazon civil airliner programme, the numbers of aircraft sold make sorry reading.

* Abandoned: Brabazon long-range piston-engined airliner; Apollo short-range turboprop airliner; Avro 693 medium-range turboprop airliner.
* Sold, 22: Ambassador short-range piston-engined airliner.
* Sold, 37: Comet 1 and 2 jet mail-carrier.
* Sold, 40 (including 28 military versions): Miles piston-engined feederliner.
* Sold, 444: Viscount short-range turboprop airliner.
* Sold, 540: Dove piston-engined feederliner.

It was scarcely a foundation to support the aircraft industry that had emerged from the war and, as the world's airline companies poured orders into Boeing and Douglas, the Conservative Government of the 1950s adopted a policy of withholding support from the British aircraft industry – partly motivated by the desire to force some necessary rationalisation. Vickers had secured a Government development contract for its V1000 (a dual-purpose military and civil transport aircraft), but this contract was cancelled in 1955 before the prototype was complete. George (later Sir George) Edwards, Managing Director of Vickers, described this cancellation as 'the biggest blunder of all'. In the opinion of many, its civil version, the VC7, would have upstaged both the 707 and the DC-8, and been in the air two years earlier. Vickers then produced the Vanguard to follow its successful Viscount. As we have seen, it was also a turboprop, powered by Rolls-Royce Tyne engines, but it could not compete with the jet-engined 707 and DC-8, nor with the French Caravelle, which appeared at the end of the 1950s.

When the Conservative Party was re-elected in 1959, it made clear its determination to force rationalisation on the industry. Duncan Sandys, the Minister of the newly formed Ministry of Aviation, stated that he wanted to see only two airframe and two aero engine manufacturers.

But rationalisation would not come without some pain and heartache. The companies boasted proud heritages and were led by strong individuals and, until the recent change of heart, the Government had spread its favours around giving the impression it wanted to sustain all the companies involved. As Charles Gardner pointed out in his book *British Aircraft Corporation*:

In 1955 . . . there was still no sign of a diminution in the number of design firms. Buggins' turn was still, somehow, enabling everyone to survive. The ultimate nonsense of ordering three V-bombers – each with its own development costs but 'split' production numbers – was helping to sustain Vickers, Avro and Handley Page. Gloster had the Javelin and a thin-wing Jaguar development, English Electric was still building Canberras and developing what was to become the Lightning, while Hawker were getting into their Hunter stride, de Havilland had the Sea Venom and Sea Vixen, Supermarine the Scimitar, while Blackburn were preparing to follow the Beverley with the Buccaneer. Saunders-Roe were developing a 'rocket-plus-jet' and Fairey were following up the Gannet anti-submarine aircraft with the FD2 Delta which, in 1956, was to carry the world's air speed record to 1132 mph . . . Armstrong Whitworth were soon to get the Argosy and there was much helicopter activity affecting Bristol, Fairey, Westland and Saunders-Roe. On the transport side, Vickers had the very successful Viscount and V1000; de Havilland were salvaging the original Comet into the Comet IV and building Herons and Doves. Bristol had the Britannia, the Freighter, the Sycamore and Belvedere helicopters and, more profitably, its Bristol engines and rockets.

A watershed occurred in 1957 when Duncan Sandys' White Paper stated that there was a definite limit to the amount Britain could spend on defence and that in future it would depend on thermonuclear power. Manned aircraft were to be phased out and replaced with missiles or 'rockets' as Sandys called them. The cancellations that followed devastated the military side of the aircraft industry and, as this represented 70 per cent of the industry's output and employed 60 per cent of its work force, the effect was profound.

Nevertheless, the overdue rationalisation of the aircraft industry finally took place, prompted in part by the decision of the Government that future military needs would centre around unmanned missiles rather than manned bombers. As a result the following groups emerged (shown below with their constituents and principal projects).

AIRFRAME GROUPS

British Aircraft Corporation
English Electric Aviation
Vickers-Armstrong (Aviation)
Bristol Aircraft
Hunting Aircraft

The principal projects of these groups were:

* VC10 jet airliner
* Super VC10 jet
* Vanguard turbojet
* Viscount turboprop
* Lightning fighter

* TSR2 strike aircraft
* Bloodhound missile
* Thunderbird missile
* Jet Provost

Hawker Siddeley Group
A.V. Roe & Co. (Avro)
Sir W.G. Armstrong-Whitworth Aircraft
Blackburn Aircraft
Folland Aircraft
Gloster Aircraft
Hawker Aircraft
De Havilland (aircraft, engines and propeller companies)

The principal projects of these groups were:

* Vulcan bomber
* Hunter fighter
* Javelin fighter
* NA-39 naval aircraft
* Gnat light fighter
* P.1127 VTOL fighter

* Comet jet
* DH.121 airliner
* Argosy freighter
* Avro 748 airliner
* Sea-slug missile
* Blue Steel guided bomb

Westland Aircraft Group
Westland Aircraft
Saunders-Roe
Bristol Aircraft's Helicopter Division
Fairey Aviation

The principal projects of these helicopter groups were:

* light transport – Skeeter; P531
* medium transport – Whirlwind; Widgeon; Wessex
* large transport – Westminster; Belvedere; Rotodyne

ENGINE GROUPS

* Rolls-Royce: Avon, Conway, RB 163, RB 141, RB 108 turbojets; Dart and Tyne turboprops.
* Bristol Siddeley Engines: Olympus, Sapphire, Viper, Orpheus, BE53 turbojets, Mamba turboprop.

INDEPENDENT COMPANIES

* Short Brothers and Harland: Britannic freighter; DC-7 light transport.
* Handley Page: Victor bomber; Dart Herald airliner.
* Scottish Aviation: Twin Pioneer transports.
* Auster Aircraft: light sporting and business aircraft.

These larger groupings were clearly going to be able to compete more strongly than their component parts could have done; nevertheless, they would find plenty of competition from overseas. As the *Financial Times* pointed out on 5 September 1960:

The most serious task facing the industry today is not only to manufacture, technically and economically, the best aircraft in the world for airline operations, but also to match, and if possible improve upon, the favourable prices and conditions of sale offered by the Americans, the French – and the Russians – who between them are the most formidable competitors facing the UK aircraft industry today. Henceforward, the financial terms offered to prospective customers will have as much, if not more, importance in winning an order than the quality of the aircraft itself.

The *Financial Times* went on to suggest that the three aircraft groupings should not compete with each other.

The two main fixed-wing airframe groups must avoid building competitive aircraft – and there is a real danger here, in that both the British Aircraft Corporation and the Hawker Siddeley Group are currently studying designs for new, small short-range jet airliners in the broad Viscount replacement category. One or the other, but not both, must enter this field, the industry must avoid having to compete with itself in world markets as well as with the foreign manufacturers.

The *Financial Times* also advocated international co-operation.

In recent months, the emergence of the Common Market, and the increasing number of link-ups between American manufacturers and their Continental counterparts have posed new dangers for Britain. If US aircraft manufactures such as Douglas, Lockheed and Pratt & Whitney can forge important financial, manufacturing and trading links with aircraft companies in France, Italy and Germany, there seems to be no reason why the British manufactures should not do the same.

Rolls-Royce was already looking at co-operation with its US competitors. The attempted collaboration with Westinghouse had ended in failure. On 18 September 1959, David Huddie wrote to Phil Gilbert:

I have just met Higgins of Westinghouse, who has reached the conclusion that they are out of the aero engine business.

They have verbally freed us to negotiate with anyone in the USA on all our engines, but before we start, could you please give me as soon as you can your legal opinion on the anti-trust aspects of our associating with either Pratt & Whitney or General Electric. All our evidence says that all future Navy or Air Force engine contracts will be awarded to one or other of these two.

Gilbert replied that there could be some anti-trust difficulties, particularly in relation to Pratt & Whitney, but said:

I would recommend that anti-trust considerations at this stage should not prevent you from exploring the possibilities of reaching the kind of agreement you desire.

Nor was Rolls-Royce immune from the pressure for rationalisation. Despite being one of the most successful of the aero engine companies – in his book published in 1957, *Britain's Air Survival,* Sir Roy Fedden wrote, 'Rolls-Royce has shown itself to be head and shoulders above – the outstanding engine company in the country' – its sales prospects and profit margins were still adversely affected by competition whether from Pratt & Whitney and General Electric in the USA or from Bristol and others in the UK.

In December 1960, Pearson was able to announce the sale of the 1,000th civil aircraft powered by Rolls-Royce gas turbine engines, and this only seven years after the first passenger-carrying aircraft powered by Rolls-Royce gas turbine engines, the Viscount, had gone into regular airline operation with BEA in April 1953. He announced:

Out of a total of approximately 1,760 civil aircraft with turbine engines which have been sold up to present in the Western world, nearly 60 per cent have been ordered with Rolls-Royce engines. Of the 1009 aircraft [with Rolls-Royce engines], 621 are British-built, 192 American, 94 French, 85 Dutch and 17 Canadian.

The surprising figure in this list is 192 American. It meant that even if **all** the other 751 aircraft, to reach the total of 1,760, were also American, their total of 943 was not completely out of sight of the British total. This lead was about to lengthen considerably. Denning and other far-sighted people in Rolls-Royce knew it and, as we shall see, they made the USA their target.

CHAPTER THREE

THE BRISTOL AEROPLANE COMPANY

THE BRISTOL TRAMWAYS COMPANY
FEDDEN
HOOKER
THE OLYMPUS
TSR2
CONCORDE
ROLLS-ROYCE SWOOPS

THE BRISTOL TRAMWAYS COMPANY

GEORGE (LATER SIR GEORGE) WHITE, was born in 1854 and, by the age of twenty, was appointed secretary of a small company that opened the first tramway line in Bristol. Joining the Bristol Stock Exchange shortly afterwards, White became instrumental in mergers that rationalised the railway and tramway concerns operating in the West Country. His tramway company, capitalised originally at £7,000, grew into an electrified operation with a capital of £1.25 million (£125 million in today's terms) by the early 1900s.

In 1909, now Chairman of the Bristol Stock Exchange and of the Bristol Tramways Company, he took a winter holiday in the South of France (you will remember from the first volume of this history that Sir Henry Royce was soon to discover the benefits of winters on the Côte d'Azur). While there he made excursions into Italy and Spain driving, as it happened, in two Rolls-Royce Silver Ghosts. He also visited a flying display at Pau and became interested enough to visit aircraft constructors in Paris on his way home, ordering both a biplane and a monoplane. At the Bristol Tramways Annual General Meeting in February 1910 he announced to his startled shareholders:

I may tell you that for some time past my brother and I have been directing our attentions to the subject of aviation, which is one hardly yet ripe for practical indication by such a company as The Bristol Tramways Company, but yet seems to offer promise of development at no distant date; so much so that we have determined personally to take the risks and expense of the endeavour to develop the science from the spectacular and commercial or manufacturing point of view.

If, as we believe, we can make the headquarters close to Bristol, we shall give our own city a prominent place in the movement nationally . . . Incidentally, I may say that we have already on order several aeroplanes of the best designs hitherto produced, with the intention to develop a British industry and make Bristol its headquarters.

The *Bristol Times and Mirror* was full of applause.

No one can truthfully say after Sir George White's speech . . . that Bristol is lacking in up-to-date enterprise. Sir George is not only abreast of the times but a little ahead of them.

The new operation, called the British and Colonial Aeroplane Company, was set up at the northern terminus of the tramways system at Filton, four miles from Bristol city centre.

As well as the British and Colonial Aeroplane Company, he also founded the Bristol Aeroplane Company, the Bristol Aviation Company and the Colonial Aviation Company to give himself maximum flexibility. The initial subscribers were himself, his brother Samuel, his son Stanley (who within a few months was appointed Managing Director, a post he held for forty-four years), Henry White Smith and Sydney Smith, his sister's sons. Henry became secretary and Sydney manager of the works.

George Challenger, son of the general manager of the Tramways Company, was appointed chief engineer and by 1911 had already designed the Bristol Racing Biplane in collaboration with the French designer, Grand-seique. Sir George sent Sydney Smith to Paris to study French methods and materials, and Smith and his colleagues returned with a licence to build the biplane and monoplane designs of the Société Zodiac, which had already produced aeroplane and airships for the Voisins and Henri Farman.

The Bristol–Zodiac biplane though elegant was not successful. Six were built and scrapped. But the Bristol men learnt from them and produced their own aeroplane, the Bristol Boxkite. By the outbreak of World War I, 300 pilots had learned to fly in Boxkites.

As well as building aeroplanes, the Bristol company founded flying schools – at Larkhill, Brooklands and Eastchurch near Chatham. Captain Frederick Hugh Sykes, a future member of the Air Council, learnt to fly on a Bristol aeroplane at Brooklands before work at the War Office. Tuition fees for a complete course were £50 (£5,000 in today's terms) with special terms for officers in the Army or Navy and for those who decided to buy a Bristol aeroplane. For those who bought a biplane the entire fee was refunded, while £25 was returned to those who purchased a monoplane. When war was declared on 4 August 1914 and the Bristol flying schools were taken over by the Government, 307 of the 644 Royal Aero Club certificates had been awarded to Bristol-trained pilots.

George White invested very heavily in his new-found love. While Short Brothers and Handley Page were founded with capital of £600 and £500 respectively, Bristol was founded with £25,000 (£2.5 million in today's terms). This was soon increased to £50,000 and by the end of 1911 to £100,000. In February 1911, the *Aero* wrote:

The B, and C.A.C. works proper would make most of our constructors green with envy so beautifully are they planned and built, being, with I think only one exception, the only shops actually designed for aircraft work. They have fine roof lights, dead level concrete floors, and all those conditions which go to turn out the best quality work.

Between 1910 and 1912, White set up complex production lines manufacturing no fewer than ten different types of aircraft. He also acquired the exclusive licence for Britain and the Colonies for the French engine, the Gnome, arguably the best aero engine available at the time. He demonstrated his aircraft not only locally but also throughout the world, sending his nephew Farnall Thurstan to India and Sydney Smith to Australia.

In the early months of the war the Bristol Scout was in action and this aircraft was followed by the Bristol F2B, a two-seater fighter with the gunner seated behind the pilot, facing aft. More than 4,500 Bristol Fighters, affectionately known as Biffs and later as Brisfits, were built – not only at Filton but also in shadow factories. The fighter carried on in service with the Royal Air Force until 1932 and was sold to the air forces of Belgium, Bulgaria, Greece, Mexico, Norway, Spain, Sweden and the Irish Free State. However, its early entry into combat was not a great success. Held back to support the spring offensive of 1917, the first Bristol Fighter squadron, No. 48, went into action over Arras in March 1917 led by Captain W. Leefe-Robinson VC against the famed von Richthofen, who wrote later:

It was a new type of aeroplane which we had not known before, and it appears to be quick and rather handy, with a powerful motor, V-shaped and twelve-cylindered. Its name could not be recognised. The D111 Albatross was, both in speed and ability to climb, undoubtedly superior.

After this inauspicious beginning in a battle in which the British pilots had manoeuvred their aircraft to give their gunners a good field of fire and in which Leefe-Robinson was shot down and taken prisoner, the Brisfit pilots realised these were the wrong tactics and began to fly their aircraft as if they were single-seaters using the front gun for the main attack and the observer's gun only as additional rear cover. This new method brought much greater success.

FEDDEN

While the Bristol Aeroplane Company had made a name for itself manufacturing aeroplanes during World War I, it had only dabbled in the manufacture of aero engines. Before he went to Canada to set up his own company, Curtiss-Reid, Wilfred Reid had designed a water-cooled, two-cylinder unit giving 65 hp. The company's decisive move into aero engines came with its purchase of the assets of an engineering company called Cosmos from the receiver in July 1920. Cosmos had been a flourishing conglomerate and included in its group was the company Brazil Straker, which, run by Roy (later Sir Roy) Fedden, had been the only company that Rolls-Royce had permitted to manufacture its engines under licence during World War I (see the first volume of this history). At the end of the war, Cosmos began shedding companies but kept Fedden and his team working on aero engines. Fedden had already designed an air-cooled radial aero engine, the Mercury, and had been given encouragement by Brigadier-General J.S. Weir, technical controller of the Ministry of Munitions, to continue his work on aero engines. In December 1918, Weir wrote to Fedden:

I need not enlarge upon the success of your work nor upon the high hope which this department has regarding the Jupiter beyond stating that it is of very great importance to the Nation, as far as aircraft are concerned, that you should press on in all haste to perfect the Jupiter, which I feel sure has a considerable future in front of it in commercial aviation.

Fedden had chosen the air-cooled, radial, approach as a clause in the Brazil Straker licence granted by Rolls-Royce had forbidden him ever to attempt a water-cooled in-line engine.

Fedden later gave due credit to Rolls-Royce. In 1936, he wrote:

Between 1915 and 1917 my own company produced Rolls-Royce, Renault and Curtiss aero engines, and I attach the highest importance to the close liaison that we had with the Rolls-Royce company at Derby. Constant liaison over these years, going into the closest details on the selection of materials, methods of manufacture, standards, fits and clearances, methods of testing etc., gave me a valuable insight into engineering, and periodic discussions with Sir Henry Royce at St Margaret's Bay, on technical problems relating to our aero engine production, were also of considerable value to me, and I can say without hesitation that without this specialised and concentrated work, it would have been very difficult to have produced a successful Mercury design.

The Bristol Aeroplane company knew of Fedden's engines but played a canny game when it came to the possibility of buying Cosmos and it was only after pressure had been brought to bear by the Government, including an order for ten Jupiters, that Bristol paid the liquidator £15,000 for the company and all its assets. The Jupiter was a nine-cylinder, 500 hp, single-row engine designed by Fedden and his team.

From the beginning relations between Fedden and his new board at Bristol were strained. In early August 1920, as Fedden moved his team from the old Cosmos works at Fishponds to Filton, Sir Henry White-Smith (Sir George White's nephew) suggested he take over some of the empty hangars and sheds around Filton House. Fedden declined the offer preferring to take his men a mile north to the far end of Filton aerodrome. He felt it was important for the pilots to appreciate the care and effort that went into the making and testing of the engines. He also wanted his engineers to spend more time in the air.

Fedden's team worked hard but by the autumn of 1921 the Bristol board was threatening to close their operation down, as they had secured no further orders on top of the one delivered to the Government. The £200,000 investment (about £10 million in today's terms) Sir Henry White-Smith had secured from the board had been almost spent. If Fedden could not improve the order position by the end of the year, the Engine Department would be wound up. He responded by taking a number of engines to the Paris Air Show and signed a manufacturing licence deal with the French Gnome-Rhône Company. The advance fee saved the Engine Department for a time and then, just as the future looked bleak again, the Air Ministry ordered eighty-one Jupiter IV engines to go in Hawker Woodcock fighters.

Over the next few years Fedden strove to make the Jupiter the supreme aero engine. Early in 1926 a Jupiter V was installed in a Bristol Type 84 Bloodhound and sealed so that any attempt to service it would have been obvious. The Bloodhound was then flown back and forth between Filton and Croydon until 25,074 miles had been flown in 225 hours 54 minutes. The engine had given no trouble and when it was stripped down in front of an Aircraft Inspection Directorate (AID) representative, everything was in perfect working order except one slightly corroded exhaust valve. This test was to secure orders for the Jupiter throughout the world including orders from the RAF for bombers, fighters and transporters. De Havilland also chose it for Imperial Airways DH.66 Hercules fourteen-seater airliners.

Unfortunately, relations with Gnome-Rhône turned sour as the French company sought ways of evading the terms of its original agreement. Two of Bristol's leading engineers, Norman Rowbotham and Roger Ninnes, had been sent out to Paris to help the French company produce the Jupiter, and through the 1920s developed derivatives that were more powerful and efficient than the original. This may have been the reason that the French company began to question the justification for the royalties. However, Fedden secured licences in seventeen other countries, which would soon be bringing Bristol £1 million a year in fees.

By the end of the 1920s Bristol was able to advertise that the Jupiter was being manufactured under licence by Gnome-Rhône in France, by W.E. Bliss Company of Brooklyn (New York) in the United States of America, by Nakajima Aircraft Company in Japan, by the Societe Piaggio of Genoa in Italy, by the Swedish Government, by Siemens and Halske of Berlin in Germany, by the Société Anonyme Saurer in Switzerland, by the Société Walter a Spol of Prague (Czechoslovakia), by the SABCA in Belgium, by the Industrija Avijonskih Motora in Yugoslavia, by the Société Anonyme des Aciers Manfred Weiss in Hungary, by the Sociedad Union Naval Levante in Spain, by the Parque de Matérial Aeronautico d'Alverca in Portugal, by the Trust d'Etat de L'Aviation of Moscow in Russia, by the Polish Government and by Bristol Engines of Montreal in Canada.

In 1927, adding to these licensing successes, the Jupiter scored two big hits at home. The RAF selected the Bristol Bulldog and the Westland Wapiti, both powered by the Jupiter. The Bulldog had beaten the Hawker Hawfinch and Hawker's designer, Sidney (later Sir Sidney) Camm, complained that the Jupiter engine supplied had been inferior to that in the Bulldog. Fedden produced the logbook and test records to show that this was not the case. The RAF bought 312 Bulldogs and 507 Wapitis; many more were exported.

The success of Fedden's air-cooled engine is made clear by this memo sent by Henry Royce himself to his colleagues in Derby on 3 June 1926:

The Jupiter may be considered to have definitely beaten the water-cooled engine, except for its one disadvantage of too much head resistance . . . Should we decide to build an air-cooled engine, as I strongly recommend . . . I think this matter is more urgent than would at first appear, owing to the loss of trade we have experienced in this country and abroad through the success of the air-cooled engine.

(In spite of this exhortation, Rolls-Royce's next engine was the water-cooled Kestrel.)

By 1930, and including the seventeen licences, 7,100 Jupiters had been produced and powered no fewer than 262 different types of aircraft – including fighters such as Bulldog, Gloster Gamecock and Hawker

Woodcock; bombers such as the Boulton Paul Sidestrand and Handley Page Hinaidi; flying boats such as the Short Rangoon and Short Calcutta; and commercial transport aircraft such as the de Havilland Hercules, Handley Page HP42 and Short Scylla.

But the Jupiter could not go on forever and Fedden knew that he must produce engines with greater power, which would almost certainly mean larger engines with more cylinders and more valves. This focused his attention on the sleeve-valve and during the 1920s his team spent thousands of hours trying to perfect the sleeve-valve system, which he hoped would replace the traditional poppet-valves. They tried in-line configurations but in 1931 abandoned them and returned to the radial. The first Bristol sleeve-valve, the Perseus, was built in 1932 and used nine Mercury-size cylinders. This is what Gordon Lewis, one of Stanley Hooker's ablest engineers, who made an enormous contribution to the development of both the Olympus and Pegasus engines and who eventually became Director of Engineering at Bristol, said of this first sleeve-valve engine:

When this engine was built it was remarkable for its elegant and uncluttered appearance, and it ran with unbelievable smoothness with a sound quite unlike that of the poppet valve engine with its reciprocating valve mechanisms. On performance grounds likewise there was no question, and comparative tests were made to underwrite the vital decision that was being taken to vest the future of Bristol engines in the sleeve valve. Reduced maintenance, low fuel and oil consumption and overall simplicity were evident advantages, but one obstacle remained that could cause the whole programme to be abandoned.

This was the inability to manufacture the main components by mass production methods in such a way that selective assembly and fitting of piston, sleeve and cylinder could be avoided.

Bristol searched for the appropriate materials and strove for the necessary manufacturing techniques. High Duty Alloys produced the low-coefficient expansion of aluminium cylinder forgings, Firth Vickers the high-expansion, steel-sleeve, centrifugal castings and Fred Whitehead of Fedden's team at Bristol perfected the process of accurate sleeve grinding.

As an insurance policy, the Pegasus and the Mercury (not a revamp of the earlier Mercury, but a brand new engine) were launched but were followed in 1935 by the Hercules, a fully supercharged engine, and in 1936 by the Taurus – these latter two both sleeve-valve engines. In 1938 came the powerful Centaurus, designed in response to Bristol's (mistaken) fear of competition from Rolls-Royce's Vulture. Apart from its success at home, 119 types of aircraft were designed outside Britain to use the Mercury and Pegasus, and more than ninety-five existing types were converted to take them.

As the 1930s progressed, Fedden was one of the few strongly advocating rearmament. As Bill Gunston wrote it in his excellent *Fedden – the life of Sir Roy Fedden*:

In 1935 there were not a few Britons who considered Fedden, with his incessant propaganda for a stronger and more widely dispersed engine industry, to be an unmitigated bore . . . One of the few who agreed wholeheartedly [with Fedden] was Air Marshal Sir Hugh Dowding, by this time Air Member for research and development . . . One day in September 1935 Fedden was visiting Dowding to argue the case for a British supply of modern variable-pitch or constant-speed propellers which for five years had been one of his hobby horses. [For the background to the setting up of Rotol see the first volume of this history.] The talk ranged far and wide; then, suddenly, Dowding said (and these were his very words): 'Look here Fedden, I think you will be relieved to learn that the Air Staff have come to the conclusion that we are going to have a war with Germany. Quite when it will come we cannot say; it may not be for 10 years. But what we must do as soon as we can is what you have been advocating, and that is to build up a vast increase in our production potential . . . What we have decided to do is to choose an established manufacturer and then try to create a replica of his production process elsewhere . . . We have chosen you. Bristol engines are essentially simple, and easily made by standard machine tools and equipment. What also counted very heavily with us was your great experience in licence agreements with many foreign firms. We are convinced that with Bristol engines of established types we are minimising the risk in launching this bold procedure.'

Fedden was given the powers to summon possible participants in the scheme. However, he soon realised that setting up the 'Shadow' scheme would not be easy. His own board at Bristol (in spite of his key position in the firm, Fedden had never been invited to become a director) were strongly against involvements. Furthermore, almost all the leading motor car firms were against it too. The one exception was Bill (later Lord) Rootes, who backed Fedden. Once the other motor firms realised the 'Shadow' scheme was going to happen they all decided that they had better be involved. By the time war broke out, Shadow factories were producing Mercury and Hercules sleeve-valve engines by the thousand.

In his book, Gunston acknowledges that Major G.P. Bulman, Director of Engine Development from 1938–41, did not agree that Fedden was a key element in the Shadow factory scheme:

Fedden took no part at all in this amazing effort of co-operation.

Gunston disagrees completely, asserting that Bulman was rewriting history.

78

He has no doubt about the contribution of both Fedden and Bristol, writing:

The astronomic growth in output schedules for Bristol engines called for enormous quantities of advanced machine tools, not only for the new Shadow plants but also for the parent company, whose works had been constantly extended since 1934, especially on the east side of the Gloucester Road. In 1936 Fred Whitehead had set up a small office in the United States, at Hartford, Connecticut [home of Pratt & Whitney]. This proved to be another of his brainwaves. Not only did it greatly facilitate the ordering and quick delivery of standard US machine tools but it also actually designed a number of special-purpose tools and other plant which American firms produced in short order. The British machine-tool industry was quite incapable of fulfilling such orders, and was in any case grossly over-extended by the nation-wide rearmament programme. Fedden estimated that Whitehead's office in America reduced the time taken to get the Shadow industry into production by more than six months. As finally constituted the original Shadow engine industry was operated by Bristol, Daimler, Rootes, Austin, Rover and Standard. The Shadow work force at the outbreak of war in 1939 was 51,850, and growing.

Just before war broke out, Fedden went on holiday to Norway. While he was away he wrote a long letter to Bristol's Managing Director, laying out his idea of how he felt the company should be reconstituted. He thought that the same structure which had existed in 1920 when he joined should now be changed. His thirty-one men had become 16,000, and the number of engines produced and delivered exceeded 26,000 with a value way in excess of that earned by the company's aeroplanes. In Fedden's view, the company should be split into two companies – or at least the Engine Department should have equal status with the Aeroplane Department.

Fedden's letter went unanswered, but just before Christmas 1939 he was summoned by the Chairman, Sir William Verdon-Smith, who said to him:

Ah, Fedden . . . I want to talk to you about the matter of your commission, paid to you on engines sold to your design. As you know it was to be in the amount of one-half per cent up to a total of £200,000 and at one-quarter per cent thereafter. In 1920 we were thinking in terms of Jupiter engines priced at about £1,000, sold in twos and threes. Now we are selling engines priced at many thousands of pounds, we are selling them by the thousand . . . When your 1920 agreement was drawn up we had no thought that it would involve such sums. The amount of your commission has reached a very high figure, and we are having difficulties with the authorities. I am afraid that we cannot go on under the present arrangement.

Fedden agreed to take a much-reduced commission and furthermore made it retroactive thereby handing back a sum in excess of £200,000 (about £10 million in today's terms). Under the circumstances Verdon-Smith offered him a new six-year contract, but Fedden wanted to know what the board proposed to do about his suggested reorganisation of the company. However, Verdon-Smith refused to discuss it beyond saying that the board had been offended by his letter. The meeting ended with Fedden saying he would work for the company for the duration of the war and for six months beyond it.

In the event, relationships had been so damaged that Fedden stayed at Bristol only until the autumn of 1942, when he was asked to leave. How could the company get rid of a man that had done so much for them, and whose ability and drive they would sorely miss?

Bill Gunston gives a balanced assessment of the faults on both sides, referring to the board as 'narrow-minded, inbred and autocratic', while saying that Fedden's 'worst enemy was unquestionably himself'.

Sir Reginald Verdon-Smith – son of Sir William, and Family Director (and later Chairman) of the Bristol Aeroplane Company – would say later of Sir Henry White-Smith's departure in 1926:

He was a strong personality in the Company, and had exercised a vitally important control over Fedden's indomitable enthusiasm. Fedden would ruin any business he ran unless he was effectively controlled, and the remarkable feature of the great years up to 1939 was that, by and large, he got a mixture of backing and restraint which he would have been unlikely to receive anywhere else ... He involved the firm in many serious difficulties, both in failing to meet over-optimistic delivery promises and in over-running cost estimates, largely due to design changes. It was this situation which led to the introduction of Rowbotham and Ninnes to bring discipline and order into engine production, whilst leaving Fedden in full charge of design and development.

Fedden thought Norman Rowbotham a toady, and Rowbotham thought Air Marshal Sir Wilfrid Freeman a bigot who favoured Hives over him. (It should be said that no one else shared this view of Freeman.) As Gunston points out, neither view was fully justified but there was no doubt that Fedden was a huge loss to Bristol. This was how Gunston described the immediate aftermath:

The traumatic shock of the departure from Bristol of Sir Roy Fedden was almost beyond belief. Few things to compare with it have ever happened in industry ... political feuding if anything increased. To make up for the sudden absence of long-term technical direction men were shuffled about in all directions but to no good effect ... One by one, most of the senior staff departed, until the fine team was dispersed. Even in the short term the departure of the 'Centaurus's father' was a disaster, and the fierce pace of development slumped – in the midst of the Second World War. In the long term the company's engine policy simply fell to pieces. Despite the great interest shown in the Orion, Rowbotham thought it would be 'sounder policy to work on lines like those of Pratt & Whitney and go for a double Hercules'.

Rowbotham instructed the design office to work on a twenty-eight cylinder, four-row engine, and an 83.5 litre engine of similar design but with twenty-eight Centaurus cylinders. However, these never got further than the drawing office, and two other engines under development, the Orion I and II, were both cancelled.

HOOKER

This was a sorry outcome for the company that had produced such aircraft as the Blenheim, the Beaufort and the Beaufighter, which had seen service in every theatre of the war save Russia, and that had produced 14,000 aircraft and over 100,000 engines during the war, with some 52,000 people employed in 150 premises throughout the country. The 1958 HMSO publication, *Factories and Plant,* showed that piston engine deliveries in the UK between June 1939 and December 1945 were as follows: Rolls-Royce, 112,183 (43 per cent); Bristol, 100,932 (39 per cent); Armstrong Siddeley, 32,868 (12 per cent); de Havilland, 10,905 (4 per cent); and Napier, 5,267 (2 per cent). The company had to wait until the arrival of Stanley Hooker in January 1949 to haul itself back into contention with other engine manufacturers – most notably Rolls-Royce.

Much of the Bristol development effort in the immediate post-war years went into the Bristol Brabazon. The Brabazon Type 1 was one of the aircraft recommended by the Brabazon Committee which, in 1943, looked at the prospects for civil aviation after the war. It was to have a transatlantic range and a high cruising speed, and was to be built by a company with the appropriate structural design experience.

What emerged was a development of the projected 100-ton bomber powered by eight 2,500 hp Centaurus engines buried in pairs within a wing of no less than a 230-feet span. With a cruising speed of 250 mph at 25,000 feet and a range of 5,500 miles, it could carry 100 passengers and twelve crew. The aircraft was of such huge proportions that a special eight-acre, three-bay hangar was constructed for it, and the runway at Bristol was lengthened and strengthened. In the process the village of Charlton was destroyed. As someone said, Brabazon production seemed to owe more to shipbuilding than to aircraft engineering. It was first flown by the Bristol chief test pilot, Bill Pegg, on 4 September 1949 but, after logging about 400 hours, was broken up in October 1953 never achieving production status.

In the middle of 1948, Rolls-Royce lost Stanley Hooker and on 3 January 1949, he joined Rolls-Royce's great rival, the Bristol Aeroplane Company.

Many have speculated on how Hooker came to fall out with Hives, who had described him as being like a son. In his book, Hooker wrote that at a private dinner in 1946, just after he was recalled from Barnoldswick, Hives had promised him he would become chief engineer. Indeed, after he returned from a sabbatical to Argentina and the USA, which had been organised by Hives, that was what Hooker expected as his next appointment. When Hooker challenged Hives with this promise early in 1948, Hives replied:

You're not ready for that job yet. You can't control your own affairs, let alone this great firm.

This was a reference to Hooker's chaotic private life and he conceded Hives's point.

In retrospect he was probably right. My marriage had irretrievably broken down, and the great strain of this, added to that of the Avon, was showing in my lifestyle and demeanour. So I went back to Barnoldswick and sulked in solitary confinement.

The Avon was certainly causing a great deal of concern. At the Rolls-Royce board meeting of 15 January 1948 it was noted that:

The situation with regard to the Avon was causing some anxiety as we were still faced with unsolved problems, and dates of delivery were rapidly approaching. The Avon was therefore being treated as of the highest priority and a great concentration of effort was being brought to bear upon it. Derby was giving active assistance to Barnoldswick.

By June, Hives had decided on drastic action and at the board meeting held on 13 June it was noted that:

The Managing Director gave a resume of the position in relation to the Avon engine, which at the moment was unsatisfactory. He informed the board that he and Mr Elliot were concentrating the whole of their efforts on the Avon engine. He had already arranged to make Elton Road the engineering headquarters of the Aero Division and all the essential staff were being moved to that office. This involved bringing back staff from Barnoldswick.

Hives and Hooker had a blazing row in June 1948, and although they made up and Hives put Hooker in charge of all research work, Hooker felt he had been demoted. He subsequently contacted Reginald Verdon-Smith, a family director at the Bristol Aeroplane Company and an old friend from his Oxford days. Then, on 30 September, Hooker told Hives he was leaving and, according to Hooker, Hives said:

I don't know what's gone wrong, Stanley. I have always treated you as my son, and you have never come into this office and asked for anything without its being granted.

It was a sad loss for Rolls-Royce, but perhaps inevitable in that Hives had to show he was the boss and would appoint his chief engineers when he deemed them to be ready. By his own admission Hooker was not ready in 1948. In the board minutes for 11 November 1948, it was noted that:

The Managing Director reported that relations with Dr Hooker had been unsatisfactory for some time and he had formed the view that it was preferable for a break to be made . . . His replacement would present no substantial difficulty as the company had plenty of available talent.

This was not quite true as was made clear by the board minute immediately above, which said:

The Managing Director emphasised that the Company was short of executive officials at the top level.

When Hooker arrived, he noted that Bristol was still producing the sleeve-valve, air-cooled radials, the Hercules and Centaurus, in considerable numbers, but was way behind Rolls-Royce in the development of gas turbine engines – although it had done some work on a complicated and heavy turboprop engine, the Theseus. In Hooker's view there was a 'ten-year lag behind Rolls-Royce'.

Hooker, of course, was determined to bring Bristol up to a pitch where it could compete with Rolls-Royce, although he admitted later that this would have been very difficult had it not been for the outbreak of the Korean War in 1950. The Government ordered Rolls-Royce to license Bristol, Standard Motors and Napier to manufacture the Avon axial engine and, at a stroke, Bristol was equipped with all that was necessary for the company to mass-produce axial engines. As Hooker wrote later:

Thus, just as did General Electric and Pratt & Whitney learn to compete with Britain by making British jet engines, so did producing a Rolls-Royce engine set Bristol on the road which, by 1960, saw them providing neck-and-neck competition with the giant at Derby.

Nevertheless, Hooker was shocked at the small amount of effort going into gas turbine engines when he arrived. His first task was to try to perfect the Proteus, the turboprop successor to the Theseus. It was originally aimed at the Brabazon II and the Princess flying boat in which it was buried in the wings and fed by ducts from the leading edge. It soon became clear that the Brabazon and the Princess were not going to be a commercial success, and Hooker and his team concentrated on getting the Proteus right for the more promising Britannia. Hooker gathered a formidable team around him including Gordon Lewis, Pierre Young and Robin Jamison, and, from Barnoldswick, Basil (later Sir Basil) Blackwell and Bob Plumb. By August 1953 they were reasonably happy with the four-engined Britannia. As Hooker said:

At last we had an excellent engine, every part of which had been redesigned on my instructions apart from the reduction gear. This was an 11:1 ratio gear with straight-tooth spur gearwheels to reduce the 11,000-rpm of the power (LP) turbine to the 1,000-rpm of the big DH propeller.

The Britannia was looking promising as a successor to the piston-engined airliners of Boeing, Douglas and Lockheed, but it was not to be. In February 1954, on a test flight with delegates from the Dutch airline KLM, a Britannia caught fire after one of the engine's input pinions, in the reduction gear at the front end of the long propeller shaft, stripped its teeth. The pilot was forced to crash-land and, though no one was hurt, the aeroplane was a write-off.

As we shall see later, the Britannia hit further snags but this one was of interest in the Rolls-Royce history in that it prompted Hives to pick up the telephone and ask Hooker if he needed any help. Later, Hooker wrote of the incident:

The familiar voice said 'I hear that you are in trouble, Stanley. Do you want any help from us?' My spirits soared as I stuttered, 'Yes, please.' 'Right,' said Hs, 'I will send down the First Eleven, who do you want?' I had the nerve to say Rubbra, Lovesey, Lombard, Haworth and gear-expert Davies. Next morning they were poring over the failure in my office. That was the last time I spoke to Hs, and I am very glad it happened. It was typical of the man that, though by this time he was calling me his 'one great failure', he never hesitated in his act of superb generosity to a competitor in trouble.

This problem was cured, but two years later a cooling problem hit the Britannia when on tropical testing for BOAC. This was to cause a further two-year delay and a cash-flow crisis that nearly bankrupted Bristol. In the end, as many major airlines waited for the faster 707, only eighty Britannias were sold. This figure was enough to save Bristol from liquidation, but not enough to put the company into a strong financial position.

The Proteus eventually established itself as a reliable and efficient engine and, as well as giving service in the air, pioneered two non-aero uses for gas turbines in propelling warships and generating electricity.

In the meantime, the Engine Division at Bristol was asked to develop a ramjet engine for the Bristol Ferranti Bloodhound ground-to-air missile. Boeing had already developed the ramjet-powered Bomarc missile and, with technical help from the company, Bristol's Thor ramjet was developed during the 1950s. Developed under the code name *Red Duster*, the Bristol Ferranti Bloodhound made the first direct hit by a British surface-to-air missile when it intercepted a Firefly target aircraft over Cardigan Bay (west Wales) in 1957. It entered service with the RAF at North Coates, Lincolnshire, in the same year using four solid-fuel jettisonable boosters for take-off and two Bristol Thor ramjets for the main propulsion to the target.

Led by Dr. R.R. Jamison, who had also come from Rolls-Royce in Derby, the Engine Division moved on from Thor to develop ramjets capable of speeds up to Mach 6. The Odin was developed for the Hawker Siddeley Dynamics Sea Dart anti-aircraft and anti-ship missile, and continued production into the 1990s, by which time 4,500 ramjet engines had been made at Bristol.

At the same time Bristol developed a small engine (the design of which had begun under Frank Owner as the Saturn), the Orpheus turbojet. Development financial support came from the Mutual Weapons Development Programme (MWDP), whose offices were very close to the Crazy Horse nightclub in Paris – much to the delight of Stanley Hooker, according to Sir Basil Blackwell.

By the end of the 1950s, the engine was available in a number of versions with thrusts varying from 4,000 lb to 8,170 lb, and could be described as the most advanced medium-thrust lightweight turbojet engine in the world. The Mark 701 powered the Folland Gnat lightweight fighter in service in India and Finland, while the Mark 101 powered the Gnat Trainer for the RAF, the Mark 803 the Fiat G91 strike fighter and trainer for the NATO air forces, and the Mark 805 the Fuji T1A, the Japanese trainer. It was also manufactured under licence in Germany, Italy and India.

Perhaps not generally known is the fact that the prototype Lockheed Jetstar used two Orpheus engines. They were certificated for civil use just in time to match the design and build of the aircraft in Kelly Johnson's 'skunk works', although production aircraft were fitted with Pratt & Whitney engines.

THE OLYMPUS

The other engine that Hooker found in the process of development when he arrived at Bristol in 1949 was the Olympus, an axial turbojet at the time more powerful than any previously built in Britain and originally designed to power bombers at 500 mph at 50,000 feet. The Bristol Engine Division was developing it at Air Ministry expense to power the Vulcan bomber. Whereas, as we have seen, the Proteus hit one snag after another, the Olympus development proceeded relatively smoothly.

Norman Rowbotham, who had become Managing Director of Bristol, appointed Hooker as chief engineer in place of Frank Owner, a long-time servant of Bristol. Hooker's comments on this move were:

I tried to be a loyal colleague to Frank Owner, and he in turn treated me with a great consideration. He was a man of great intelligence, well read, a brilliant pianist, possessed of a good sense of humour and a great command of English, both written and spoken. Yet he was capable of acting like an ostrich and putting his head in the sand, though he must have known the day of reckoning would come when the giant Brabazon and Princess would need their Proteus engines. Meanwhile, this abysmal engine sank ever deeper in the morass, failing its compressor blades, turbine blades, bearings and many other parts, even at totally inadequate powers well below 2,000 hp.

Designed initially to produce 9,750 lb thrust, the Olympus had reached 11,000 lb by the time it flew in a prototype Vulcan in September 1953 and by the mid-1950s it was giving 13,500 lb. In 1957, the Mk 201 version of the Olympus 6 reached 17,000 lb in the Vulcan B.2. At that point the Air Ministry switched its support to Rolls-Royce's Conway and instructed Avro to redesign the Vulcan to take the Conway. Rolls-Royce had been successful in selling the Conway to Boeing as an option engine on the 707, whereas Bristol had failed to secure any commitment to the Olympus from Boeing. Rolls-Royce was able to argue strongly to the British Government that it made sense to use the Conway in the Mark 2 Vulcan to achieve commonality with the Victor Mark 2 and broaden the base for the sale of the engine to airlines buying the 707 and DC-8.

In Gordon Lewis's view, Dr. A.A. Griffith did not help the cause of the Olympus [and why should he? After all, he worked for Rolls-Royce, Bristol's chief rival] by telling the Air Ministry that a two-shaft engine would not operate above 36,000 feet without stalling.

Bristol resisted this move and said it would develop the Olympus at its own expense. The Ministry officials assumed that this would mean a more expensive engine, but Bristol said it would still sell it at the same price as the Conway. Hooker wrote later:

In the event the Olympus 200 went like clockwork . . . Production rose to the occasion [Hooker had been very critical of Bristol's production capabilities] and churned them out at a keen price. The Conway, on the other hand, became more and more expensive, partly to pay for changes to suit the civil DC-8 and 707-420, of which a handful had been sold, and partly because the production price was much

The Bristol Fighter, this one with a Rolls-Royce Falcon
engine, was described by German fighter ace, von Richthofen,
after he had fought against it over Arras in March 1917,
as 'a new type of aeroplane which we had not known
before, and it appears to be quick and rather handy,
with a powerful motor.'

ABOVE AND OPPOSITE: The Bristol Aeroplane Company
founded flying schools at the same time as it
was building aeroplanes. These photographs were
taken in the early 1920s.

Roy, later Sir Roy, Fedden, a brilliant engineer and leader of men, but also extremely single-minded, and certainly not *one of us* as far as the directors of the Bristol Aeroplane Company were concerned.

OPPOSITE: In 1918, Fedden was urged by Brigadier-General J.S. Weir, technical controller of the Ministry of Munitions, to 'press on in all haste to perfect the Jupiter, which I feel sure has a considerable future in front of it in commercial aviation.' Weir was right, the Jupiter was enormously successful not only in Britain but also overseas, where Fedden secured licences in eighteen countries. Here is a civil Mk 10 FBM Jupiter (TOP) and a military Mk 7F Jupiter (BOTTOM).

A Cosmos Jupiter engine in a Bristol Badger.

A Hercules-powered Beaufighter. Under Fedden, Bristol
perfected the sleeve-valve engine and, by the outbreak
of war in 1939, was turning out Mercury and Hercules
engines by the thousand.

The Bristol Blenheim. Originally called the 142, it first flew on
12 April 1935 and was probably the most advanced aeroplane
in Europe. At the time it was faster than fighters and within weeks
the Air Ministry had ordered 150. However, by 1939 both the
British and German fighters could fly at much higher speeds.

The Bristol Brabazon. Recommended by the Brabazon
Committee, much of Bristol's efforts in the immediate
post-war years went into this aircraft. Powered by eight
2,500 hp Centaurus engines buried in pairs within
a wing of no less than a 230-feet span, it was built in
a special eight-acre, three-bay hangar. The runway
at Bristol was lengthened and strengthened, and, in the
process, the village of Charlton was destroyed.

Servicing a Centaurus 20 in the huge wing of the
Bristol Brabazon.

Gordon Lewis, who made an outstanding contribution to
Bristol engineering in the post-war years.

The Bristol Britannia. It looked promising as a successor to the
piston-engined airliners of Boeing, Douglas and Lockheed.
However, several difficulties, including a cooling problem while
on tropical testing with BOAC, caused delays, so that only 80
were sold as the world's airlines waited for the faster Boeing 707.

After designing the Dart and the Tyne at Rolls-Royce,
Lionel Haworth moved to Bristol where he made a major
contribution on the Olympus and the Pegasus.

The Bloodhound entered service with the RAF at North
Coates, Lincolnshire, in 1957, using four solid-fuel
jettisonable boosters for take-off and two Bristol Thor
ramjets for the main propulsion to the target.

TSR2 (Tactical Strike and Reconnaissance), born out of a
cancellation – the project to produce a supersonic bomber to
replace the V-Force, the Victor, Vulcan and Valiant – it suffered
cancellation itself. Gordon Lewis would say: 'Much could be
written on the persecution of the TSR-2, and one can only
wonder at the ineptitude and strange political motivation
of the Labour Government.'

Sir Reginald Verdon-Smith. Realising the need to have a strong British aero engine company, he was happy to sell Bristol to Rolls-Royce providing an understanding could be reached with Bristol's partner in the Bristol Siddeley Group, Hawker Siddeley.

higher than expectation . . . we went on to produce the 300-series engine rated at 20,000 lbs, more than double the original design figure, by adding a zero-stage to increase the mass flow. These became the standard engines of the RAF's Vulcan B.2 force, which stayed in use 25 years to 1983. These were the trusty machines, which bombed the runway at Port Stanley in what were by far the longest bombing missions in the history of air warfare.

TSR2

The TSR2 (Tactical Strike and Reconnaissance) was born out of a cancellation, that of a project by Avro to produce a supersonic bomber to replace the V-Force – the Victor, Vulcan and Valiant. However, this was doomed as soon as Minister of Defence, Duncan Sandys, produced his 1957 Defence White Paper. According to Sandys, big bombers had no future and therefore the RAF, if it was to have any traditional bomber role at all, needed a smaller tactical aircraft able to attack at twice the speed of sound and at a low level to avoid the enemy's defence mechanism.

Vickers and English Electric, with some prodding from Government, began talking in 1958 and eventually, on 1 July 1960, merged as the British Aircraft Corporation taking in the Bristol Aircraft Company at the same time. It drew up a detailed specification for a tactical, strike, reconnaissance aircraft, which became know as the TSR2. What emerged over the following two years was a fully automated machine able, through radar in its nose, to follow at heights down to as low as 200 feet the undulating shapes of an enemy landscape, well underneath the scan of any radar system, at the same time transmitting back to base pictures of the territory over which it was passing. Its range would be 3,000 miles, armoury nuclear or conventional, with a speed around 750 mph at low level and Mach 2 – 1,350 mph – at altitude. As Arthur Reed put it, in his book, *Britain's Aircraft Industry*:

In comparative terms with the military aircraft, which the industry had been turning out during that era – the Hunters, Canberras, Swifts, Javelins, the V-bombers – TSR2 was as advanced as were the Spitfire and Hurricane over the wire and fabric biplanes which they superseded.

One of the demands made for the TSR2 was that it should be able to fly at its maximum Mach 2.2 for a full forty-five minutes. This would mean a total redesign of Bristol's Olympus, the engine chosen, in high-temperature materials. Furthermore, the engine would need a lot more thrust as well as a reheat system (afterburner) capable of being fully modulated over the entire range of augmentation instead of being a mere on/off device, as had previous British afterburners.

In order to compete for the engine contract, Bristol Aero Engines had also been forced into some rationalisation and, in April 1959, had merged with Armstrong Siddeley Motors of Coventry to form Bristol Siddeley Engines Limited, jointly owned by the Bristol Aeroplane Company and the Hawker Siddeley Group. The company had already reorganised itself in 1956 into Bristol Aircraft, Bristol Aero Engines and Bristol Cars with the Bristol Aeroplane Company remaining as the holding company.

Hooker thought the demands for performance were excessive and wrote later:

Also demanded was an operational radius of 1,000 nautical miles. I met the Vice-Chief of Air Staff, Sir Geoffrey Tuttle . . . I asked him, 'Geoffrey, why do you insist on a 1,080 mile range? It is clearly a number carved out of the sky. Do you realise that the final 100 miles will cost you something like a million pounds a mile for engineering development?' (This was certainly an underestimate) . . .

Do not think that I, or anyone at Bristol, did not do our best to produce good engines for TSR2. In this programme even the Olympus did its best to bite us, when one blew up during ground running under the Vulcan flying test-bed.

Hooker is being somewhat economical with the truth. According to Arthur Reed:

Testing of the Bristol Siddeley Olympus 22R engines produced the first major technical crisis. Four blew up at intervals, three on test beds, one slung under a Vulcan bomber, and although vibration of the low-pressure shaft was established as the cause, [Roland] Beaumont had to make the first flight from Boscombe Down . . . on 28 September 1964, without the essential modifications made and in the knowledge that during the climb through the first 1,000 feet after take-off, when full power was essential, either one of the two engines, each asserting 15 tons of thrust . . . could explode, destroying in one go both of them [he and his co-pilot], the TSR2 aircraft and the whole project.

This and other technical problems caused the aircraft and engine development to run late as well as hugely in excess of the budget. By 1964 there were many voices in Whitehall and the media calling for its cancellation. Apart from challenging its own cost, many in the Labour Party opposed it on the grounds that it perpetuated Britain's membership of the nuclear club.

When Labour returned to power (after a gap of thirteen years) in October 1964, Prime Minister Harold Wilson ordered a review of aerospace projects. Indeed, in February 1965 he told the House of Commons that the research and development bill for the TSR2 had risen to £300 million (£5.4 billion in today's terms), way above its original estimate. On an order for 150, each aircraft would cost £5 million, 25 times as much as the Canberra it was replacing and about as much as a pre-war battleship. [This was disingenuous, as overall costs had risen by about four times since before the war.]

The axe came in the 1965 Budget. It was a mighty blow for the British aircraft industry and, of course, to Bristol. Gordon Lewis would say later:

Much could be written on the persecution of the TSR-2, and one can only wonder at the ineptitude and strange political motivation of the newly elected Labour government of the time. The calamity was compounded at Bristol by the almost simultaneous cancellation of BS.100 and the P.1154, which was still some way off first flight, but the real bitterness of feeling at Bristol was for the TSR-2. The remains of this superb and potentially formidable aircraft can be seen at Duxford as the surviving memorial to great engineering and doctrinaire politics.

However, Bristol at least had the Olympus, the engine destined for Concorde SST (supersonic transport).

CONCORDE

Concorde's development began in 1956. By then it was clear that Britain had fallen behind the USA in the development of subsonic jet airliners. BOAC's lack of interest in the Vickers VC7 had killed Government support for the V1000 project, or probably the cancellation of the V1000 meant that the VC7 option for BOAC was removed and it seemed that Britain's best chance was to leapfrog and produce the world's first supersonic airliner. The Government asked the Royal Aircraft Establishment to carry out research and make a proposal. It formed the Supersonic Transport Aircraft Committee (STAC) and in its final report in March 1959, STAC recommended an SST to cruise at Mach 2.2 or 1,450 mph, with transatlantic range. It was going to be a daunting aircraft for Britain to develop alone and many were in favour of finding a partner.

There had been some examples of European collaboration during the 1950s but these were limited. However, by the early 1960s attitudes were beginning to change. George Edwards saw internationalisation as the logical conclusion to the rationalisation proceeding apace in the British aircraft industry at this time. Air Commodore F.R. Banks was in favour of a 'whole-hearted link-up', which, in his view, should have started at the end of the war. However, others felt that it would not be easy. Sir Reginald Verdon-Smith said:

We have arrived at the stage where international co-operation is necessary. But to say that it is necessary does not mean to say that it is easy. All concerned in these international projects find them very complicated and it is said that the engineering involved is simple compared with the political and administrative problems.

The Government, frightened by the potential cost, which it thought might be as high as £100 million (about £1.8 billion in today's terms) – although, in the event, it was to cost about twenty times as much! – was certainly keen to share the costs. The Treasury favoured the USA, but the Americans chose a more ambitious and expensive project, and attention turned towards the French who responded positively to British overtures.

Design teams from the British Aircraft Corporation and the French company, Sud Aviation, began to work together. After much negotiation (the French favoured a medium-range aircraft while the British felt that a trans-atlantic capability was absolutely essential), a joint programme got under way. On the engine side, Bristol and SNECMA, the French engine company, came together more easily and reached agreement to develop an engine based on the Olympus. The two companies had worked together in the 1950s. Arnold (later Sir Arnold) Hall, then the Managing Director of Bristol Siddeley Engines, said later:

We went to talk to my friends at SNECMA to see what they felt about it all. The result of these conversations was that we concluded that, if at any time there were an aircraft project, the two companies would work together on the engine.

Hooker said that Bristol and SNECMA enjoyed the happiest of relationships.

Their two top engineers, Michel Garnier and Jean Devriese, quickly became BSEL's friends, and I cannot recall a single technical difference that was not settled by a single short meeting.

Bristol's relationships with SNECMA went back a long way. As we have already seen, Fedden had signed a licence agreement with the Gnome-Rhône Company, and Rowbotham and Ninnes had worked with the French company in the 1920s and 1930s. After the war all the important French aero engine companies except Hispano-Suiza were amalgamated into SNECMA and in 1948 the French company had approached Bristol to license the Hercules as a replacement for its 14R (which Ninnes had developed for it in the 1930s). Meanwhile, Hispano-Suiza built the Rolls-Royce Tay, and its uprated version, the Verdun, under licence. Relations between SNECMA and Bristol became close.

The official work (and money) was split 60 per cent to Bristol and 40 per cent to SNECMA. Bristol was responsible for the basic engine, and the French for the new and complex reheat jet pipe, thrust reverser, noise suppressor and the convergent/divergent final nozzle, which was essential for supersonic operation.

It became necessary to increase the thrust of the Olympus engines for Concorde because the aircraft grew heavier, as new aircraft usually do in development. Increased weight meant more fuel and larger engines, which in turn meant more fuel. More fuel meant a larger wing, which meant more weight. This spiral was exacerbated in the case of Concorde because there could be no reduction in the Paris-to-New-York capability. The initial weight of 326,000 lb increased to 408,000 lb and the Olympus engines would have to achieve more thrust to cope.

The designers at Bristol gave the Olympus 593 a zero-stage on the front of the LP compressor, to pump more air with a higher pressure ratio, as well as a redesigned turbine with air-cooled rotor blades to allow a higher operating temperature. This achieved a thrust of 30,000 lb, but it became clear that 36,000 lb would be necessary. Bristol persuaded SNECMA to incorporate partial reheat (afterburning) in the jet pipe, which would give a 20 per cent thrust-boost at take-off. Nevertheless, there were drawbacks – extra complication in the variable propelling nozzle, possibly higher fire risk and, certainly, extra noise (although more thrust would inevitably mean more noise however it was produced). There was some reluctance to use reheat but, according to Hooker:

SNECMA produced a superb reheat and nozzle system, and today, though the captain usually informs the passengers when he is switching on reheat to start his transonic acceleration through the once-feared sonic barrier, it is all a non-event and Mach 2 is as quiet and smooth as subsonic flight.

At 50,000 feet the Olympus 593 gave 10,000 lb of thrust but as, at its cruising speed, each pound of thrust was the equivalent to 4 hp, the total horsepower of Concorde was no less than 140,800. The Concorde engines weighed seven times as much as Whittle's engine, but gave twenty-five times the thrust at three times the speed, with a much lower specific fuel consumption. Alec Collins, a long-serving Rolls-Royce engineer, said of Concorde's engines:

The most impressive thing is that they still have the highest overall efficiency of any jet engines in service today. Fuel consumption is around 30mpg/passenger, which at 1320 mph is very impressive and is probably better than that of the cars that most of Concorde's passengers drive at about one twentieth of the speed.

However, Concorde threatened to be even noisier than the already very noisy Boeing 707 and Douglas DC-8. Hooker appointed E. Ffowcs Williams at Imperial College, London, to take charge of his Noise Research Department. He, in turn, gathered other experts around him and their research enabled SNECMA to design an exhaust system that, with little performance and weight penalty, reduced noise levels – although not to that of subsonic jets.

Noise became a big factor in the opposition that built up to Concorde. The sonic boom was exploited to the full, even though it resembles distant thunder. As a result, Concorde has never been allowed to cruise at Mach 2 over land.

Hooker gave credit to Pierre Young (whose French mother and childhood in France meant that he was bilingual) for achieving a successful propulsion system, and also Lionel Haworth of whom he said:

Haworth had been a friend [at Rolls-Royce in Derby] since 1938 and I was thrilled when he threw in his lot with us at Bristol. He did many enormous tasks in perfecting the design of the Olympus and Pegasus, bringing to bear equal proficiency in mechanical engineering, aerodynamics, vibration, material properties and accurate estimations of weight and cost. I shall never forget his eloquence in explaining to Pierre Sartre [chief designer at Sud Aviation] how to solve the unprecedented problem of installing the very long and rigid engine in the extremely long nacelle fixed to the highly flexible wing, which in any case varies in dimensions according to how hot or cold it gets in flight. He was a pupil of Rubbra's, and there is no higher praise than to describe him as Bristol's Rubbra.

As with any new leap forward in technology or design, Concorde encountered many problems – not least of which was its rapidly escalating development cost. From an original estimate when the agreement between France and Britain was signed that the development cost up to 1970 would be between £150 and £170 million, the figure mounted steadily. By July 1964 it was £275 million, largely because of the change from medium to long haul. But that was just the beginning. Through the late 1960s and into the early 1970s, the development costs rose inexorably until they broke the £1,000 million (about £15 billion in today's terms) barrier before development was complete.

It should be remembered that the US Government and aircraft industry spent $1,000 million *not* to have an SST when the Boeing 2707-300 1,900 mph, supersonic 250-seater was abandoned in 1971.

Dr. Archibald Russell (known to everyone as 'Russ') – who, with Dr. Bill Strang and Mick Wilde, had produced the Bristol 198 proposal for an aircraft powered by six Olympus jets under the wing during the 1950s, thus virtually assuring that the Olympus would be the engine selected for Concorde – carried in his wallet throughout the 1960s a piece of paper that grew grubbier and grubbier as he showed it to all and sundry – including the media. It said:

All-up weight, 385,000 lb – total development costs £1,000 million.

Partly as a result of these escalating costs and partly as a result of Britain's poorly performing economy in the 1960s – a decade that, even though it often looked like a 'golden economic age' of low inflation, low unemployment and relatively high growth, was actually punctuated by economic crises (including a sterling devaluation in November 1967) – Concorde's future was constantly in doubt. As soon as the Labour Party returned to power in October 1964, a White Paper (dubbed the Brown Paper after George Brown, the Minister for Economic Affairs) was published stating that the Government would carry out a 'strict review' of all Government spending to relieve the strain on the balance of payments and release funds for more productive purposes by cutting out expenditure on items of low economic priority, such as 'prestige projects'. Few doubted that one of the 'prestige projects' was Concorde.

On the Sunday before the Paper was presented to the House of Commons, word went to Paris instructing the British Ambassador, Sir Pierson Dixon, to tell the French Foreign Minister, Couve de Murville, of the British Government's wish to cancel the Concorde project. Unfortunately, neither Dixon nor de Murville could be found. (As it happened, Dixon had gone shooting with the Chairman of Hispano-Suiza, the French aircraft engine company). Nevertheless, word of the British plans reached President de Gaulle. When the British Aviation Minister, Roy Jenkins, visited Paris just after publication of the White Paper to negotiate the terms of the cancellation, his French counterpart, M. Jacquet, under instructions from de Gaulle, informed him that there would be no cancellation and that Britain would be taken to the international court if necessary. The interview lasted less than half an hour.

In spite of all its dramas, rows in Parliament, political rows with the French, protest groups in the UK and USA, crises in design and escalating costs, Concorde was a huge technical achievement. Charles Gardner summed it up well in his book *British Aircraft Corporation*.

The immensity of the Concorde achievement is still not really appreciated outside the limited circle of those who are in the business of aviation. The problems, which faced the designers, were mountainous. A supersonic airliner not only had to cruise beyond the speed of sound, it had, for aerodynamic reasons affecting range and economy, to cruise at twice that speed i.e. at Mach 2. It also had to produce this performance not for the ten-minute dash of a fighter (which was then pampered by a maintenance crew until its next short flight, maybe several days later) but for three or four continuous hours per sortie, with at least two such flights a day up to 2800 hours to 3000 hours a year. The comfort of the conditions inside the passenger cabin had to be indistinguishable from those inside any conventional jetliner, the safety, reliability, handling and the ability to face engine failure at take-off or en route, plus docility at lower speeds round the airfield, had to be in conformity with the highest standards of British and American certificates of Airworthiness. The engine intakes and airflow had to function, by variable geometry, equally well at low speeds at ground level as at Mach 2 at over 60,000 feet, and the whole aircraft had to be able to operate anywhere in the world and to go on doing so, without excessive maintenance costs, for ten years or more.

Nevertheless, the sad truth is that significant sales were not achieved. Environmentalists in the USA mounted a huge campaign against it, as they had against the Boeing SST. OPEC (Organization of Petroleum Exporting Countries) quintupled the cost of fuel in 1973/4 so that the price of a barrel of oil, $1.50 when Concorde was conceived in 1956, rose to $30. And initially there were virtually no routes for Concorde to fly. South Africa was not possible because there was no politically viable refuelling stop available in black Africa. The route to Australia was held up by problems with the authorities in Singapore and Malaysia. And in Tokyo, students would not allow even conventional traffic into the new airport. In the USA, Government hearings held up flights to Washington until 1976 and to New York until 1977.

ROLLS-ROYCE SWOOPS

The Government and its, or rather the taxpayers', money had been important to the aircraft industry from its birth, and we have seen how aircraft and aero engine manufacturers flourished or struggled, with or without help from that money. By the time Harold Wilson's Labour Government came to power in October 1964 there were those, including Wilson himself and a number of his Cabinet colleagues, who felt that the industry had absorbed too much of the country's engineering resources without giving much in return. With the economy struggling it was not long before aerospace projects were being cancelled. First went the AW681 and P.1154, quickly followed by the TSR2. The replacements were to be the F-111 from General Dynamics in the USA, the Phantom from McDonnell also in the USA, and more Buccaneers from the Blackburn Aircraft Company, already in service. The F-111 was later cancelled but the Phantoms, re-engineered to take the Rolls-Royce Spey 201 engine, were ordered. One Government-funded project, which did survive, was Hawker Siddeley's vertical take-off fighter, the P.1127, or Harrier, powered by Bristol's Pegasus engine.

Shortly after the 1964 election, the Government set up an enquiry chaired by Lord Plowden, the Chairman of Tube Investments (now TI), which was asked to examine 'the future place and organisation of the aircraft industry in relation to the general economy of the country'. This committee, which reported in December 1965, was critical, blaming the industry for a series of poorly planned projects and failure to develop aircraft suitable for the world market. Nevertheless, it did not want the industry to be allowed to decline, suggesting that, with its highly skilled labour force and negligible import propensity, it was 'exactly the sort of industry on which Britain should concentrate'. The main problem was the low production runs; the solution, collaboration. It did not see the USA as a likely partner in view of its size and strength. France was the most obvious partner, sharing as it did many of the same problems as the UK.

This was music to the ears of the Government, which had already embarked on co-operation with France. In May 1965, agreement had been reached on two military projects. The first was a supersonic attack fighter/trainer, the Jaguar, with the French company Breguet, supported by BAC, taking the lead on the airframe, and Rolls-Royce, supported by Turbomeca of France, taking the lead on the Rolls-Royce Adour engine. The second was the sophisticated multi-role combat aircraft, the AFVG (Anglo-French Variable Geometry). The lead contractor was BAC, supported by the French company, Dassault. Bristol Siddeley, already working with SNECMA on the Olympus for Concorde, would co-operate on the engine. As it turned out, one project, the Jaguar, with sales of 600 aircraft, worked well, while the AFVG failed as it quickly became clear that Dassault was not going to subordinate itself to the British and broke away in 1967 to develop its own variable geometry aircraft. Britain was forced to find other partners which, two years later, it succeeded in doing. In 1969, Britain, Germany and Italy agreed to develop the Multi-Role Combat Aircraft (MRCA), later called the Tornado. Some 800 were ordered, and BAC benefited enormously in the 1970s and 1980s.

As well as commenting on the military side, the Plowden Committee also expressed views on the civil aspect of the aviation industry. The three major airframe manufacturers – BAC, Hawker Siddeley and Sud Aviation – were all considering replacements for their BAC 111, Trident and Caravelle, and again the Plowden Committee urged collaboration to take advantage of the new, more powerful, bypass turbo engines becoming available. In the USA, Boeing was developing its 747, the 'Jumbo', and Lockheed and McDonnell Douglas (McDonnell and Douglas merged in 1967) were beginning work on what eventually became the three-engined, wide-bodied airbuses, the TriStar, and the Douglas DC-10 – although they started as twin-engined aircraft to meet the specification of Frank Kolk, the chief engineer of American Airlines.

After Plowden, both British and French Governments encouraged the idea of consortia to produce a smaller, wide-bodied airbus, and the German aircraft companies Bölkow, VFW (formed from Wezer, Focke-Wulf and Heinkel), HFB, Messerschmitt and Dornier grouped together to join the French and British. Designs were called for and at the end of 1966 the A300

Airbus, a twin-engined aircraft capable of carrying 300 passengers, was chosen. Development of the airframe was split 37.5 per cent Sud Aviation, 37.5 per cent Hawker Siddeley and 25 per cent to the German consortium. On the engine front, the French favoured the Pratt & Whitney JT-9D, already chosen by Boeing for the 747. Agreement had been reached for Bristol Siddeley and SNECMA to manufacture this engine under licence. Bristol, like Rolls-Royce, had been studying high-bypass-ratio fan engines but could not take the risk of the large investment necessary to develop such an engine. Rolls-Royce was outraged by this possibility. It was already very disappointed at losing out to the JT-9D on the 747, and it was not going to be squeezed out of its *home* market as well.

It was probably the clinching argument in Rolls-Royce's decision to buy Bristol Siddeley Engines in June 1966, although it was inevitable at some stage that the two remaining British engine manufacturers should come together to compete against the US giants, General Electric and Pratt & Whitney.

Rolls-Royce had for some time been studying and carrying out research and rig work into the optimum configuration of the next generation of large subsonic transport engines for short- and long-range aircraft to replace the Conway. As a result, Rolls-Royce had decided to build a demonstrator engine and, within a relatively short period, expected to be able to put forward an engine both for the European Airbus and any similar US aircraft.

Rolls-Royce was fairly confident that Bristol had not been carrying out similar work and was therefore not in a position to compete. This made Rolls-Royce all the more concerned when it read press reports that Bristol, to compensate for its lack of knowledge and experience in this field, was seeking an association with Rolls-Royce's competitor in the large subsonic field, Pratt & Whitney.

As Rolls-Royce saw it, such an association would consolidate on a permanent basis the existing link between Pratt & Whitney and SNECMA and, in effect, provide Pratt & Whitney with a firm foothold right inside the British and European market. If this association led to an engine for the Airbus, it would really be a Pratt & Whitney engine as neither Bristol nor SNECMA possessed the experience or expertise to develop such an engine.

Rolls-Royce felt it could not stand idly by and allow this to happen, and one way in which the company could prevent it was to take over Bristol. With its hands pretty full in developing the RB 211, it was not an ideal moment – though at this stage the Rolls-Royce directors could not foretell how much of their resources would be tied up in developing that engine. That aside, a take-over or merger made sound industrial and commercial sense. By this time, the UK manufacture of large- or even medium-sized airliners had dwindled to practically nil. In the future, customers were going to be either in the USA or part of a European consortium. Size would matter in every way, whether concerning research, design, development or manufacture. The pressure and logic for the two companies to come together was virtually irresistible. There were, of course, many practical details to be resolved but, with goodwill, these should not present any major obstacle to the creation of a coherent and powerful aero engine design and manufacturing company.

A confidential Rolls-Royce memorandum of early 1966 laid out the reasons for merger.

* Sir Reginald Verdon-Smith is in favour of a single-engine company and is prepared to sell Bristol to Rolls-Royce provided an understanding reached with Bristol's partner in Bristol Siddeley Engines, Hawker Siddeley Group.
* A duopoly inevitably leads to playing off with deterioration of margins.
* Ministry of Aviation strongly in favour to avoid difficulties of sharing out projects particularly when Europe involved.
* The removal of the continual threat [from Bristol] of a link-up with Pratt & Whitney.
* Control of total capacity, which would, in the event of contraction, give some control of where and when.
* Reduction in costs through rationalisation.
* Avoidance of Buggins' turn next approach by Government in award of contracts.
* Link-up of technical teams and manufacturing expertise and facilities.
* Rationalisation of sales and service.
* Industrial gas turbines and rocket motors need benefit of amalgamation.
* Solve problems of joint projects with SNECMA and of engine for variable sweep fighter.
* Bristol would benefit from Rolls-Royce's knowledge and experience of operating at high turbine inlet temperatures for Concorde engine.
* Finally, there should be no personality problems.

The 'Reasons for merger' section of a letter written on 9 September 1966 by Sir Reginald Verdon-Smith, the Chairman of the Bristol Aeroplane Company, stated:

After a successful record as a manufacturing organisation for nearly 50 years, your Company has, since the mergers within the aircraft industry in 1959 and 1960, derived its profits mainly from its 50 per cent ownership of Bristol Siddeley Engines Limited (the other 50 per cent being owned by Hawker Siddeley Group Limited), partly from its own operations in this country and Canada, and partly from other investments including British Aircraft Corporation Limited, Westland Aircraft Limited and certain other companies. In collaboration with Hawker Siddeley Group Limited, your Company has successfully developed Bristol Siddeley Engines Limited as a major unit in the British aero-engine industry, its principal competitor in the United Kingdom being Rolls-Royce Limited.

In 1959 Bristol Aero Engines had merged with Armstrong Siddeley Motors to form Bristol Siddeley Engines, with the Bristol Aeroplane Company and the Hawker Siddeley Group each owning 50 per cent of the shares.

The Siddeley Autocar Company had been formed in 1902 and the Deasy Motor Car Manufacturing Company in 1906. Captain H.H. Deasy, a founder member of the RAC, had sponsored the Deasy company. In 1909 J.D. Siddeley joined Deasy as general manager of the Deasy company and the cars became known as the J.D. Siddeley-Type Deasy cars. In 1912 the name was changed to the Siddeley Deasy Motor Car Company. As with many such companies, World War I brought the opportunity for Siddeley Deasy to establish itself as it was flooded with orders for cars, ambulances and several types of military vehicle and subcontract work on aero engines to Royal Aircraft Factory designs. The company also became involved in the manufacture of aero engines. Before the war, the shipbuilding and engineering company Beardmore in Glasgow had taken a licence to manufacture the Austro Daimler aero engine. This engine was progressively developed with the involvement of the ubiquitous Frank Halford and a colleague, T.C. Pullinger, the Managing Director of Arrol Johnson. In this form it was known as the BHP – Beardmore Halford Pullinger. Siddeley Deasy were asked to manufacture the engine and, while doing so, improved it and renamed it the Puma. By the end of the war the company was producing this engine at the rate of 700 per month.

In 1919 a merger was arranged with Armstrong Whitworth, from which came the Armstrong Whitworth Aircraft Company and Armstrong Siddeley Motors. The company returned to motor car production and introduced the Siddeley Six. However, it continued to build aero engines. As at Bristol, Armstrong Siddeley concentrated on air cooling. Major F.M. Green joined the company from the Royal Aircraft Factory at Farnborough and became chief engineer. He brought with him the preliminary designs of the projected RAF8 fourteen-cylinder, 22.4 litre, air-cooled radial engine. This engine was developed and became the 24.8 litre Jaguar. The Rolls-Royce Eagle and Falcon engines had been superseded in the early 1920s by the Napier Lion which was now overtaken by the Jaguar. However by the late 1920s, the Jaguar was itself overtaken by the superior and simpler Bristol Jupiter.

Armstrong Siddeley continued to develop air-cooled engines culminating in the Cheetah which was widely used in training and liaison aircraft during the Second World War.

World War II brought concentration on aero engines, though tank gearboxes, based on the pre-selector gearbox for the Armstrong Siddeley car, and torpedo engines were also produced. Armstrong Siddeley was an early entrant into the gas turbine field. In 1936, under Colonel Fell, the chief engineer, a study was made of a German combined steam and gas turbine aero engine project. However, there was little follow-through until, in 1938, the company undertook the manufacture of the components of an RAE turbo-supercharger design that employed mechanically independent contra-rotating discs along similar lines to the CR1 gas turbine that Dr. A.A. Griffith was working on at Rolls-Royce. In the early part of World War II, the Ministry of Aircraft Production decided Armstrong Siddeley should pursue the development of gas turbines and assist Metropolitan-Vickers with its F2 axial-compressor single-spool jet engine and, in 1942, the ASX, an axial-compressor single-spool jet engine of 5:1 pressure ratio and a target thrust of 2,600 lb was designed. This engine was developed as the Python turboprop, the first turboprop in the world to go into production though not to fly. That honour went to the Rolls-Royce Trent. In the Westland Wyvern it was the first to enter military service. It was succeeded by the Mamba and Double Mamba turboprops and the Viper turbojet, first introduced as an expendable engine for the Jindivik target drone, and later developed into a successful long-life engine for piloted aircraft, both civil and military.

The largest of the turbojets developed by Armstrong Siddeley was the Sapphire, which powered the Handley Page Victor B Mk 1 bomber, all the marks of the Gloster Javelin fighter, and the Hawker Hunter Mk 2 and Mk 5 fighters. The Sapphire was also produced in the USA as the J65, which powered five different production aircraft and many prototypes.

In the post-war years Armstrong Siddeley also produced rocket engines – the Snarler, the Screamer, the Gamma and the Stentor – and diesel engines, initially of its own design and then of the Maybach-type under licence. The company also introduced a new range of motor cars.

Further rationalisation within the industry had taken place in 1961 when Bristol Siddeley bought the de Havilland Engine Company and Blackburn Engines, both of which had been formerly operating within the Hawker Siddeley Group. As we saw in Chapter Two, the de Havilland Aircraft Company had been founded in 1921 and produced the famous light aero engine Gipsy series originally designed by Frank Halford. Six-cylinder Gipsies powered the Comet Racer for its record-breaking England–Australia flight in 1934.

In 1941 de Havilland made an early entry into jet propulsion and produced the Goblin engine, the first gas turbine to pass a full military Type Test, and subsequently the Ghost, which was the first turbojet to be approved for civil operations. These engines were produced under licence in a number of countries and were used in the Vampire, Venom, SAAB J21R and J29. The Ghost also powered the Comet 1. Additionally, de Havilland manufactured rocket engines and its Sprite was the first to obtain a military Type Test certificate. In 1954 it produced its first axial turbojet, the reheated Gyron of 30,000 lb thrust.

The Blackburn Aeroplane and Motor Company had entered the aero engine field in 1934 when it bought the Cirrus-Hermes Engineering Company, which produced the Cirrus engine – another Frank Halford design. In 1952 Blackburn signed an agreement with the French company, Turbomeca, enabling it to produce the Turbomeca range of small gas turbine engines. The Palouste, Artouste and, later, the Cumulus, were developed for use as airborne auxiliary power units for large aircraft. Turboshaft developments of these engines were the Nimbus and Turmo, both of which had applications in the helicopter and air-cushion craft fields. The Nimbus powered the Westland Scout/Wasp helicopters and the SR-N2 Hovercraft, while the Turmo powered the Vickers-Armstrong VA-3 Hovercraft.

In his letter, Verdon-Smith continued:

The aero-engine industry is a highly competitive one in which it is essential to operate internationally. This demands large units capable of dealing adequately with the size and complexity of the problems presented and commanding sufficient resources to face the capital expenditures and the research and development cost necessary to keep in the forefront . . . Apart from the benefits, which will arise from rationalisation, a merger will mean that, instead of two units in world-wide competition with each other . . . there will be created one of the largest organisations of its kind in the world.

In the long term, the acquisition of Bristol was beneficial to Rolls-Royce, Bristol and the country, as it allowed Rolls-Royce to build an aero engine company capable of competing on a world scale with General Electric and Pratt & Whitney. In the short term, it served to compound the financial problems that were to overwhelm the company at the end of the 1960s. As R.A. MacCrindle and P. Godfrey put it, in their investigation for the Department of Trade and Industry:

We have made calculations which indicate that the profits earned by The Bristol Aeroplane Company Limited and Bristol Siddeley in the years 1966–1969 were, in aggregate, less than the interests and dividends which became payable by Rolls-Royce as a consequence of the acquisition. Not until 1969, when the Rolls-Royce dividend for the year was reduced to six per cent, did the Bristol companies' contribution to group profits exceed the relevant interest and dividends.

And, it has to be said, Rolls-Royce moved very slowly to grasp the rationalisation possibilities of the merger. The acquisition was completed on 7 October

1966, and three days later the Gas Turbine Policy Co-ordinating Committee met for the first time. From Rolls-Royce came Sir Denning Pearson, Sir David Huddie, Fred Hinckley and Adrian Lombard; from Bristol were Hugh Conway, Stanley Hooker and Brian Davidson. In his report to the board on 1 November, Pearson defined the object of this committee as:

to resolve policy issues which arise with two operating units of the company serving the same market and to decide on changes for more effective and economical working by co-ordination of activities or by other means.

By the middle of 1967, four subcommittees had been set up to deal with co-ordination in the areas of marketing, engineering, advanced engineering and experimental manufacture. And these subcommittees were powerful in the sense that they consisted of the principal directors of both the Aero Engine Divisions in Derby and Bristol. However, nothing happened. Bristol Siddeley retained its individual identity within the group. During 1968 and 1969 the various divisions began to take notice of each other, but Bristol Siddeley and the Derby Aero Engine Division were still competing on certain projects.

Some efforts were made to rationalise planning, accounting and personnel policies but it was not until November 1969 that, faced with the rapidly deteriorating financial position in the company, real impetus was given to a rationalisation of resources. Certain main board directors with divisional responsibilities were asked to examine specific areas for rationalisation and cost saving. Even so, Rex Nicholson felt obliged to write to Sir David Huddie in December saying, *inter alia*:

If a meaningful rationalisation of our manufacturing resources is to be achieved, somebody has got to be given the authority to make decisions which will be accepted. To have proposals escalated to a discussion between the proposer, the Chief Executive and the manager of the present facilities, makes nonsense and imposes an intolerable load on the Chief Executive in that he has to make a judgement based on whose views he is going to support.

I believe that if any progress is to be made in this field, the Chief Executive must state clearly that he is giving the authority to one person to make decisions and his decisions will be final, subject naturally to question at any time by the Chief Executive but not in the role of an arbiter.

Nicholson's suggestion was not taken up perhaps because it was felt that such an executive could not be spared. Nicholson himself, for example, was completely involved in production of the RB 211. The logical answer would seem to have been the appointment of a group managing director for gas turbines, but this was not done, perhaps because there was no obvious

101

candidate. It was not until the middle of 1970, after David Huddie's departure that such a position was created and Hugh Conway appointed to fill it.

Some rationalisation gradually took place but, as late as 13 October 1970, one of the directors reported to the board:

To the extent that co-ordination could be achieved at working level a great deal of progress has been made in many fields. However, wherever co-ordination significantly affected divisional interests progress has been slow, owing to the preoccupation of senior officials with the problems of their own divisions.

In the course of interviewing the many former employees of Bristol for this book, the author regularly met with the suggestion that Bristol was just as likely to take over Rolls-Royce as *vice versa*. A glance at the minutes of the board meetings at Bristol during 1966 suggest that this was extremely unlikely and, indeed, given Bristol's precarious financial position, would have been impossible. At a meeting of the directors of Bristol Siddeley Engines, held at the company's London office in Mercury House, Knightsbridge on 19 July 1966, the minutes read:

The cash position had worsened since the beginning of the Ministry's financial year and the forecasts for July and August showed no improvement. Continual delay in paying suppliers was causing considerable dissatisfaction and there were some threats of delay or rejection of the Company's orders.

At the next meeting on 20 September, the minutes read:

The cash position continued to deteriorate; and some £4 million [£72 million in today's terms] due to suppliers was now being withheld.

At the board meeting held at Patchway, Bristol on 1 December, the minutes read:

The cash position remained serious . . . Mr Burns said that the position had been eased somewhat by a loan of £2.8 million [£50.4 million in today's terms] from Rolls-Royce.

We would have to conclude that far from Bristol being able to take over Rolls-Royce, Bristol needed to be rescued and, under the circumstances, Rolls-Royce paid too much. As we shall see, they could not have chosen a worse moment to do so.

CHAPTER FOUR

THE RB 211
FIGHT FOR THE ORDER

'SHOULD ACCEPT TOTAL INVOLVEMENT'
THE HISTORY OF LOCKHEED
DEVELOPMENT OF THE RB 211
DECISION NEAR
'SUCCESS!'

'SHOULD ACCEPT TOTAL INVOLVEMENT'

NEITHER THE BRITISH AIRCRAFT manufacturers, by this time represented by the newly formed British Aircraft Corporation (BAC), nor Rolls-Royce itself were enhancing their reputation in the United States of America in the mid-1960s.

Phil Gilbert, Rolls-Royce's legal adviser in the States, wrote an internal memorandum on 20 September 1965 and his exasperation is clear.

Tom [Bowling] said that with Ken Crooks and others there had been worked out a system for integrating the service organisation. Basically, I gather that this would mean that service will really be conducted out of Derby with Rolls-Royce of Canada and indeed Rolls-Royce Inc pretty well working into the integrated set-up. He said that this was made necessary by reason of the really disastrous events of the last several months in which Rolls-Royce had virtually acquired for itself a poor reputation in this country. This stems largely from the great shortage of parts . . . He asked me what I thought and I said the Rolls-Royce reputation here was seriously deteriorating. I said I for one thought they should accept total involvement characterised by a true commitment and continuity. I said I did not see how in the long run they could exist without the US market, particularly in connection with relationships with US aircraft constructors.

Bowling and Gilbert were very pessimistic about the sales prospects for BAC in the US market. Bowling had been trying to convince American Airlines that Rolls-Royce was ready to deliver, 'but that not even Rolls-Royce had the slightest notion of when BAC would deliver the first airplane'.

By 1966, Phil Gilbert had become convinced that Rolls-Royce was not going to be successful in the USA unless it increased its commitment to the market. On 21 April he wrote a long letter to Fred Hinckley, the head of the International Division in Derby, and laid out no fewer than fourteen actions he felt were necessary for Rolls-Royce to win new business. His first action was an establishment of Rolls-Royce with 'the *visible* appearance of *commitment* and *continuity*'. He went on to say that commitment was not shown 'by sporadic visits by high-ranking executives'. He felt that John Waite's presence in the USA for most of the preceding month was a good start and 'it is eminently clear that his presence here has led to all kinds of contacts, discussions, etc. which would not otherwise have occurred if Rolls-Royce had continued to try to operate out of Derby or Montreal'.

Gilbert's second action point was a much more daring one. He wanted Rolls-Royce to make dramatic price reductions. In 1961 Rolls-Royce had begun development of a large new two-shaft civil turbofan demonstrator, the RB 178. This engine, which had a bypass ratio of 2.3:1, first ran in July 1966 delivering 28,000 lb thrust. Gilbert wanted Rolls-Royce to try to offer an engine based on it at $50,000 to $100,000 under the price offered to Boeing and Pan Am, and presumably under the price offered by Pratt & Whitney. In view of what was to happen four years later, this could only be described as a risky strategy.

Gilbert wanted a sustained effort to be made on Douglas 'to capitalise on all the advantages, which you can offer'. He wanted Rolls-Royce to offer a price demonstrably lower than Pratt & Whitney's, and for Rolls-Royce executives to live with Douglas 'to the same extent that Pratt & Whitney undoubtedly does'. He felt that Rolls-Royce was not exploiting the quietness of its engines and recommended the giving of noise level guarantees. He wrote:

I would recommend that these guarantees be brought to the attention, either through press leaks or otherwise, of the FAA [Federal Aviation Administration] and all other parties. The climate of opinion is plainly right for this.

Gilbert went on to suggest Rolls-Royce grasp the nettle of inflation guarantees as well, and build in escalation keyed to US indices, as 'any customer here is prepared to accept US inflation, but UK inflation is to him a great unknown'. US customers should also be offered dollar rather than sterling prices. He concluded:

I believe that such a program might very well force a review of the whole Boeing and Pan Am tentative decision on the Pratt & Whitney engine, and could cause others to slow down commitments to Pratt & Whitney. It would, I think, vastly improve your long-range prospects with Douglas. Failing this . . . it seems to me only reasonable to assume that Pratt & Whitney will tie up the 747 and then the jumbo-twin, and the European airbus, and the other airlines and Douglas would be likely to follow along. This, I am sure you will agree, should not be allowed to happen. I think it lies within your power to prevent it . . . The risks are great, but I think that the requirements for a successful long-term future indicate that they should be taken. Nothing could be worse, I think, than for Rolls-Royce to try to remain uncommitted and to think it can effectively compete with Pratt & Whitney by sporadic effort. Nothing could be less promising than an attempt to sell [to] TWA and other airlines and Douglas on terms no more attractive than those already offered to Pan Am and Boeing and no more attractive than those already offered by Pratt & Whitney.

In May 1963, Adrian Lombard presented to the main Rolls-Royce board the case for starting work on a new engine. It was to replace the Conway, by this time only just competitive with the latest engines from Pratt & Whitney. Rolls-Royce had been studying the possibilities of such a new engine for some time and in July 1961 had put together a brochure describing two versions of a large turbojet engine called the RB 178. The brochure described a two-shaft engine of high bypass ratio, giving a take-off thrust of approximately 25,000 lb. Rigs were built so that the necessary aerodynamic development work could be carried out.

Meanwhile, further development led Rolls-Royce to believe that a three-shaft engine would be a more robust and elegant design, giving much greater thrust than the RB 178. It was invited by both Boeing and Douglas to quote for a 'big' engine with a high bypass ratio for a longer-range, wide-bodied aircraft that would have both passenger and freight application. Rolls-Royce began development work on a three-shaft engine.

At the same time, in the USA, both Pratt & Whitney and General Electric were forging ahead in their development of high bypass ratio engines. They were greatly assisted by Government funding for demonstrator engine programmes, to produce an engine to power the C5A military transport aircraft, and the contract to produce this engine was secured by General Electric in 1965 when the US Government ordered the TF39 engine.

Coincidentally, the new Jumbo jet, the 747, was being developed by Boeing, and Rolls-Royce lost out to Pratt & Whitney in the competition to be the first engine supplier when Pan Am ordered the 747 specifying the Pratt & Whitney JT-9D engine. The Rolls-Royce board came to the conclusion that the company must produce a world-beating 'big' engine if it was to maintain its position as one of the three major aero engine manufacturers.

Forecasts suggested that the future for aero engines in the civil market would be for engines up to 50,000 lb thrust and that most would be sold in the USA. The alternatives for Rolls-Royce were bleak. It was expected that only the Spey would contribute significantly to turnover after 1970. On 26 February 1967, charts presented to a main board subcommittee showed sales of existing engines would run down from £58.9 million in 1969 to £3.5 million in 1975, while sales of spares would decrease from £36.5 million to £31.9 million in the same period. Without a new engine, Rolls-Royce would drop out of contention with Pratt & Whitney and General Electric. The committee decided that firm offers for the RB 207 and the RB 211 should be submitted for both the European Airbus and the Lockheed TriStar respectively.

Rolls-Royce had already sought written assurance from its financial advisers, Lazards, that it would be able to raise the necessary finance, should it obtain a contract from Lockheed. Lazards would say later:

Based on Rolls-Royce's own financial estimates for the five year period 1967–1971, Rolls-Royce should have no difficulty in arranging suitable short- or long-term finance in the amounts envisaged as required to complete the engines for the L.1011 and for other major engine projects in the UK and overseas on which the company was already engaged or to which it was committed.

THE HISTORY OF LOCKHEED

The Lockheed Aircraft Corporation was founded in 1916 by the Loughhead brothers, Allan and Malcolm, to exploit their aeroplane, the F-1, a flying boat. It was originally called the Loughhead Aircraft Manufacturing Company, but Allan explained that it changed its name.

The true pronunciation of our name (Loughhead) is Lockheed, but so many people called it Loghead that we decided to legalise the phonetic spelling.

The brothers were soon joined by John K. Northrop, a self-taught engineer, later to become world-famous as the prime mover in the concept of the Douglas DC-1, a father-figure in the design of the famous DC-3, and ultimately as President of his own manufacturing company.

Although the company sold its F-1 flying boat successfully to the US Navy, the harsh economic conditions immediately after World War I forced the company to suspend its operations in 1920. Malcolm Lockheed left and subsequently set up his own operation to make four-wheel braking systems for motor cars. This company, Lockheed Hydraulics, grew into a worldwide outfit as a subsidiary of the Bendix Corporation.

Meanwhile, Allan was backed by the same Californian businessman, Fred

Keeler, who had helped his brother, and a new Lockheed Aircraft Company was formed in 1926. Northrop was brought back from the Douglas Aircraft Company and they developed the Vega powered by a Wright J-5, a 225 hp engine with nine cylinders. The aircraft broke many records and sold so well that the company moved its operations from Hollywood to the site nearby at Burbank, where it expanded and expanded from the mid-1930s onwards.

However, before those happy days arrived, the company was once again plunged into crisis by the economic depression of the early 1930s. Lockheed went into receivership in 1931. It was saved from extinction by a Harvard-educated businessman, Bob Gross, who paid $40,000 for Lockheed's assets in association with his brother Courtland, Walter Varne and Lloyd Stearman. They brought in a Massachusetts Institute of Technology-trained aeronautical engineer, Hall Hibbard, who designed, at Bob Gross's request, a small twin-engined transport aeroplane. After a number of teething troubles that nearly sank the company again, this new aircraft, called the Electra, proved a huge success. (It was a British Airways Electra that flew Prime Minister Neville Chamberlain from Munich to Heston in September 1938, for him to wave the piece of paper and declare 'peace for our time'.)

As we know, it was not to be 'peace for our time', and it was the ensuing world war that turned Lockheed from a small struggling company into a giant corporation. Even before war broke out in Europe, Lockheed sold the Hudson bomber to the Air Ministry in Britain, which signed a contract on 23 June 1938 for 200 plus as many as could be delivered by December, up to a maximum of 250. Early in the war, the Hudsons carried out reconnaissance work for Royal Air Force Coastal Command. Later, as bombers, they took part in air strikes in Norway, at Kiel and in the Dieppe raid. They served as dive bombers, skip bombers, medium-range bombers, submarine busters, escort fighters and ambulance ships. Prime Minister Churchill told President Roosevelt:

They helped to save our necks.

Manufacturing other aircraft as well – for instance, the Ventura, the Lodestar and Boeing-designed B-17s – Lockheed's workforce grew from 2,500 to 94,000 by the end of the war. Bob Gross said:

Altogether during the war years, nearly 20,000 military aircraft of assorted types rolled out of our factory doors.

Following the war, Lockheed continued to compete with Douglas and Boeing. Each company enjoyed its successes – Douglas with the DC-4, DC-6, DC-6B and ultimately the excellent DC-8, then Lockheed with the Constellation followed by the Electra, and Boeing with the Stratocruiser and the superb 707. By the early 1960s there were three projects where all three were competing – the US Air Force's proposed long-range, king-sized logistics transport; the Supersonic Transport (SST) project; and the US Navy's advanced design, long-range patrol aircraft. Boeing secured the contract to develop the SST, and Lockheed the United States Air Force's logistic transport, and while Lockheed was still working on the Navy's requirement, its design engineers conceived the idea of a commercial twin-jet airliner. As we shall see, Douglas was also working on its version of a new jet liner.

DEVELOPMENT OF THE RB 211

In spring 1966, Frank Kolk, the chief engineer of American Airlines, called on Rudy Thoren, his equivalent at Lockheed, and discussed the idea of Lockheed building an aeroplane that American could operate as a kind of 'shuttlebus' on its busy La Guardia, New York–Chicago route. As it turned out, because the La Guardia runways were relatively short and, in part, built on piles into Flushing Bay, the aircraft required a large-area wing and numerous tyres. As a result, the larger wings could hold more fuel and the 'shuttlebus' became more than that: it could fly across the whole of the USA. Of course, Lockheed was interested in Kolk's ideas. From Burbank, Kolk flew on to call on Douglas, which was also interested. The race to build the 'Airbus' had started.

Kolk was originally thinking of a twin-engined aircraft but when Lockheed talked to other airlines – it would need more than one airline to be interested to justify the investment – the consensus was in favour of a tri-jet. For example, Eastern Airlines, whose longest non-stop leg on its routes at that time was from New York to San Juan – about 1,800 miles and mostly over water – leaned towards the tri-jet. Nevertheless, it liked the idea of a wide-bodied aircraft with a high-density seating potential; a large aircraft but not as large as the 747. TWA was also interested, even though it was heavily committed on the 747. The 747 was too large and uneconomical on its Chicago, Kansas City and St Louis routes to the West Coast. However, TWA did not like the two-engine idea. If one engine cut out over the Rockies, FAA safety limits would require an engine of 55,000 lb thrust for drift-down procedures to emergency fields, and no such engine existed at that time.

As a result, in June 1967 American Airlines decided in principle in favour of a three-engined rather than a twin-engined aircraft and saw Lockheed as a potential supplier. On 23 June, in response to a request from Lockheed, Rolls-Royce offered the RB 211-06 engine generating 33,260 lb of thrust on a 90 degrees Fahrenheit day. The engine was a high-bypass engine and included new features not found in any previous Rolls-Royce engine.

The first was the three-shaft concept, the second the extensive use of composite materials such as 'Hyfil', and the third the use of an annular combustion chamber in a high-pressure engine. The company's only experience of such features was confined to the limited running of the RB 178 demonstrator and to the development work on the three-shaft concept incorporated in the original Trent engine. The engine had run in July 1966 and, after early problems, had achieved a thrust corresponding to 93 per cent of the 28,500 lb for which it was designed. However, serious mechanical defects were revealed and furthermore the project had run £218,000 in excess of the approved budget. Nevertheless, in retrospect, it was a great pity that the development programme was suspended, as the RB 178 had a number of common features with the RB 211, particularly in regard to the high-pressure module. A number of Rolls-Royce engineers were to say later that some of the problems encountered on the RB 211 would have been solved two years earlier if the RB 178 development programme had continued. Sir David Huddie later described the termination of the programme as 'one of our greatest mistakes'.

Apart from the new concept and materials, Rolls-Royce was offering a thrust much higher than anything it had achieved before and furthermore it was offering to put an engine *into service* in 1971, about four years away. It was an ambitious offer. Nevertheless, the technological advances were such that Rolls-Royce was confident of winning orders, and somehow it would have to meet the programme.

What were the technical advantages of the RB 211 over its rivals?

First, they were inherent in the three-shaft design and the composite materials used in portions of the engine. The three-shaft engine provided simplicity of design with translation into favourable maximum parts cost and performance guarantees. The modular construction meant ease of installation and repair, using modern advanced techniques, and permitting the fullest use of the rapidly developing methods of flight and ground diagnosis of trouble. It was a completely new approach allowing overhaul of the parts of the engine that needed it, rather than the removal of the whole engine as previously.

The relatively modest turbine inlet temperatures gave assurance that the components at the 'hot end' of the engine, which normally required lengthy and expensive development, would have the lives and durability required for a successful commercial engine. Furthermore, engine performance growth was assured in a straightforward way that would match the anticipated requirements for aircraft growth after the start of service.

The composite materials provided advantages in weight and parts costs, which would be critical aspects of short-haul operations with the increased number of flight cycles per thousand hours flown.

THE MAGIC OF A NAME

A combination of these three technical advantages allowed Rolls-Royce to give *guarantees* on maximum parts cost, specific fuel consumption, noise, smoke and performance, which it believed could not be matched by any other engine manufacturer.

Phil Ruffles, Director of Engineering and Technology at Rolls-Royce in 2000, worked on the RB 211 from its earliest days in late 1963 and recalled later that it had extremely ambitious design objectives compared with the Conway. It aimed to reduce fuel consumption by 20–25 per cent with low installed drag, and to reduce engine noise at sideline, flyover and approach conditions by 10–15 PNdB. The aim was to devise a 'smokeless' engine and minimise unburnt hydrocarbons, carbon monoxide and oxides of nitrogen, and to evolve a robust mechanical arrangement enabling high pressures and blade speeds, operating in a high-temperature environment. Most important, the aim was to provide the potential for thrust growth and achieve a substantial reduction in specific weight while achieving good flight response, control and handling characteristics as well as tolerance to intake flow distortion. These ambitious targets were to be achieved with an engine that was easy to build and strip, and that would be cheaper because it would have fewer parts and would not compromise on reliability or safety.

Ruffles said:

The scale of advance was enormous. The RB.211 was the largest turbo fan built by Rolls-Royce, being roughly double the thrust of the Conway. It had a fan diameter which was almost double that of the Conway and a fourfold increase in total airflow. The bypass ratio increased from a mere 0.70 to nearly 5.0 at cruise.

Although the diameter of the engine increased dramatically, the length was five inches less. I might say that there were many times when we wished that the length of the compressors and combustor had been a bit longer. Five inches would have gone a long way to easing many of the problems. However, the most challenging advances were in cycle pressure ratio (up from 17.3 to 25.5), turbine entry temperature (increased by 150° centigrade with an increase of 90°C in combustion inlet temperature) and weight.

Added to the technical advantages, Rolls-Royce was prepared to offer commercial advantages. First, it quoted firm prices for several years and guaranteed levels of escalation for initial orders. Rolls-Royce also felt confident enough to amortise its launch cost over a substantial programme. Thanks to this and the low recurring costs inherent in the simplicity of the design, Rolls-Royce felt able to quote prices substantially lower than the competition. Furthermore, the company offered to organise long-term finance at low rates of interest for all its launching customers.

An added advantage was what was called 'the total propulsion system concept'. Rolls-Royce saw it as logical that the engine manufacturer should have primary responsibility for the entire propulsion system, and offered its facilities – for testing complete power plants and for an endurance flight programme for an airline – well in advance of aircraft certification. This programme would not be satisfactorily carried out if the engine pod was subcontracted by an airline to another manufacturer.

Finally, Rolls-Royce offered prospective customers evidence of its experience and capability. By the end of 1967 its five basic gas turbine engines had achieved over 70 million engine hours in commercial operation. Fourteen million engine hours had been accumulated using air-cooled turbine blades and high turbine inlet temperatures. Several breakthrough features had been pioneered by Rolls-Royce that had become accepted as common practice for subsonic commercial jet aircraft. The most notable were the start of the continuous gas turbine commercial operation with the Dart engine, the introduction of the bypass, or fan, principle with the Conway engine, and the introduction of air cooling for turbine blades, permitting the use of higher turbine inlet temperatures.

Rolls-Royce itself was an extremely substantial operation employing 84,000 people, 73,000 of whom were concerned with gas turbine engines. Some 43,000 employees would be committed to the RB 211 engine programme. Furthermore, the programme for the US Tri-Jet allowed more time from first engine run to the major milestone dates – that is first flight, delivery of first production engine, engine certification and delivery of first series of aircraft to operators – than was achieved on the original Conway and Spey engines.

Rolls-Royce realised that there were arguments being put forward for a 'buy US' policy and sought to counter them. On the 'balance of payments' front it argued that both the USA and UK Governments had a policy of encouraging strong international trade. From the UK point of view it was essential to earn dollars to maintain some sort of balance in a situation where, for many years, the USA had sold more to the UK – and indeed to the rest of the world – than it had bought. In the short term, Rolls-Royce would be making considerable investment in the USA in setting up repair facilities, spare parts depots, training facilities and other after-sales services. This would be consistent with President Lyndon Johnson's expressed desire to stimulate foreign investment.

Conscious that Britain as a whole had acquired something of a reputation for irresponsible trades union behaviour, Rolls-Royce emphasised its own good record in the field of labour relations.

One of the factors helping Rolls-Royce to sell the three-shaft engine to Lockheed was its quietness. On 17 June 1966, Rolls-Royce, understanding that people around the world – especially those living near the increasingly busy international airports – were demanding that governments force the world's airlines to reduce noise pollution, issued a press release on the quietness advantages of its new three-shaft engine. *Inter alia*, it said:

The low noise level of this new Rolls-Royce turbofan created great interest at the recent Conference of Commonwealth Airlines, and provided ample evidence of the importance of noise to airlines. The Conference was told that this new generation of three-shaft turbofans would reduce engine noise considerably, without loss of performance or significant increase in engine cost . . .

While the basic design of the new turbofan engine will produce significantly lower noise levels in its own right, the three-shaft design can be used to reduce noise still further. One of the problems in current aircraft is noise when approaching to land. In this condition, with the three-shaft engine, it is possible to slow the fan down and still further reduce the noise the engine makes on the approach. This feature is only possible with a three-shaft engine.

And on 16 August the *New York Times* ran a story on its front page under the headline 'Government seeks curb on jet noise', which noted that President Johnson had called for a broad attack on the problem (of jet aircraft noise). This gave Rolls-Royce a further opportunity to emphasise the quietness of its three-shaft engine.

In the offer to Lockheed of 23 June 1967, firm prices were quoted for the engine and accessories while prices for other equipment, such as pods and the power plant items of thrust reverser and acoustic material, were given as indications only. The quotation was on Rolls-Royce's normal terms – 15 per cent payable before delivery and 85 per cent on delivery. Deferred payment terms with an increased price were offered. The price quoted for an installed engine was $527,800 (£188,500 or about £3.77 million in today's terms).

Rolls-Royce projected an initial level of sales of 1,989 engines for the Lockheed L.1011 and a further 1,300 engines in a twin-engined aircraft. After an internal competition in Lockheed to select a name for the aircraft, TriStar was chosen and this became its official name. These forecasts included sales for a more powerful or 'stretched' version, which would require a further £20 million of development costs. At the same time, Rolls-Royce planned to pursue the development of the RB 207 aimed at the European Airbus at a cost of £49.5 million to be shared with European partners.

While this was happening, Rolls-Royce was having discussions with Allison in Indianapolis regarding the possibility of joint development and production of the engine for the US airbus projects. This would make entry into the US market easier in that it would remove objections concerning the balance of payments. It would also share the heavy development costs. However, these discussions ended in July 1967 when General Motors decided that its subsidiary, Allison, would not become involved in the RB 211 project. The official reasons why Allison had not been able to give Lockheed a firm proposal were that Rolls-Royce had not given the necessary details to Allison in time and also that Rolls-Royce making an independent proposal compromised a joint programme. Some speculated that decisions took so long to come from General Motors with regard to aero engines, which was a very small part of their business, Rolls-Royce felt it must pursue its own bid. Whatever the reasons, Rolls-Royce was on its own.

At a main Rolls-Royce board meeting on 18 July 1967, a product profit plan for the RB 211-06 and RB 207 was presented and discussed. Certain assumptions were made including the development and production of a stretched version of the RB 211-06. Another assumption made was that there would be sales for a twin-engined aircraft. The forecast was for sales of 3,289 engines to be used in 510 L.1011 aircraft and 500 twin-engined aircraft, both assuming a 30 per cent spare engine sales requirement. The total, spread over ten years from 1970 to 1979, would be 1,160 engines for the initial L.1011 and 687 for the twin, followed by 829 engines for the higher-thrust L.1011 and 613 for the twin.

The forecast was based on the premise that there would be only one engine manufacturer and one aircraft manufacturer supplying this market. It was a dangerous premise but not out of line with the consensus of opinion among airlines and the aircraft supply industry at the time. Total launch costs were estimated at £91.4 million (over £1.8 billion in today's terms). The forecast profit before charging interest was £264 million, or 19.2 per cent of gross sales of £1.375 billion. If such sales and profits were achieved, no one could dispute that Rolls-Royce was still in the big league.

After discussion of this plan the board, not surprisingly, gave its authority for final bids to be made for the RB 207 to power the European airbus and the RB 211 to power the US airbus and possibly the BAC 211. The board said it was essential Rolls-Royce should be the prime engine contractor, and also requested a financial review to show the effect of supporting the dual programme and the alternative of supporting only the RB 211 programme.

DECISION NEAR

The jet age in civil air transport, which began in the mid-1950s, brought great prosperity to the airlines, especially in the USA. The jet engine, helped by improved wing design, increased the cruising speed of civil aircraft by more than 50 per cent. The cost of a seat per mile flown, the key economic gauge for airlines, was halved by the larger and larger jet aircraft, even though they consumed much more fuel in absolute terms. And by the mid-1960s further technological strides meant that ever-larger aircraft, bringing greater savings, could be produced.

This was the background to the airlines' encouragement of the aircraft manufacturers to design and build new 'wide-bodied' aircraft. For the Lockheed management, led by Dan Haughton, it was a way back into the civil market. The company had endured severe problems in the late 1950s with its Electra, which had suffered five crashes, two of them caused by the aircraft disintegrating in the air. This had soured Lockheed's view of the civil market and for much of the 1960s it concentrated on military business. However, this market was not without its problems and the Lockheed board wanted to re-enter the civil arena.

Haughton, who was to feature strongly in the life of Rolls-Royce for ten years from the mid-1960s to the mid-1970s and who, in the words of Sir Ralph Robins, the current Chairman of Rolls-Royce, was to 'teach us a great deal', was described by *Fortune* as:

. . . a lean six-footer with a florid face, thinning hair, and a ready smile. One friend describes his manner as one of 'red-dirt southern courtliness' – nothing affected, nothing high-flown, and yet unmistakably gentry.

His father had been a small farmer and storekeeper in the backwoods of Alabama, and Haughton had worked on the farm and down a coal mine to pay for his time at Alabama University where he graduated in Accounting in 1933. He had joined Lockheed in 1939 when it was still a small company, and was soon spotted by Robert and Courtland Gross and eventually appointed Chairman. At the same time he promoted Karl Kotchian to President. Haughton was extremely hard-working, arriving in the office at 5 a.m. and working flat-out with three secretaries. When asked about his punishing schedule he replied, 'I sleep fast.'

On Tuesday 12 September 1967, *The Wall Street Journal* reported that Lockheed had announced that its proposed airbus could be in service by 1972. It would re-enter the airliner business with the Model 1011, which would carry between 227 and 300 passengers, cruise at 600 mph and have a range, when fully loaded, of about 2,000 miles. Haughton said that negotiations with airlines would begin shortly, around a price of $15.6 million. He did not reveal the minimum number of orders required to begin production but estimated there was a worldwide market for 800 of the 1011s. He anticipated competition, especially from Douglas, which was known to be showing airbus designs to airlines. Boeing, with its commitment to the 747 Jumbo and its short-range 727, already in production, was not seen as a likely participant in the airbus field in the near future. The *Journal* surmised that the British–French airbus design was in trouble because 'of an immediate lack of interest on the part of the European lines that were looked to for the minimum number of orders needed for production'.

A Lockheed design official, William Magruder, said that an engine for the 1101 had not been selected. General Electric, Pratt & Whitney and Rolls-Royce were all in competition for the contract. The design called for three turbofan engines, two to be slung on the wings with the third to be housed in the aircraft's tail section.

In Walter Boyne's book on Lockheed, *Beyond the Horizons*, he describes the vicious competition in the aircraft industry:

A complex, intense struggle ensued, in which the competition between airlines overlapped the competition between the aircraft and engine manufacturers. In a rational world, an agreement would have been reached in which all of the airlines would have selected one aircraft using one type of engine. This would have provided the airlines with the lowest cost means of transportation and the winning engine and aircraft companies a viable product from which a profit might be made. Even the losing aircraft and engine manufacturers would have benefited, for they could have cut their losses and turned to other products.

The RB.211 seemed to have been designed specifically for the L-1011, in terms of weight, power and dimension. Gerhard Neumann, General Electric's combination super-engineer and super-salesman, tried to convince Haughton to use the GE CF6. He argued that the Rolls-Royce was a 'paper engine' and that the claims being made for it were impossible to accomplish within the proposed schedule. For his part, Haughton deemed the CF6 to be too long for placement in the rear fuselage; the L-1011 would have to be redesigned to accommodate it.

On 16 September, Rolls-Royce wrote to aircraft manufacturers explaining that the RB 211 was offered with a fan blade manufactured from a composite material known as 'Hyfil', but to give reassurance the company was pursuing a parallel development programme on an alternative fan blade constructed of solid titanium. Another letter on 13 October said that if a titanium blade was used, the engine price would increase by $9,520.

Seven days later, in response to a request from Douglas, Rolls-Royce provided details of an engine of 35,400 lb thrust, designated the RB 211-10. At the same time, Lockheed was also requesting greater thrust. And by this time, the Rolls-Royce sales effort in the USA aimed at both the aircraft manu-facturers and at the final customers, the airlines. In the autumn of 1967, as negotiations with Lockheed reached a critical stage, David Huddie took up permanent residence in New York. At the same time, technical and sales representatives were flying back and forth to the USA constantly, and there was a continuous stream of requests back to Derby on technical changes and the financial implications.

On the price front, some relief was forthcoming when in November the pound was devalued from $2.80 to $2.40, some 14 per cent. On 6 December,

115

Rolls-Royce quoted Douglas prices 5 per cent lower than those previously quoted, which still gave an extra 9 per cent to Rolls-Royce. However, on the technical front Rolls-Royce had to contend with both Douglas and Lockheed yet again demanding more thrust. And this difficulty was compounded by intense pressure from the US engine manufacturers, Pratt & Whitney and General Electric. One of the arguments they used was the adverse effect on the US balance of payments if a Rolls-Royce – that is, foreign – engine was specified for the US airbus.

To counter this argument, Rolls-Royce had been discussing with both Douglas and Lockheed the possibility of arranging an offset order for the aircraft from the UK, and Rolls-Royce and its advisers concluded that machinery could be set up whereby Douglas or Lockheed could be guaranteed a certain volume of sales outside the USA. A company called Air Holdings Limited was found to be willing to agree to purchase from Douglas or Lockheed a number of aircraft fitted with Rolls-Royce engines. It was arranged that Air Holdings should order fifty aircraft but, at the same time, it was arranged that every sale achieved by the aircraft manufacturers outside the USA would reduce Air Holdings' obligation by the number sold. Air Holdings was to pay a deposit, which would be forfeited if their order was not taken up.

Air Holdings did not want any aircraft itself. Nor did it have any reason to put itself at risk. Rolls-Royce therefore indemnified it against any loss, a risk of no less than $30 million, or about £12.5 million [approaching £200 million in today's terms]. Finance for the deposits would be provided by Rolls-Royce, which would also bear the necessary interest charges. Air Holdings would receive a fixed fee of $250,000 plus expenses.

This was later described as a 'curious series of arrangements'. Nevertheless, there is evidence that the Air Holdings offset order may have been vital. Lockheed knew that the US Government was extremely concerned about the balance of payments effect of Lockheed buying a foreign engine and, in January 1968, Dan Haughton broached the idea of an offset arrangement. In London, Lord Poole, Chairman of Rolls-Royce's financial advisers, Lazards, approached Sir Myles Wyatt, Chairman of Air Holdings. It also happened that Lord Poole was closely associated with Lord Cowdray's interests which, through Broadminster Nominees, owned 8.1 per cent of Air Holdings. By 6 February sufficient progress on an offset deal had been made for Lockheed to present it to the US Government. For Lockheed it was an insurance policy to make sure it retained US administration support. On 9 February the Transportation Department replied that it would 'look with some favour' on the plan. Nevertheless, the airlines were wary of placing orders that would bring them criticism for damaging the balance of payments and costing America jobs. The Air Holdings deal may have been critical. As

David Palmer, the New York correspondent of the *Financial Times*, wrote once the Lockheed contract was secured:

It looks now as if the careful work done by Lockheed on the offset deal with Air Holdings may have been the deciding factor in winning it along with TWA's and Eastern's loyalty. The airlines wanted Rolls, but they also wanted a face-saver. Lockheed had it. By all reports, Mr James McDonnell [by this time, McDonnell had merged with Douglas] had come round to believing that in the circumstances it would be better to buy American despite the fact he is said earlier to have preferred the Rolls engine. If this is true then Rolls not only won the contract for itself, it also won it for Lockheed.

In early 1968, General Electric reduced the price of the engine it had offered to Lockheed and Rolls-Royce felt obliged to make a similar concession. On 14 February, it offered a 3 per cent discount in respect of all firm orders before 30 September 1968. In addition, the 3 per cent discount would apply to options – granted on or before 30 September – to airline customers for aircraft for their own initial use, up to a total for each customer equal to its continuing firm orders placed on or before that date. The major airlines were advised of this reduction.

And it was not only Rolls-Royce that was suffering from price competition. Lockheed itself was being forced to meet competition from Douglas and Boeing. Dan Haughton, the Lockheed Chairman, told Huddie that on 19 February he proposed to ask his board to give approval for a massive reduction in the price of aircraft ordered before 1 April 1968. On 22 February, Huddie wrote to Haughton saying that Rolls-Royce would make a reduction of no less than $75,000 per ship set – that is, three engines – in respect of engines purchased for aircraft ordered by US airlines on or before 1 April 1968. Huddie said that if price was to win the competition then the aircraft price reduction should be large enough to ensure victory.

The first airline order was forthcoming on the day of the Lockheed board meeting. It came from American Airlines and went not to Lockheed but to Douglas for the DC-10.

According to Walter Boyne:

American also favoured the Rolls-Royce engine [we shall see below why it did not say so], as did United, TWA and Eastern. Delta preferred the GE engine, as did McDonnell Douglas. Lockheed interpreted American's preference for the Rolls-Royce engine as a strong sign that they also preferred Lockheed. Haughton did not know that Kolk, American's Chief Engineer, and the 'father' of the wide-body program, was disappointed that Lockheed had decided not to build a twin-engine aircraft and was now determined that American would buy the Douglas aircraft. It was not that the DC-10 was superior to the L-1011; most engineers, pilots and passengers will confirm that the opposite was the case. Kolk, and American, melded their disappointment with Lockheed and their confidence in Douglas into a decision to buy the DC-10.

Haughton's continued belief that Kolk favoured the TriStar was fostered by American's entertainment of successively lower bids from Lockheed. In the hardball game of aircraft sales, American was using Lockheed merely as a foil to force the price of the DC-10 down, even though it never intended to buy the TriStar. So similar were the aircraft, and so intense was the bidding, that the price variant on a roughly $15 million airplane was on $200,000 – just over one percent.

On February 19, 1968, Haughton sought to ice the deal by making a last minute significant price concession to George Spater, American's president. To his mounting horror, he, and his chief finance man, Frank Frain, were kept waiting outside Spater's office for almost two hours, before he was informed – by an American vice-president, Donald J Lloyd Jones – that American had just picked McDonnell Douglas. The purchase was for twenty-five aircraft and an option for twenty-five more. It was both a blow to Lockheed and an apparent calculated insult to Haughton, who was outraged.

His immediate reaction was to go to American's competitors and offer an even greater price reduction; he would sell the first L-1011s to the first buyer at $14.4 million, a price based on anger and desperation and not on practical economics.

Following this order the other airlines pressed for greater thrust and Rolls-Royce offered the RB 211-18 with a thrust of 40,600 lb on an 84-degrees-Fahrenheit day for initial service. This was a level of thrust it had expected to achieve in the second year of service. This offer was clearly something of a gamble but one which they felt they had to make.

As the screw was tightened further and further with more thrust for less money, the day of decision was looming. And, of course, Rolls-Royce's competitors were not standing idly by without putting up a fight.

General Electric, in a last-ditch attempt to win the order, turned to the politicians. Senator Frank J. Lausche and Representative Robert Taft Jr., both from Ohio (where work on the General Electric engine would take place), formed a group of six Senators and six Congressmen to lobby on its behalf. They said that the Rolls-Royce engine did not have the potential power development of the General Electric model (Rolls-Royce had no problem countering that argument). Lausche argued that awarding the engine contract to Rolls-Royce would cost Ohio 10,000 jobs and $7.2 billion of income over fifteen years, while Taft produced figures showing that the balance of payments would be $3.8 billion better off if the contract went to the USA.

The Blackburn Buccaneer with Rolls-Royce Spey engines.
Already in service, more were ordered alongside the F-111
from General Dynamics and the Phantom from McDonnell,
both in the USA, while the British AW681, P.1154 and TSR2
were all cancelled.

The Avro Vulcan bomber, powered by four Bristol Olympus engines, was the most successful of the RAF's V-Force bombers. Having never flown a combat mission, the Vulcans were recalled during the Falklands campaign to take part in the attack on Port Stanley.

OPPOSITE TOP: Rolls-Royce developed the AR 963, effectively a Spey, which came close to being the launch engine for the successful Boeing 727. Later, UPS modified some of their 727s to take the Rolls-Royce Tay, a derivative of the Spey.

OPPOSITE BOTTOM: The TF41 Spey, jointly developed with Allison, powered the A-7 Corsair fighter bomber which saw active service in the Vietnam war.

Concorde, powered by four Bristol Olympus engines.
A wonderful example of what could be
achieved by co-operation between those old rivals,
the French and the British.

Sir Stanley Hooker, following his successful career at Bristol,
returned to Derby to help solve the problems of the RB 211,
and, during the 1970s, led a delegation to China to secure a
licensing deal for the Spey.

An RB 211-06 engine with Hyfil fan blades. Its launch was
authorised in July 1967 and it went on to win the Lockheed
order in March 1968.

A VC10 was borrowed from RAF Transport Command to test
the RB 211. The aircraft was converted at Hucknall to take one
RB 211 to replace two Conway RCo42s on one side. The first
flight took place in March 1970.

The Lockheed TriStar, with its three Rolls-Royce RB 211-22B
engines, just before its maiden flight at Palmdale, California,
on 16 November 1970.

However, the US Government was not swayed by these arguments and told the group on 20 March that it could not take sides in such a commercial transaction.

Rolls-Royce could point to the assurance given by John Conner, the US Secretary of Commerce, to Douglas Jay, UK President of the Board of Trade, in a letter on 9 April 1965 in which he wrote:

We wish to assure Her Majesty's Government that limitations on imports have no place in the United States program for improvement in the balance of payments. Our policy remains one of expanding world trade. Import restrictions would be inconsistent with our current endeavours to get an expansion policy adopted by all the trading nations.

Specifically, in regard to aircraft purchases, we expect American firms to contract for deliveries strictly on the basis of normal commercial considerations. We are not aware of any attempts by American aircraft firms or American companies to apply balance of payments considerations in sales or purchases of aircraft.

Rolls-Royce had stipulated that it could not cope with both the Douglas and the Lockheed programmes, but in March 1968 it did not have either. However, to avoid the possibility of the Rolls-Royce engine being selected by both airframe constructors, so-called 'go-ahead' provisions were laid down.

In the case of Lockheed, the provisions stipulated that firm go-ahead orders for the L.1011 aircraft with the RB 211 engine should have been received by 15 April 1968 from three out of four of the airlines – American, United, Trans World and Eastern. If it became clear by 31 March that this condition would not be achieved, Lockheed and Rolls-Royce would discuss whether to extend the deadline.

Richard Turner, currently a main board Director of Rolls-Royce, was involved in the negotiations with Lockheed between 1966 and 1968 in his capacity as personal assistant to David Huddie. He recalled that neither Lockheed nor Douglas was prepared to launch its tri-jet unless they had commitment from three of the four major airlines – TWA, Eastern, United and American. This would have meant that whichever aircraft manufacturer was the loser would not proceed and the field would be left clear for the other. In the middle of March 1968, American Airlines, keen to gain the credit for launching the new tri-jet, called a press conference to announce its decision, which was that it had chosen the DC-10 to be powered by the Rolls-Royce RB 211.

However, before the press conference was held, the US President, Lyndon Johnson, announced that he had appointed the President of American, C.R. Smith, as Secretary of State for Commerce. Under the circumstances, American felt it would be embarrassing to announce that it had chosen an

THE MAGIC OF A NAME

aero engine for its new aircraft made by a foreign company. At the press conference American announced that it had chosen the aircraft, the Douglas DC-10, but had not yet made up its mind about the engine.

This announcement was to have enormous repercussions – for Douglas, for Lockheed and for Rolls-Royce. Douglas had leapt ahead in the race to capture three of the four major airlines. Lockheed now had to secure the other three. For the moment Rolls-Royce could only wait and see what happened, or that is what the company thought initially. But Dan Haughton of Lockheed was nothing if not a fighter. He went to TWA and Eastern and asked if they would commit themselves if he changed his launch conditions to two airlines instead of three. They said yes, but enquired whether Rolls-Royce would be prepared to do so. After board discussion the company decided it was prepared to do so.

'SUCCESS!'

So Rolls-Royce waived its go-ahead provisions with Lockheed when Lockheed informed Rolls-Royce that it had decided to proceed with the L.1011 programme as soon as letters of intent to purchase that aircraft had been executed with Trans World and Eastern. On 29 March 1968, Lockheed informed Rolls-Royce by letter that, as the go-ahead condition for the L.1011/RB 211 programme had been met, the company had decided to proceed and that Rolls-Royce would indicate *its* willingness to proceed by signing and returning a copy of the letter. On the same day, orders for the Lockheed L.1011, the TriStar, by Trans World Airlines and Eastern Airlines for a total of ninety-four aircraft (forty-four from TWA and fifty from Eastern), together with the fifty for Air Holdings, were announced and on the same day Lockheed placed an order with Rolls-Royce for 150 ship sets of the RB 211-22.

With every justification the British press went mad with delight:

Success! Rolls-Royce win Britain's biggest export order.

ROLLS-ROYCE TRIUMPH – 'Battle of Britain' victory acclaimed.

The Observer waxed lyrical:

You've just pulled off the deal of a lifetime; you now know that Rolls-Royce can afford to stay in the aero-engine race for at least another decade; you've brought continuing prosperity in Derby and the prospect of new jobs in Belfast; and you look like helping Britain's balance of payments to the tune of £1,000 million. Now you know what Sir Denning Pearson, chief executive of Rolls-Royce feels like this week-end.

120

The *Financial Times* recorded the enormous sales effort and meticulous attention to detail that had secured the order.

Beginning in September 1966, Rolls-Royce personnel began a round of 230 journeys across the Atlantic which were to cost more than £80,000 in air fares and expenses alone. At any one time there were 20 members of Rolls visiting the US as part of the sales drive – the undramatic process of talking not just to top men, but talking to men from the grass roots as well, right round the country. [This is what Phil Gilbert had advocated and Rolls-Royce carried out his advice.]

One of the results of Rolls-Royce's 'cold analysis of past failures' [Huddie's words] was to make contact with technicians and engineers at every level of decision making, and to anticipate their problems. Each aircraft manufacturer and each of a dozen or so interested airlines were bombarded with specifications for the RB.211, which make a pile over two feet high. Each document was bound in black with the individual company's name embossed in gold. [The present Chairman, Sir Ralph Robins, ran a Special Project group precisely to co-ordinate this activity. David Huddie said later, 'When I was away inter-departmental affairs didn't get done so I put Ralph on to it and boy, did he do it well!']

In the House of Commons on 1 April, the Minister of Technology, Anthony Wedgewood-Benn, said:

With permission, I wish to make a statement. Hon. Members will already know that Rolls-Royce Ltd has been successful in gaining a most valuable order in the United States to supply RB.211 advanced technology engines for the Lockheed Airbus.

This order is of particular importance. It constitutes a foothold in the American civil aircraft market far bigger than anything which we have achieved before. It is of special value to the British economy, and is, above all, an outstanding encouragement for the skills and technology of British industry . . .

We can be justly proud of the company's achievement in designing the RB.211 engine and in winning the order in the face of determined competition.

The House is aware that the Rolls-Royce RB.207 engine, which is of the same advanced technology design, has been adopted for the European Airbus. This aircraft should be complementary to, rather than a rival of, the American market. It will be of shorter range and more economical, and hence should be better suited to many airline routes, particularly in Europe. The securing of the RB.211 order by Rolls-Royce does not lessen in any way our support for the European Airbus and our determination, in association with our French and German partners, to do all we can to ensure that this aircraft meets airline requirements and can thus be a commercial success.

It was a moment of triumph for Huddie, who had long seen the importance of widening Rolls-Royce's horizons and breaking into the all-important US market, for Pearson who had backed him, and indeed for all those in Rolls-Royce who had worked all hours to develop the engine and for those who had helped Huddie to convince the US buyers that here was a breakthrough engine and that Rolls-Royce could make and deliver it.

CHAPTER FIVE

THE RB 211
COMPLETING THE ORDER

STRONG SALES PRESSURE
DEVELOPMENT-ORIENTATED DICTATORS
A MEETING EACH MORNING
ONLY AN EXTRAORDINARY EFFORT
A LONG AND COMPLEX PROCESS
THE WHOLE COUNTRY WAS STUNNED
WHAT WENT WRONG?

STRONG SALES PRESSURE

ROLLS-ROYCE HAD WON THE ORDER. But could it deliver? First it had to resolve the Douglas situation because Douglas might well order the RB 211 as well. Rolls-Royce had already advised Douglas and Lockheed that it could not support two programmes and the obvious thing to do was to withdraw its offer to Douglas. However, David Huddie, ever the salesman, felt it was impossible to treat Douglas this way, and a review was undertaken at the end of which he and Pearson decided that, if it was necessary, additional experimental capacity could be found and more Rolls-Royce personnel could be redeployed.

Later, there was fierce criticism of this decision on the basis that if Douglas had decided to proceed with the RB 211 engine, the load on Rolls-Royce, which proved to be burdensome even with one programme, would have been substantially increased. It was an early indication of the underestimating of the task that lay before the company. In the event, Douglas did not order the engine.

And it was going to be a big job. We will come to the financial aspects later. The engineering task was formidable in the extreme. We have already seen some of the promises made to secure the Lockheed contract. And the pressure did not ease once the contract was secured. One of the engineers

123

was to say later that he felt conscious of being 'beaten up' with regard to the engineering guarantees being required.

On 11 March 1968, even before the order was won, John Perkins, Commercial Director of the Derby-based Aero Engine Division, wrote to David Huddie saying:

Just a short note, prompted by your query and that I believe of Mr Kendall [Director of Sales and Service under Perkins] in our telephone conversation last week, on the extra cost and risk of the new stage 2 RB.211 rating.

I sense that you and others in New York, under obvious extreme sales pressure, are impatient and critical of what you believe is a conservative commercial outlook back here. The truth will, I believe, ultimately be shown to be very much the opposite. The latest engineering early growth proposals were, in my opinion and observation, not very well considered – they were generated quickly under strong sales pressure, as almost a last desperate throw, and in the absence of yourself, Sir Denning Pearson and Mr Hinckley [former senior engineer in the Merlin era, by this time Director of the International Division] were not really subject to a sufficiently critical review. I was myself somewhat critical of the risk element, and would have been more so had I known as much as I do now; at the same time and with the urgency, I restricted any action to attempting some minimum commercial protection. Although you are close to the action, this sometimes causes a lack of perspective, putting together all the various reports we have here I would have seriously doubted whether another $10,000 per engine would really be a major factor in the ultimate decision. We may have done some damage to our credibility – a lot more if the truth were known.

We now have an engine, which appears to make more optimistic assumptions on component efficiency than earlier designs, while the cumulative recent engineering experience of other projects (Trent, Adour, etc) is completely in the opposite direction ie well below expected component performance. [This Trent, the RB 203, was the first three-shaft engine Rolls-Royce built. Its take-off rating was similar to early Speys, but it was very much lighter. It was seen as a Spey replacement. Development work on this engine was stopped in 1970 to ensure concentration on the RB 211. It should not be confused with Rolls-Royce's present Trent family of engines.] In my opinion a considered general management view of this position would anticipate a serious increase in engineering and production costs, and foresee a real risk of failing to achieve performance in time and the associated massive rework costs. This is clearly a matter of judgement, there are no specifications or costs today to back it up; but supporting the argument I have already heard from very senior engineering levels that we ought to scale up the HP system.

If this letter has a practical message, it is 'don't give any more away'. There is already a risk you may win a Pyrrhic victory.

It was a prophetic letter, but at the time it infuriated Huddie who replied:

In your position, and considering the state of the present competition and the decisions which you know to have been taken, I honestly would not have sent your note [of] 11 March 1968. I do not see what good you think it could have done after the event . . . Anyone who has ever lived through one of these competitions before will know that in the last phases decisions have to be taken without full information, and in the circumstances we have to do the best we can and then loyally abide by the decision taken. There will be time enough for the critical appraisal after the event if we win, and no need for it if we lose. There is very little more time left, and outstanding answers and technical changes simply have to be effected quickly.

As the months passed in the late spring and early summer of 1968, the engineers at Rolls-Royce began to express their concern. On 20 May the Derby-based Aero Engine Division Director of Experimental Manufacture, Les Buckler, reported to the Divisional board that the dates for the redesigned RB 211 engine gave concern, *particularly as it was not yet known what the redesigned engine configuration would be like*. On 5 July he said to the board:

It is difficult at this stage to be factual about the -22 situation next year, but the fact remains that designs have gone back six weeks or so [which] means that the May 1969 date for the 8th engine is becoming more and more unrealistic.

And on 1 August the RB 211-22 chief designer, Don McLean, wrote to the Derby-based Aero Engine Division Director of Engineering, Ernest Eltis:

RB.211-22
Major Tasks in Design
The following items represent areas in which major design activity must take place urgently in order to give us an engine which has a satisfactory degree of integrity and improved prospects of achieving the guaranteed performance level. It is a list that I have discovered by critically reviewing the engine and it is highly probable that several more will be found following a more extensive review of the engine design. In addition to this list, there are other problems for which we have known solutions and for which action is already in hand.

He then listed twenty-six components.

DEVELOPMENT-ORIENTATED DICTATORS

The truth was that both Lockheed and Rolls-Royce knew when the contract was signed that the RB 211-06 design was not adequate to meet the performance specified in that contract and soon after the contract was signed

Lockheed requested a change to the mounting of the engines. The proposal was to move, from the rear of the engine core to the top of the fan casing, the point at which the engine thrust was transmitted to the airframe. Lockheed believed that this would enable a substantial reduction to be made in the overall weight of the aircraft.

Rolls-Royce agreed to make this change, without any extra charge, even though it knew it would paradoxically increase the weight of the engine itself. Furthermore, it meant a great number of design changes involving the deployment of a considerable amount of engineering effort. The fan casing had to be strengthened and, because the load-bearing parts were to be different, all the drives to the wheelcase, all the external pipework and all the fairings over the core of the engine would have to be changed. All in all, the change in the engine mounting proved to be very costly.

In those months after the Lockheed contract was signed, it became clear that the project work on the RB 211 carried out *prior* to the signature had been inadequate. Many people were later to blame the organisational changes made in 1965 which, in their view, led to a situation whereby not all the company's best design talents were released from other projects to be deployed on the RB 211. This means we must look back to what those organisation changes were.

In 1964 the board asked the management consultancy firm, McKinsey, to analyse the company's structure and recommend any changes it thought necessary. Previously, all the designers of the Derby-based Aero Engine Division were responsible through the various levels of seniority, to the chief design engineer. These designers had been responsible for the designs of all aero engines under development. The development process had been in the hands of development engineers responsible to the chief development engineer.

The McKinsey report recommended a complete change whereby the new organisation would be project-orientated and project groups were set up, each headed by a chief engineer. The more experienced specialist designers were put into staff design groups within the technical services area with the idea that they should apply their experience over all projects. McKinsey warned that:

Project managers (the chief engineers on the various projects) should have appropriate design and development experience and must not be permitted to become design or development orientated dictators.

In spite of this warning the senior project engineers who became line-responsible to the chief project engineers did find themselves cut off from the experience of the staff designers, and in the design and development of an engine the development function began to dominate.

126

The loser was the chief designer, previously on an equal footing with the chief development engineer and enjoying greater independence of thought and action. Promotion had been earned through design experience and this experience had been gained by small groups of junior designers being managed by more experienced section leaders who specialised in particular areas of engines. The section leaders had, in turn, been supervised by more senior and experienced men. The inclusion of all designers in one management group had made it easier to bring their accumulated experience to bear on particular problems.

The change to the project organisation had its advantages, not least in breaking up the fiefdoms that had become established under the old system. However, it also broke up the method by which the young designers gained experience. Furthermore, there was no structured way in which experience gained was analysed and recorded to be used by future generations. Everything depended on the close relationship between junior and senior designers, and this relationship had been changed. The overall effect was to downgrade the role of the designer, subsequently causing the resignation of the chief design engineer of the Derby Engine Division, Fred Morley, who moved to Leavesden to take over small engine development.

The effect on the RB 211 programme was that the best Rolls-Royce designers were not all working on a project that was vital to the future of the company. John Bush, the engineering director, was to say later:

. . . I have mentioned that a lot of the details of the installation were still being resolved in April 1968. We needed from then on a concentration in my opinion of the best or among the best designers that we could lay our hands on. At that time we were involved in the Trent engine [the RB 203 – see earlier explanation]. We had work to do on the Adour; we had work to do on the Phantom engine. [What he meant was – on the Spey engine for the McDonnell Douglas Phantom.] I was very conscious of the fact that in my own particular design group I had not got a fair proportion of the good designers. Whenever I asked for more design capacity what happened was that generally speaking the lower standard of designers from the other groups were taken off or spun off to go into the RB.211 design organisation . . . If I were to look for one single cause for our problem I would say that it was lack of senior design experience on the 211 at that time and the fact that up until that time and even after the competition had been won I and many of my senior people were called upon very often to go back to the United States to discuss with Lockheed the specification and also to do sales tours of the airlines.

Further pressure was applied by the bringing forward of the delivery of the RB 211-22. In order to ensure prompt delivery it was essential that Rolls-Royce's Production Department should, from an early date, design and develop the necessary manufacturing equipment and tooling for the production of the engine components. Furthermore, it was considered essential that manufacture of the components should begin while development was still in progress in order to build up stocks for the future assembly of engines. As a result, the drawings of the RB 211-22 had to be released to the Production Department long before the RB 211-22 engine made its first run. This put additional pressure on the design team. One senior engineer said later:

The RB.211 production release process forced designers to complete compatible schemes to mate up with other schemes with which they were not satisfied and to dissipate part of their energy in doing that.

But while these general difficulties were making themselves manifest, how were specific developments progressing? We have already noted the Hyfil fan blade and the reassurances given to Lockheed about parallel development of a titanium blade. During the early summer of 1968, Hyfil blades were being tested in Conway engines in a VC10 aircraft. Initially the test seemed satisfactory, but a bird ingestion revealed a weakness in Hyfil in relation to impact. Further development work was put in hand, but in July the Director of Advanced Engineering, Lindsay Dawson, wrote to the Divisional board:

HYFIL
The bird impact problem on Hyfil blades is serious. It is possible that making the blade able to withstand reasonable sized birds may be a problem of making the material absorb sufficient energy. This is difficult with a material which fails with low elongation.

We have reached the stage where there is little doubt that Hyfil fan blades will run satisfactorily on engines on the test bed. It will be sometime before we know if we can make them sufficiently impact resistant. I would like to intensify the work on the hollow titanium blade, which is the only way of making a wide chord blade as an alternative to Hyfil. I regard a wide chord blade as extremely desirable. With a correctly designed programme, we should be able to make a hollow titanium blade and Hyfil blade interchangeable using very similar engine casings. This would help the problem of what to instruct on Production.

Tests began on the original RB 211-06 in August 1968 and soon revealed that further redesign would be necessary. Without it, the required specific fuel consumption would not be achieved nor would the required thrust be obtained within the limits of the planned turbine inlet temperatures. As this redesign proceeded, Rolls-Royce began to feel pressure building up from Lockheed and, in January 1969, made a significant change in the administration of the RB 211 project. Tom Metcalfe, a member of the Derby Aero

Engine Division board, was appointed RB 211 Programme Director. Huddie made it clear that his responsibility was one of co-ordination – 'The responsibility of line officials for the successful achievement of their part of our programmes is in no way changed by the appointment of Programme Directors' – but nevertheless, the appointment made it clear to Lockheed and everyone else the level of emphasis now given to the project.

Lockheed was also pressing constantly for up-to-date information and for access to Rolls-Royce's engineering work. This caused John Bush to complain to the Divisional board that:

I am still of the opinion that the continuous pressure from Lockheed for up-to-date status information and complete visibility of our engineering work is taking up too much time of the senior engineers involved in the project.

This pressure to meet deadlines both from the Production Department and Lockheed meant that the designers sometimes released schemes before they had considered and studied them properly. It was probably the leading cause of the high level of modifications on the RB 211. Giving priority to time-tables rather than quality and excellence was a complete reversal of Rolls-Royce's traditional methods.

Early in 1969 it became clear that further substantial redesign was still necessary. Phil Ruffles showed the author a Test Report from January, which stated:

* Seizure of LP spool during initial attempts to start engine, LP module removed
* Strip revealed distorted blades due to severe foul with seal segments
* Turbine module from Engine 3 fitted to resume testing
* After 19 hrs 40 mins, several Hyfil fan blades were found damaged
* Hyfil assembly removed and Titanium assembly fitted to resume testing
* Surge at 36,000 lb thrust after 21 hrs 34 mins (day temperature: –7°C)
* Engine rejected from test due to seized HP spool
* Strip examination revealed HP turbine blade foul with seal segments
* Thermal deterioration of the flame tube
* Best performance to date!

In addition, redesigned fan casing, modified because Lockheed had wanted to alter the engine mounting, was found to be unsatisfactory and needed a further redesign. Nevertheless, by the end of January David Huddie was cautiously optimistic and wrote to the main board:

The RB.211 presents a formidable task for the Division and it will take all our efforts to produce a working engine on time. The engine is not yet anywhere near its performance but we are planning to run an engine at 40,600 lb by April 1969, and something very close to the specific fuel consumption by September 1969.

There are several serious mechanical problems but I do not think we have yet seen anything which, with an all-out effort, cannot be solved in the time.

There is no doubt that the closest liaison between Engineering, Experimental Manufacturing and Production is going to be required to produce engines to the right specification for flight and service which will work well.

Just over a month later, on 3 March 1969, it was reported to the Divisional board that bench-running of the first six RB 211-06 engines had reached 375 hours. A thrust of 36,000 lb had been achieved. However, there were problems not only with the fan blade but also with the intermediate-pressure compressor and high-pressure compressor, as well as with the combustion chamber. Additionally, the engine weight was 4 per cent above specification. Concern was now being expressed about meeting the timetable, and a design change board was set up to speed up and approve design changes.

Four days later the Divisional Design Management Committee produced a report based on its investigation of the activities of the project design teams. *Inter alia*, it said:

There is considerable evidence to suggest that we are short of design effort, both in quantity and quality.

The RB.211 project started off with a smaller proportion of the best available designers than any other, then current project. The task of correcting this has involved the RB.211 project supervision in a long and continuous battle.

The programming and progressing of design work has absorbed considerable time on the part of our best technical brains. This is a misapplication of ability to the detriment of design supervision.

Staff Designers are allocated on engine sections such as Compressors or Turbines, to a particular engine project. Communication between projects then depends on Staff Designers talking and working together. There is evidence that this is not happening.

At about the same time, it was decided to build four RB 211-22 engines to an interim standard. It had originally been intended to convert the RB 211-06 engine to the full RB 211-22 standard, but there had been so many alterations this was deemed impractical. It was also decided to introduce into the development programme a further three engines to the full RB 211-22 standard, which meant that the development programme was increased by the including of an additional seven engines.

In early May 1969, Tom Metcalfe reported to the Divisional board that all seven RB 211-06 engines were running and that take-off thrust to the RB

211-22 standard of 40,600 lb had been achieved on 23 April at an air temperature of 7.5 degrees Centigrade. However, a turbine entry temperature of 1,600 degrees Kelvin had been required to achieve this thrust, whereas the specified entry temperature was 1,372 degrees Kelvin. Although the Divisional board was told of this problem the main Rolls-Royce board was told only that the required thrust had been achieved.

A MEETING EACH MORNING

By August 1969 the situation had reached the point where John Bush requested that every engineering change should be approved either by himself or by the chief engineer, and agreed by the general manager of Experimental Manufacture.

It was agreed at a programme meeting held on 18 August that a meeting should be held each morning to review the situation. Bush also requested that all personnel engaged on the stretched version of the RB 211 should be switched to working on the RB 211-22. Lockheed and Rolls-Royce had reached the conclusion that progress on the Hyfil fan blade was sufficient for them to stop further development on the back-up titanium fan programme and at the same meeting it was confirmed that this work had stopped.

The Divisional board met again on 1 September and Ernest Eltis, who, following the death of Adrian Lombard, was effectively in charge of Engineering, presented a report, which said:

We are very concerned by the slow progress in achieving a satisfactory engine standard and I believe the situation is now becoming sufficiently critical to require exceptional measures to ensure a timely recovery.

One year after the first engine run on the RB.211-06 we have completed just over 1020 hours, 350 hours less than planned.

The 8th engine, the first with fan case mounting, long fan duct and accessories on the fan case, but with only some of the RB.211-22 performance modifications, ran slightly behind schedule at the beginning of August. However, in the short time of running accomplished before it failed, we found that only half the expected improvement in performance had been obtained . . .

The main reasons for the performance deficiencies on current development engines lie in the turbine section. Apart from the performance losses due to excessive gas leakage in the turbine area we have now identified flow separation in the outer annulus of the two HP gas producer turbines. The consequent substantial loss in turbine power results in mismatching of the fan in relation to the gas producer section, thus aggravating the basic performance loss due to the turbine.

While the development engines are running in this mismatched condition their operation is unrepresentative, probably hiding other problems which will be revealed

131

when we get the matching right, as well as producing irrelevant problems while the matching is so far out.

With the short time remaining, it is clearly vital that we recover from this situation urgently. Although the compressor performance seems to be reasonably satisfactory, and 5 second accelerations have now been demonstrated on the HP [high pressure] spool, we are still suffering from lack of surge margin at low speed on the HP compressor, causing starting difficulties, as well as from lack of surge margin on the IP [intermediate pressure] compressor. However, some of this will probably go away when we get the overall performance and therefore the matching correct . . .

The combustion chamber, although still far from satisfactory, is gradually being improved both from the point of view of cooling and traverse.

Good progress has been made with the work to achieve a satisfactory Hyfil fan blade and although there is again a great deal more to be done, we have recently come to the conclusion that the titanium backup programme can be stopped. We have already obtained Lockheed's agreement to this decision and are now in the process of informing the customer airlines, consistent with our undertakings . . .

We shall need jointly to apply all our ingenuity to ensure that our available resources are organised and used so that we can overcome the present unsatisfactory technical situation on the engine while containing development expenditure to the best of our ability and achieving the required production factory cost and engine weight.

It is clearly necessary to find a way to achieve this while at the same time intensifying our technical sales effort in the USA, Europe, Japan, etc. However, it has been obvious for a long time that our sales position would be greatly improved if we had an engine running today of which we could be proud. For this reason alone, not to mention the disastrous consequences to our customers and to the Company if we fail to achieve what we have sold, we must ensure that the engineers responsible for the programme and for the support of this programme are from now on protected as much as possible from any activity which does not directly contribute towards getting the engine and the production release standard right.

Since there are frequent occasions when airlines insist on seeing senior engineers and specialists concerned with the project and do not accept officials from other departments as adequate – however excellent their past engineering experience may have been – we have decided to make special appointments (full time or part time as appropriate) in the 211 Project Group and in key specialist groups in Engineering, whose main responsibility will be, albeit with an Engineering title, to fulfil that part of the technical sales effort which the other departments at present responsible are not capable of performing.

During the autumn, the problems continued and it was becoming clear that very slow progress was being made towards achieving the required performance. As a result, Tom Metcalfe decided that modifications to achieve reliability and performance should have a higher priority than those designed to achieve weight reduction. And at a main board meeting on 16 September, David Huddie at last made all the directors aware that all was not well, when he said:

Meeting our programme on the RB.211 now represents a formidable task and we are putting the programme on an emergency basis. The running of engine No. 8 was disappointing in that we are 11% away from the cruise specific fuel consumption we have to achieve. To put this in perspective there is every reason to believe that Pratt & Whitney were more than this adrift with their first flight engines in the 747.

We have some good indications that our deficiencies are largely in the turbine area and know of many lines to follow but we cannot yet put on paper a specification, which would meet our contract commitments.

At the divisional meeting held on 3 November the Director of Engineering reported that there was no significant change in the serious position of the RB 211. Three more RB 211-22 engines had been made available for testing, but the actual amount of testing had been insignificant because of a serious fault in the combustion chamber and a vibration problem on the high-pressure turbine blade. The test-running hours were now well behind programme. The Director of Engineering sent a report to divisional directors, which said that lessons to be learnt were:

Instead of building 7 RB.211-06 engines, which we knew in March 1968 to be of an obsolete standard, we should have built only 2 or 3 and cut the running programme accordingly during the first year of development until RB.211-22 engines could become available.

We should not have given Production such an extensive and detailed drawing release as early as September 1968. Much of this was worthless. Instead, we should have used the available manpower to accelerate the work on cost saving, weight saving and performance schemes.

We seriously considered both these steps at the time but decided that the adverse impact on Lockheed and our airline customers by such a drastic curtailment of the published programme was unacceptable.

Recognising that the most important contribution that can be made towards reducing development costs is better design, it should be remembered that:

(a) The RB.211-06 design was substantially complete in October 1967

(b) Most of the big decisions affecting the RB.211-22 design were made early in 1968 when the senior engineers responsible were heavily engaged in, at the time, essential activities in support of the technical sales effort and technical liaison with Lockheed and Douglas.

Although we have, in the last few years, been making great efforts to strengthen the design and stress departments, really experienced designers and engineers are few and far between.

It is a very long time since we designed and developed a brand new commercial engine. The last was the Spey (first run 1960). Even then it represented a less drastic step than the RB.211 since the Spey design leaned heavily on the RB.141 and Conway.

At the main board meeting on 18 November 1969, David Huddie reported that to deliver the RB 211-22 engines to specification, in time to correspond with the delivery of aircraft to passenger-carrying standard, was a formidable task, which should be just achievable with an all-out effort. He also reported that two highly experienced engineers, formerly based in Derby, Fred Morley and Lionel Haworth, were being seconded onto the project on a part-time basis. As we have seen, Morley had left his Derby position unhappy with the organisation changes put in hand in 1965 to go and run the small engines operation at Leavesden. Haworth had left Rolls-Royce and joined Bristol Siddeley before Bristol had been acquired by Rolls-Royce.

On 28 November, Tom Metcalfe, the RB 211 Programme Director, wrote to Huddie saying:

In a nut shell, I am convinced that unless we take some very special steps right away, we will not have an engine that works at all, we will be seriously late in deliveries and eventually we will be the cause of delay to the aircraft programme and the aircraft deliveries to the airlines.

The problem lies in Engineering – and they do have some difficult problems, – but in spite of the fact that all concerned in Engineering are working flat out, the plain facts are that we are not making positive technical progress anything like fast enough. Indeed in certain areas we are little further forward today than we were 15 months ago when we first ran an engine, and we are almost unable to run the -22 engine at all today and that is just 9 months away from delivering the first Production engine. I believe that, based on any of our previous new engine programmes, we should now have been able to run at least, say, a 25 hour test at full thrust.

If one analyses the things that are stopping us there are not all that many of them. The combustion chamber and its poor traverse and the HP turbine blade and nozzle guide vane are probably the main items stopping us today. Clearly performance needs continuous and vigorous attention, even to get to the point where we can achieve the thrust at anything like reasonable temperatures. One can also foresee that the hyfil fan blade and the fan casing will be stoppers before long for rather different reasons. But until we can get running engines, and for a reasonable length of time at the right speeds and temperatures, we will not be able to discover those other serious problems which surely must be just under the surface today and will stop us when we do get around to discovering them.

134

I believe we have plenty of bodies on the job in Engineering – probably too many. It seems that what we are lacking is enough top-class senior experienced development talent. We do have Mr Bush, Mr Hare and Mr Mayall, all first class, experienced and down-to-earth development men – but we obviously need more of them.

Some minor relief arrived just before Christmas 1969 when Lockheed agreed to reschedule the delivery programme. In mid-January at a meeting in Derby between Lockheed and Rolls-Royce representatives, Lockheed expressed its dissatisfaction with Rolls-Royce management, particularly in relation to the RB 211 project. At the same time it was agreed that it would be necessary to redesign the inner casing of the combustion chamber, which would also mean changes to production parts. In addition, concern was expressed about reconciling engineering and production requirements with regard to engine cooling. Furthermore, it was suggested that the titanium fan blade back-up programme should be restarted as there were continuing problems with the Hyfil fan blades. All of this precluded any intensive endurance testing, which was so vital for the development of any engine.

We shall come later to the financial implications of the development of the RB 211 with all its attendant delays and frustrations – suffice it to say here that the intense financial pressure from which Rolls-Royce was suffering by early 1970 caused it to abandon a vital research programme. Although Rolls-Royce was as technologically advanced as its competitors in almost every area – indeed in many areas, it was ahead of them – it had been accepted that the company was significantly behind General Electric in high-temperature technology. Even by January 1970 the RB 211 was still unable to run at temperatures higher than those achieved on the Spey. Rolls-Royce had set up a programme to master the technology for developing the engine for application in both the civil and military spheres. This involved a high-temperature demonstration unit which first ran in February 1969. As financial pressure increased, Rolls-Royce felt it could no longer afford the programme and, early in 1970, shut it down. The Director of Advanced Engineering, Lindsay Dawson, was to say later that he felt that a continuation of the programme would have been a great help in solving the turbine blade cooling problem on the RB 211-22. It was a re-run of the RB 178 demonstrator unit saga.

ONLY AN EXTRAORDINARY EFFORT

On 2 March 1970, Ernest Eltis presented a report to the Divisional board which showed that significant technical problems remained unsolved. The main ones were as follows:

* The combustion chamber, which had been responsible for two engine failures since the last meeting.
* The HP turbine blade, which had been responsible for three engine failures since the last meeting.
* Inadequate turbine sealing, which had been responsible for poor performance and one failure since the last meeting.
* Fan flutter.
* High-pressure nozzle guide vane cooling.

Because of these problems there was still no engine fit for meaningful endurance running.

And later in the month, confidence in the Hyfil fan blade was badly shaken by a major fan failure on an open-air test bed at Hucknall. There was a primary root failure of at least one Hyfil blade and the impact of debris from this primary failure caused other blades to fail in rapid succession with the resulting imbalance leading to breaking of the fan shaft. The fan disc and the attached piece of shaft flew out of the engine over a distance of 60 yards. Confidence was so severely shaken that in May the Hyfil blade was abandoned. The substitution of the titanium blade meant not only extra expense but also extra weight. In addition, it meant that the containment ring, designed to prevent any broken fan blades being ejected, would have to be made of stainless steel rather than aluminium. This subsequently meant more cost and more weight.

On 28 May 1970 Rex Nicholson, a main board Director, who, largely in response to pressure from Lockheed, had replaced Tom Metcalfe as the RB 211 Programme Director, wrote to David Huddie saying:

There can be no question that there is a grave shortage of design effort to deal with, to cite only three requirements, weight reduction, gas generator fairings and thrust reversers . . . I can see no alternative but to employ the resources necessary in an endeavour to recover the programme, which is, at the moment, in grave jeopardy and also to meet the commitments which we have undertaken, for example – weight reduction.

To compound the weight problem, Lockheed was also suffering similar difficulties with the airframe and began to press Rolls-Royce for more thrust, requesting that take-off thrust be increased to 42,000 lb by March 1972. Rolls-Royce had been planning to reach that level by September 1973.

By July 1970 the situation was so bad that we can only use the word 'crisis' to describe it. In June, David Huddie had collapsed from exhaustion and was forced to rest. In the event, he never returned to work for Rolls-Royce. At a Divisional board meeting on 6 July, the news was almost universally bad.

Tom Metcalfe reported that there was now only a very slim chance that Lockheed could be supplied on time with engines with the required thrust. Eltis and Bush agreed with him. Rex Nicholson reported that launch costs had been reassessed at £148.5 million and could be more. This was £20 million more than had been calculated as recently as April and way, way above original estimates. Furthermore, he calculated that the agreed selling prices were hardly above cost.

Three engines for the first aircraft were due for delivery between 15 August and 1 September 1970, but the engine to be tested on the modified VC10 flying test bed was six weeks late and the VC10 would not fly again until mid-August. Production was due to start building engines on 18 July, but there were still many problems. The combustion chamber, nozzle guide vanes, fan casing, gas generator cowlings and fan reversers had all been subject to continuous changes and, finally, Eltis reported that they had still not established the most effective and economical way of containing a titanium fan blade in the event of a blade failure.

By the autumn, Rolls-Royce was faced with the dilemma of whether to produce engines for Lockheed to meet the deadlines or concentrate on testing to solve all difficulties on the way to a final solution. On 1 October, Bush wrote to Nicholson stating the difficulties:

Since the flight Engines came on the scene, development testing has to all intents and purposes ceased. As an example, Engine 8, which is a key engine on performance, last ran on September 4th. All that was required for its rebuild was the fitment of another HP turbine, which was available then. Only today is the engine going to run – a slip of at least 3 weeks – and it would not have run today but for my personal progressing of its rebuild . . .

Nevertheless I am convinced that only an extraordinary effort on development engine testing is going to prevent us from failing completely to meet our commitments to Lockheed . . .

The Programme will not be met because we will be unable to define a standard of engine which would get anywhere near achieving the required performance. Production would be able to build engines but their Batch 3 and Batch 5 engines would give little or no more thrust than 23, 24 and 25 simply because there would be very little or no differences between them, and this because we would not have done any development testing to enable us to establish any changes resulting in performance improvements.

By the autumn of 1970, as we shall see, the company's financial difficulties were severe, and drastic steps had been taken. Ian (now Sir Ian) Morrow, had been appointed to the board as Deputy Chairman and was running a committee that replaced Sir Denning Pearson in his role as Chief Executive.

Morrow came with a big reputation as a distinguished management accountant. He had been a founder member of the British Institute of Management, a past president of the Institute of Cost and Works Accountants and vice-president of the Institute of Chartered Accountants in Scotland. Pearson had then resigned as Chairman to be replaced by Lord Cole, a former chairman of the detergent and cosmetics manufacturer, Unilever. Some people were bemused by the appointment of someone who had no experience of engineering, let alone the aero engine industry.

Unfortunately, the news from Derby was still not good. The first L.1011 was due to fly in November and this was to be the beginning of an intensive flight programme aimed at full certification by the following November. However, it was reported to the Divisional board on 29 September that the combustion chamber fitted in the first flight engines was expected to last for only 35 hours. The first engines were expected to produce 35,000 lb thrust, but tests suggested that the engines would not operate satisfactorily at the temperatures required to give this thrust.

There were some chinks of light. Apart from the problem of high-temperature running, the engine had generally proved to be extremely rugged although the experienced engineers fully expected new problems when really intensive endurance flight testing began, as was the case with all new engines when they first flew.

The first flight of L.1011 took place as planned on 16 November 1970. However, at a Divisional board meeting on 30 November, the Director of Engineering, Ernest Eltis, warned that the achievement of meeting the first flight deadline should not give rise to a false sense of security. There were still many problems, notably fuel consumption and performance. Nicholson and Bush flew out to visit Lockheed in California in the middle of December (as we shall see, Lord Cole, Ian Morrow and Hugh Conway, by this time group managing director, Gas Turbines, were also at Lockheed talking about finance). They told Lockheed about the thrust deficiency, but that Rolls-Royce expected to achieve 38,500 lb by April 1971.

A LONG AND COMPLEX PROCESS

As we move towards the horrendous climax of the initial RB 211 contract with Lockheed, it is essential to look at the financial position of Rolls-Royce and the impact of the development of this new engine.

Even before the RB 211 began to have a serious impact, the financial return earned by Rolls-Royce had begun to decline. Sir Denning Pearson had said as long ago as July 1961:

Our best estimates show that neither in volume nor profit will the turnover of the years ahead be sufficient to carry present launching costs while maintaining the ability to undertake profitable new projects as they arise. The point has therefore been reached when the company will be prevented from achieving the fullest exploitation of the position which has been won in the world aero engine business . . .

No project is started without the most careful scrutiny of its commercial prospects and possible profitability; but no matter how successful any one project may be, the time which elapses between the commencement of expenditure and the delivery of a sufficient quantity of engines and spare parts to recover the launching costs may be as much as 15 years. The length of this period is a reflection of the highly competitive prices which limit the rate at which launching costs can be recovered from sales revenue. This cycle of expenditure and recovery means that large sums of money are unproductive for a considerable time and that the total amount unrecovered will clearly increase as new projects are started and will only reach a peak when the revenue being derived from current projects is enough to provide development funds for new projects. We confidently look forward to this situation but it would be optimistic to expect to see this happening for several years bearing in mind that it is only some six years since the first major private venture project was started.

From 1961 onwards the reported results of the company were put on a different accounting basis whereby some of the research and development costs were carried forward over a number of years rather than written off in the year in which the money was actually spent. This was a perfectly acceptable accounting practice and is constantly used throughout the world by companies engaged in research involving high expenditure and long development lead times. Some do it while others charge the expenditure in the year the money is spent. Over the years 1961–69, Rolls-Royce decided to adopt a gradual method of writing off its research expenditure. It did not make any difference to the actual level of expenditure – the money was still spent whichever method of accounting was used – but it made a difference to some people's perception of the balance sheet. Over this nine-year period the change meant that reported profits were £11.7 million higher than if the previous method had still been used.

From 1960, Rolls-Royce became increasingly dependent on borrowed money, beginning in the difficult year of 1961. In that year £11.9 million was raised by a debenture issue and the bank overdraft nearly doubled from £11.3 million to £21 million. The proportion of funds from borrowed money, as opposed to capital and reserves, more than doubled from 22 per cent to 46 per cent. And the ratio stayed high throughout the 1960s – not that Rolls-Royce was different from other companies in the aviation industry.

The Plowden Report said:

Much of the industry's capital has been obtained by borrowing. In 1961, 47 per cent of the physical assets was financed in this way, compared with four per cent in industry generally.

The development of *any* new aero engine was already very expensive and risky before the RB 211 came off the drawing board, and we should consider how the development of a highly sophisticated engine could consume money at a prodigious rate. It goes without saying that development of an aero engine was, and is, a long and complex process.

The process begins with the preliminary phase during which the design is proved by the testing of what are known as the development engines. During this phase, the engine is gradually improved to the specification required by the customer. The customer's specification can also change as his airframe modifications may additionally necessitate engine modifications. Following the development phase, prototype engines are built for the flight-testing programme leading up to final certification. Throughout all stages, improvements and modifications are made continuously and the whole process can take about seven years. And the development process continues even after the engine is in service, sometimes to remedy faults that are only exposed through extensive flying hours, sometimes to extend parts life and reduce warranty costs, sometimes to increase the thrust or reduce the weight of the engine to maximise its sales potential, and almost always to bring the engine, which has been proved safe for flight, up to the customer's specification.

In the design phase, the main cost is the salaries of the designers and any modifications merely meant new drawings. (This was true in the 1960s; in 2000 there are virtually no drawings as everything is defined electronically as indeed are modern aircraft – for example, Boeing has called its 777 the 'paperless airplane'.) However, once the development phase begins, costs escalate rapidly, and changes at this stage, when metal is being cut, are much more expensive. In this development period, the main costs are the salaries of the designers and development engineers, and the material and manufacturing costs of components, engine building, test facilities and instrumentation – most of which are expensive on aero engines.

The engine consists of several thousand parts and even a relatively small part can take a department of up to fifty people several months to complete. As well as the engine, other accessories and power plant components will also be under development either in-house or at specialist suppliers, including control systems, thrust reversers and pods. Mock-ups of all these have to be made (again, not in 2000 where virtual reality suffices) and rigs set up to test them. Much of the test equipment – such as vacuum pits in which fan blades are rotated to destruction – is sophisticated and highly specialised.

Once the development starts, the engineers are constantly juggling with several requirements: thrust, specific fuel consumption, weight, time between overhauls and production costs. An improvement in one can often be at the expense of deterioration in another.

We have already seen the technological leap forward of the RB 211 and as the development phase rolled along, the pressures of time and specification – which was altered several times – became increasingly onerous. We have also seen how problems that arose took longer than forecast to solve. Such problems were not unusual and any experienced engineer would have expected and would, given time, have been confident of solving them. However, the timetable was extremely tight and, because of the new technological boundaries being breached, solutions to these problems were taking much more time to evolve. As a result, a fairly relaxed approach gradually gave way to a frenetic effort to produce engines to specification on time.

The initial relaxed approach, both to solving problems and to the financial risks, may have stemmed from a feeling of over-confidence following the success of developing the Spey on time and within the forecast budget. A main board member was to say later:

We launched the Spey engine in about 1962 with its various marks in subsequent years and we forecast that the launching cost of that engine would be £30 million approximately. It was quite remarkable, so remarkable in fact that we even had teams going round in this country and in the United States saying, 'Look, this is the way to calculate the development cost of an engine, and it works. We know.' All the evidence was that we did know, we then introduced or started to introduce, an engine called the Trent, which was never fully developed to the stage where we were getting very satisfactory performance from it. Its importance was that it was the first three-shaft engine. Up to the stage when it was abandoned, everything about that engine had been absolutely right – performance, cost, the lot. We were aware of this. Nobody failed to tell us that this had happened. It undoubtedly gave a great deal of over-confidence . . . Looking back, it looks like a dream world and one wonders how one lived through it, but at the time there was this short but good history, very high confidence in the engineering field, and there were people there who had clearly done quite remarkable things and one expected them to go on doing them.

The effect of this frenetic effort on Rolls-Royce's finances was calamitous. In January 1968, when the contract negotiations with Lockheed were moving towards a climax, the main board Finance Committee was warning of lower profits in the years ahead and of the company's increasing reliance on short-term borrowing. It concluded:

This is not a situation which any of us can accept. It means that short term borrowing would outstrip any prudent level of borrowing and with disappointing profits there would be no means of raising additional equity.

We have seen how competitive the bidding to secure the contract became and the result was not only a significant uprating of the engine from the RB 211-06 to the RB 211-22, but also a price that reduced the total for a ship set of engines by $86,000. In addition, Rolls-Royce offered a discount of 3 per cent, the equivalent of $76,000 per set, for all orders placed before 30 September 1968. To compound the competitive worries, Douglas decided to proceed with their wide-bodied aircraft and contracted General Electric to supply the engines. This reduced the number of airlines likely to buy the L.1011 from Lockheed.

This was a crucial decision by Douglas. Many in the industry believed that the market for such aircraft was only large enough to give one aircraft manufacturer sufficient business for it to climb the inevitable learning curve of manufacturing a new aircraft and eventually make a profit. If two manufacturers were to split the market, neither would win sufficient orders to reach that profitable stage. Furthermore, while they were competing for the orders, they were likely to offer extra price incentives and offer more, usually expensive, extras to try to win airlines' business. Of course, it was in the interests of the airlines to have more than one supplier for any type of aircraft they wanted to buy. First, it would give them this competitive situation and allow them to feel confident that they were not paying any more than was absolutely necessary. Needless to say, they became extremely adept at playing off the manufacturers against each other. Second, it provided an insurance policy in case one of the manufacturers ran into difficulties and was clearly not going to deliver on time – or even at all.

As we have seen, development costs soared. Furthermore, the forecast profitability on each sale once the engines were made dwindled to nothing. Indeed, by 1969 it was clear that, at the originally negotiated price, Rolls-Royce would lose money on every sale. In a report to the Divisional board in mid-1969, Tom Metcalfe predicted that the RB 211-22 project – from conception to manufacture – would lose approximately £16.3 million; a sum in contrast to the preceding projection, which had indicated a potential profit of £46.9 million. This was a negative difference of no less than £63.2 million (about £1.1 billion in today's terms).

By this time, the company was in a terrible dilemma. As Metcalfe put it later:

The difficulty . . . was that any steps one could reasonably take to limit or reverse the adverse financial trends would inevitably prejudice the achievement of the technical objectives of the contract specification in the time available. So one had an almost insoluble problem, in that it was necessary to deploy large engineering effort which cost money in order to overcome the specification deficiencies that were beginning to become apparent, without which one could not comply with the conditions of the contract and would be open to all sorts of redress if one was materially deficient or significantly deficient on the technical specification, plus the fact that the design changes we referred to earlier to try and get out of those technical problems were themselves causing escalation both to the development cost and to the production factory cost, and it seemed that this was a box it was almost impossible to get out of.

By late 1970 when the Government had been made aware of the seriousness of the situation, the financial numbers were far removed from anything originally forecast or envisaged. In the figures given to Government in September 1970, research and development expenditure had increased from the original estimate of £89.3 million to £170.3 million, and the estimated production cost of an installed engine had increased from £153,655 to £237,452, which was greater than the £230,375 selling price.

In the autumn of 1970, an extra £60 million from Rolls-Royce's bankers was negotiated. However, and this proved to be critical, the availability of the £60 million was to be conditional on the completion of a report by the accounting firm, Cooper Brothers & Co., regarding the long-term position and financial structure of the company. Lord Cole would say later that it had not been a condition in the negotiations but was added afterwards. To a degree, it was an academic argument because events overtook the existing situation, with each new analysis and forecast presenting a blacker and blacker picture.

When the 1971–75 group forecast became available, it was apparent that even the extra £60 million would be inadequate. The estimated 1971 production cost of an installed engine had risen to £281,000, which amounted to £44,000 more than the estimated cost used in the £60 million negotiation and £128,000 more than the original estimate. It was estimated that the provision for production losses should be raised from £45 million to £90 million. Furthermore, estimated launch costs had risen from £170.3 million to £202.7 million and, finally, there was no hope of producing an engine to Lockheed's specifications within the timetable. The programme would have to be extended by six to twelve months.

By 21 January 1971, Rolls-Royce had prepared forecasts incorporating this programme extension. The forecasts showed that the cash requirement had grown by £47 million to £107 million (over £1.6 billion in today's terms) higher than facilities originally available. There was a £28 million contingency figure, which on the record thus far might prove a conservative figure. The figures did *not* include a provision for damages liable because of the delay. These were estimated speculatively at £40 million to £50 million.

Indeed, it was this estimate of potential damages plus the fact that money

143

would be lost on every engine sale that precipitated the final agonising crisis. It seemed that only a re-negotiation of the contract terms with Lockheed could save the day.

When Dan Haughton arrived in London on 2 February, the position was put before him at a lunch at which cold salmon was served. Karl Kotchian, his deputy, said later that when Haughton telephoned him to tell him the news he said:

The weather and food were cold and the news wasn't so hot either.

Haughton was asked to agree to a revision of the terms, whereby Lockheed would pay £100,000 more per engine, and also for an extension to the programme. At this stage, Prime Minister Edward Heath was involved and Haughton told him that he did not have the authority, nor Lockheed the resources, to agree to either new condition.

By this time it was clear that Rolls-Royce was technically insolvent as it was unable to pay its debts as they fell due (the banks had reduced its over-draft facilities from £50 million to £40 million); its future cash requirements were £170 million in excess of the £70 million borrowing facilities normally available, and the loss of resources committed to the RB 211-22 project, with the liability for damages that might arise due to the delays in meeting the programme, were likely to exceed the net tangible assets of the company.

The Government was particularly frightened by the open-ended commit-ment of unknown liabilities for damages. The official White Paper, *Rolls-Royce Ltd and the RB.211 Aero Engine*, recorded:

Rolls-Royce's liabilities were large, hurriedly estimated, in part unquantified and unquantifiable. The extent of these liabilities would depend on whether the RB.211 were completed or cancelled. If it were cancelled the Government were advised that the liabilities could have run into hundreds of millions of pounds. If the RB.211 went on, the company thought they would need at least an additional £150 million to cover their expected cash flow deficiency (£110 million) and to provide for claims by Lockheed and the airlines for delay (estimated at £45–50 million). No great reliance could be placed on the company's estimates which had been prepared hurriedly, and which they themselves stated would be subject to considerable amendment as more information came to light. In the circumstances reported to them and having regard to the magnitude of the figures Ministers considered that, whether the RB.211 project was stopped or continued, it would not be a responsible use of public funds to assume a very large unquantified commitment either by supporting the company with funds which it had no prospect of repaying or by the Government taking the company over and thereby making itself responsible for all the company's debts and obligations.

THE WHOLE COUNTRY WAS STUNNED

Once the Government had said it would do no more, the inevitable consequence was receivership and on 4 February 1971, a date seared into the brains of every Rolls-Royce director, manager and employee, Rupert Nicholson of Peat Marwick Mitchell & Co. was appointed receiver.

The whole country and, of course, all the Rolls-Royce employees, were stunned. Although there had been a great deal of comment in the press over the previous six months, under such headlines as 'From bad to worse at Rolls-Royce', 'Can Morrow save Rolls-Royce?', 'Can Britain afford a Rolls?', 'Rolls cash Crisis' and 'Cabinet battle on Rolls', hardly anyone had thought it possible the company would be allowed to go bust.

From Rupert Nicholson's point of view it was a question, as always, of assessing the assets of the company and securing the best price for them. There were plenty of people willing to say their piece but first, if the assets were not to deteriorate, the business had to be kept going. On the first day of the receivership, events showed that Nicholson would need to take control quickly. Two of his lieutenants were sent to Derby but on arrival they found that a manager from Cooper's Insolvency Department was already there and had organised the shutting of all factory gates at 8 a.m. This had the unfortunate consequence that no work was done that day and that important supplies were turned away.

It was made clear to Nicholson that the Government would buy the aero engine business, and he received a call from Sir Ronald Melville, the Permanent Secretary at the Ministry of Aviation Supply, asking that supplies to Government departments and airlines would not be interrupted and assuring him, in view of the Government's intentions, that the creditors would not suffer in any way as a result. Lord Carrington, Secretary of State for Defence, was apparently annoyed later that this assurance was given without reference to him but nevertheless agreed in principle.

On his first day at Derby, Don Pepper, the Rolls-Royce personnel director, arranged for Nicholson to meet two representatives of the unions. As the Government had already said in its announcement that there would inevitably be redundancies, Nicholson was expecting a hostile reaction, but the two union representatives said that labour relations had always been good at Rolls-Royce and that he could count on their co-operation. This meeting was followed by a 'leisurely' (Nicholson's word) lunch with Hugh Conway, who gave Nicholson his views on the cause of the collapse. After lunch Nicholson met a number of 'second rank Rolls-Royce staff' (his words) who showed some resentment. He said later:

They said openly, Rolls-Royce is not bust; this has been engineered by the government, or someone.

Indeed, there was an undercurrent of this refusal to face facts and this may explain the telegram said to have been sent from Derby on 4 February, saying that the RB 211 was on that day giving the performance standard required under the contract. In fact, it was not until March that this stage was more or less reached.

A full performance run (of engine 10011 of batch 3) began at 6.00 p.m. on 3 February 1971. Eltis sent the results of this test run by telex to Conway in London at 9.30 a.m. on 4 February. The telex stated that the engine had achieved 38,500 lb thrust at 1,475 degrees Kelvin on a standard day against a commitment to provide a thrust of 38,000 lb by 12 February.

The contract said that the thrust would eventually have to be 40,000 lb so perhaps they were playing with words. The aim of Eltis's telex was to show that the engine was responding to all the modifications.

Nicholson was soon receiving conflicting advice about the RB 211 contract. He met Lord Carrington on 9 February, who was 'brusque' (Nicholson's word) in telling him that he must repudiate the contract. However, on the same day he met Phil Gilbert, who, as we have seen, had been enormously helpful to the company over the years as its USA legal adviser, and who, according to Nicholson:

. . . was most insistent that the RB.211 contract was not onerous and that Lockheed and the American airlines would be helpful and reasonable, if asked. If, however, I defied them I was laying the company open to immense damages and I should expect no mercy.

The next day, Nicholson telephoned Dan Haughton and told him that he was, with Government authority, 'maintaining capability to produce the RB 211 engine'. In other words, research and development, especially testing of the engines, was moving full-speed ahead. Haughton confirmed what Gilbert had said. He wanted 'engines, not damages' and he was willing to give Nicholson all the help he could, in the form of a large technical team, if he would take on the RB 211.

Five days later, Nicholson flew to the Coventry, Glasgow and Bristol factories with Dan Haughton in one of Rolls-Royce's HS.125 aircraft. Sir Denning Pearson met him at the steps of the aircraft at East Midlands airport and added his plea that the contract should not be repudiated. At the same time, Nicholson met Secor Browne, the chief of the American Aeronautics Registration Board, who had been sent over by President Nixon on a fact-finding mission. (In the last hours before declining to save Rolls-Royce, Edward Heath had been in touch with Nixon discussing the implications of receivership.)

Nicholson and Browne had breakfast together at Duffield Bank House and Nicholson said later:

The talk was most reassuring to me as it was a completely different picture to the one given by HB Doggett and the others or even by Hugh Conway. In this sense, it fitted in with what Dan Haughton was saying, that although they could adapt to the General Electric engine, if they must, they would prefer the RB.211 engine even if it cost them more than the contract price. I was left with the impression that there was appreciable leeway in price, before coming up to the price of the General Electric engine. He explained high bypass ratio and triple spool and contrasted General Electric and Pratt & Whitney engines with sketch illustrations: he also spoke of fuel consumption and noise – all favourable to RB.211. This was all extremely useful independent background for me in the negotiations in the weeks which followed.

Three days later, Nicholson met the senior directors of Rolls-Royce, Lord Cole, Ian Morrow and Hugh Conway, along with Dan Haughton and a number of his senior men. Haughton again pressed them to avail themselves of Lockheed's expertise. He and his colleagues were adamant that it would be preferable to get the RB 211 right, rather than switch to Pratt & Whitney or General Electric.

While Lockheed was trying to convince Nicholson that there was a great future for the RB 211, the Government seemed determined to undermine that future. Frederick Corfield, Minister of Aviation Supply, argued in the House of Commons that continuance of the engine was not essential to Britain remaining in 'the big aero engine league'. Furthermore, he did not believe there was 'an enormous potential market' for the engine, which was 'to a large extent behind engines of equivalent power and performance', by which he meant Pratt & Whitney's JT-9D and General Electric's CF6. Corfield, a pig farmer by profession, was out of his depth. The ultimate success of the RB 211 showed that he did not know what he was talking about.

Meanwhile, Lockheed, itself not in a strong financial state, was desperate to find a solution, and Haughton ordered a complete reappraisal of the three engines that it had considered four years earlier. The Lockheed engineers again compared the GE CF6 and the Pratt & Whitney JT-9D with the RB 211, and yet again it was clear that the RB 211 was superior. Furthermore, a change would be time-consuming and expensive. Haughton was told that a change would cost around $100 million and delay delivery to the airlines by a further six months.

Haughton was trying to juggle his airline customers and his twenty-four banks, and at the same time convince the Nixon administration that a failure of the TriStar programme would be bad for the USA. One of his airline customers wavered. Delta signed a letter of intent to buy five DC-10s, but also made it clear it would still buy the twenty-four TriStars ordered, if the current problems could be overcome.

On 10 May it was announced that Rolls-Royce's assets would be transferred from the receiver to the Government and that a new company called Rolls-Royce (1971) Ltd. would operate from midnight on 22 May.

And on 11 May a new contract was signed between Rolls-Royce (1971) and Lockheed. This followed two days of meetings at which the four leading airline customers – Trans World, Eastern, Delta and Air Canada – expressed their satisfaction with the project. Under the new agreement the price of each engine went up by £110,000 to £460,000, and penalties for late delivery were annulled. The great Rolls-Royce enterprise was saved. All that skill, experience and expertise built up over many years could remain as a team and be used effectively. The alternative hardly bore contemplation.

However, the British Government's insistence, if it was to back the RB 211, on US Government guarantees for the $250 million bank loans needed by Lockheed to complete the programme, almost scuppered the whole deal. After some hesitation, President Nixon made it clear he supported the guarantees and would ask Congress to pass the necessary legislation. Whether he was successful in persuading them we shall see in the next chapter but, in the meantime, we have to consider what brought the great company of Rolls-Royce to this impasse.

WHAT WENT WRONG?

To many people, Rolls-Royce going bankrupt was akin to Great Britain going bankrupt. How could the country's premier engineering company, the company that had produced the Merlin engine which won the Battle of Britain, the company that had pioneered jet transport, the company that had developed an engine that took business from under the noses of the US giants General Electric and Pratt & Whitney and made 'the best car in the world', be allowed to fail?

Many had their views and the Government set up a Department of Trade and Industry investigation under Section 165(a)(i) of the Companies Act 1948. The two inspectors – R.A. MacCrindle, QC and P. Godfrey, FCA – interviewed all the relevant people and produced a comprehensive report in 1973.

In their conclusion, they summarised what they felt were the causes of the crash. First, they made the point about Rolls-Royce's reputation.

Down the years, Rolls-Royce had built for itself a worldwide reputation second to none. The words 'Rolls-Royce' had passed into daily use and were synonymous with a high standard of workmanship. In the public eye, the company enjoyed a splendid and exceptional status. It was also regarded by the man in the street as being 'as safe as the Bank of England'.

The report analysed both the technical and the financial challenges faced by Rolls-Royce as it attempted to enter the 'big engine' market in competition with General Electric and Pratt & Whitney. (It should be borne in mind that Rolls-Royce had lost out in the competition to power Boeing's 747.) On the technical front, the report stated categorically:

The step forward in technology that Rolls-Royce was proposing to take was big; bigger on any view than steps which it had taken in its previous civil engine developments. The amount of research and experience available to it relevant to an engine such as the RB.211 was limited, comprising mainly –

(i)   RB.178 demonstrator programme which had been terminated in 1966; the demonstrator engine ran for only a few hours and showed serious mechanical defects
(ii) development work on the three shaft concept in the Trent engine which had run for a limited time only
(iii) design and development work on the RB.211-06 which had not run.

In addition, it must be remembered that in the course of the negotiations, the engine had moved from the -06 rating earlier proposed to the -22 rating eventually sold, with some 6,000 lb increase in thrust. This increased thrust was to be achieved without any significant increase in the overall dimensions of the engine, with the consequence that the increased thrust had to come mainly from operating at higher temperatures than those originally planned. What would have been a difficult enough task on the original specification had become a very much more formidable task on the specification called for by the contract.

The inspectors also noted that the engineering directors on the main board should have appreciated that the technical hurdles to be jumped were formidable and that the timetable was extremely tight – especially in view of other difficulties and commitments. They wrote:

Four other points are germane to a consideration of the technical matters. First, the unexpected death in July 1967 of Mr A Lombard, then director of engineering. By all accounts he was a brilliant engineer. In addition he was able to provide the dynamic leadership that brought forth the optimum results from the team of highly qualified men who formed the DED [Derby] engineering department. His leadership was not replaced. Secondly, the RB.211-22 was the first major civil project to be undertaken after the reorganisation of the engineering department and the creation of project groups. This reorganisation contributed to a situation where the invaluable experience of senior design personnel was not allocated to the RB.211, notably in the early critical stage of design. Thirdly, at the time that Rolls-Royce was taking on the Lockheed contract the engineering department was heavily committed to the Spey Phantom, Trent and Adour engines; and BED [Bristol Engineering Department] was engaged on the Olympus (for Concorde), Pegasus and other engines. Fourthly, the stress office was generally understaffed, was short of experienced men, and was therefore not able to cope with the sudden massive inflow of work which was bound to arise from a development such as the RB.211-22.

As the timetable became tighter and tighter due to the inevitable setbacks in developing a brand new engine, the financial implications became more and more serious, and in their reaction to this situation the directors were criticised by the inspectors, who wrote:

On the basis of contemporary long term forecasts the -06 programme committed some 30 per cent of the existing net worth of Rolls-Royce. The -22 programme increased the commitment to 60 per cent. All the evidence leads us to the conclusion that to those Rolls-Royce directors with wide experience in the exercise of engineering judgement examples of runaway overspend on advanced technological projects should have been trite knowledge. With such huge amounts involved that experience would normally have dictated that the forecasts should be viewed as being subject to substantial margins of error. But we have not found any evidence to indicate that the board considered the risks inherent in the Lockheed contract in relation to the existing net worth of the company. Nor are we satisfied that the board sufficiently considered the susceptibility of this type of project to crippling overspend or the implications of failure in the task. It is true that it must have seemed to those in Rolls-Royce that had they not concluded this contract the future for the company and its employees would have looked black indeed, for in truth no attractive alternative was available or was likely to become available. The decision to conclude the contract on the terms offered, despite the firm limit on the financial aid promised by HMG, was a decision which it would have required much courage to oppose. But the risk inherent in the decision was alarming, and in the ultimate such risks are borne by shareholders. It is their funds that the decisions of directors may put in jeopardy.

The inspectors continued:

With all the advantages of hindsight we feel bound to record that in our judgement the chances of failure, and above all the consequences of failure, should it occur, were not adequately probed. The whole future of the company was hazarded on one contract. It is not always wrong for a business to put all its eggs in one basket. Nor is it always wrong for a business to undertake a speculative project. But for a business to combine the two is to court disaster. In the case of the RB.211 this was not adequately appreciated. The evidence which we have heard suggests that the management of Rolls-Royce never seriously thought when the Lockheed contract was signed that, if the worst came to the worst, the company could lose sums so large as to force it into bankruptcy.

LEFT: Geoff Wilde – involved in most of the key design
developments at Rolls-Royce in the post-war years.

RIGHT: Fred Morley, originally based at Derby, had moved to
the Small Engine Division following the McKinsey-inspired
rationalisation of the mid-1960s. In 1969, Huddie brought him,
as well as Lionel Haworth, back to Derby to help overcome the
development problems on the RB 211.

Adrian Lombard originally came to Rolls-Royce via the
takeover of Rover's operation at Barnoldswick. In the late 1950s
he became Director of Engineering for the Aero Engine Division.
In 1963 he presented to the Board the case for starting work on
a new engine to succeed the Conway. His untimely death in
1967 robbed Rolls-Royce of his strong leadership at a critical
point and created a vacuum which took time to fill.

RB.211-Modular construction

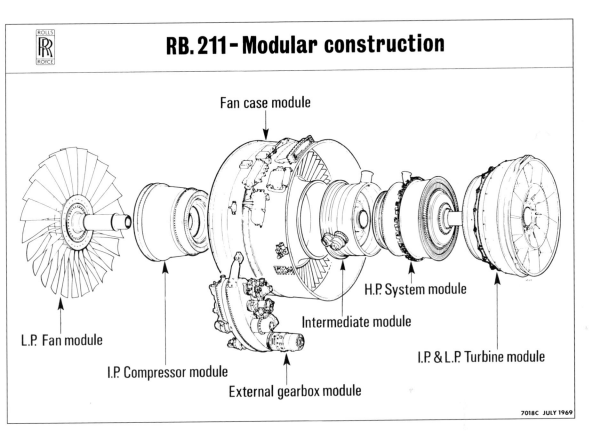

Fan case module

L.P. Fan module

I.P. Compressor module

External gearbox module

Intermediate module

H.P. System module

I.P. & L.P. Turbine module

7018C JULY 1969

A fundamental benefit of the RB 211 has been
its modular construction.

The Lockheed TriStar, powered by three Rolls-Royce RB 211-22B engines, takes off on its maiden flight at Palmdale, California, on 16 November 1970. It has been said that it was so quiet that the sound of the engines was drowned out by a diesel fire-truck and some thought the engines had failed.

David, later Sir David, Huddie. He knew that Rolls-Royce must be successful in the USA if it was to remain in the big league of aero engine manufacturers. He could see that the three-shaft RB 211 engine was a world beater and he sold it successfully to Lockheed and to leading US airlines. A great engineer who inspired nothing less than hero-worship, he broke his health in trying to make sure Rolls-Royce supplied the RB 211 to Lockheed on time and to the demanded specification.

Jim, later Sir Denning, Pearson. Like Huddie, a man of vision
who concurred with Huddie's view that Rolls-Royce must break
decisively into the US market. Perhaps he spent too much time
in Derby and not enough at Westminster 'tramping the corridors
of power' or in the City raising the funds necessary to see a
mould-breaking engine through its expensive design and
development stage. He was determined that the Rolls-Royce
receivership should only be seen as a failure due to lack of
finance rather than lack of engineering know-how.

Dan Haughton, the President of Lockheed who started work at 5.00 a.m. each day, and employed three secretaries. Sir Ralph Robins said, 'He taught us a great deal'. Others felt he provided the only leadership at Rolls-Royce as well as at Lockheed in the dark days after the collapse in February 1971. A friend in the USA described his manner as 'red-dirt southern courtliness – nothing affected, nothing high-flown and yet unmistakably gentry.' He is seen here with, left to right, David Huddie, Rex Nicholson (Production Director), Hugh Conway (Managing Director of Bristol) and Sir Denning Pearson.

Minister for Aerospace and Shipping, Michael Heseltine,
with Ralph Robins on his left and Peter Thornton (Secretary,
Aerospace and Shipping, DTI) below him, by an RB 211
engine at the Lockheed facility at Palmdale, California, in 1972.

And Sir Denning Pearson accepted the blame:

Sir Denning Pearson frankly accepted, before us, that the board did not sufficiently examine at the time the risks which the company faced and the consequences which would arise should the company's projections in a notoriously difficult field turn out to be wrong. The technical assessment of the task proved to be very far out, and the contingency margins allowed were grossly inadequate. In the event the job proved to be so much more difficult and costly than had been foreseen that by the autumn of 1970 it was plain that Rolls-Royce had insufficient funds to enable the task to be completed.

Pearson was very keen that the collapse should be seen as a financial one and not one caused by technical failure or the incompetence of Rolls-Royce engineers. It was a tragic conclusion for Pearson. He had seen the necessity for Rolls-Royce to tackle a world, rather than just a British, market and that meant breaking into the USA, both with an aircraft manufacturer and the airlines. His career had been a sparkling one – as we saw, it was he who realised more than anyone including Hives that the future after the Second World War was in the civil more than the military business – and this was a horrendous and largely undeserved denouement.

The inspectors accepted that forecasting not only development cost but also manufacturing cost on a brand new engine was extremely difficult, and the consequences for getting it wrong were severe.

We have seen that in the case of the RB.211 the design specification at the date of the Lockheed contract was far from final. With an engine which was required to embody so many advances on the existing art things could not be otherwise. In circumstances such as this one cannot hope to form a realistic view of eventual production costs, and it is idle to pretend otherwise. Estimates of production cost were based in part upon materials and concepts which were untried. When in the event they were tried they were not always successful. The effect on production cost was dramatic. As time went on the estimated costs of production of each engine went up and up, until in the end these costs themselves well exceeded the sale price of the engine.

The inspectors were also critical of the management of the contract.

Men of skill abounded, but that is not enough to ensure that a venture of this sort is tackled in the most efficient way. The optimum concentration of available talent had to be brought into operation, and the maximum effort inspired. The venture could not hope to be tackled in the most effective manner if there were not imposed from the outset strong control from the top. When the Lockheed contract was obtained the managing director of the division was Sir David Huddie. He was a fine engineer

151

and a man of great ability. But it must be recorded that for much of the critical time during his tenure of office when a strong driving force was particularly necessary in the division it was not there. Sir David had led the team which negotiated the contract in the US, and this had kept him away from his duties in Derby for many months. After the contract was concluded he continued to find that the project made many demands upon him which obliged him to be away from Derby. In particular he and a number of his senior engineers found it necessary to travel abroad frequently for the purpose of discussion with potential customers for the aircraft of which the RB.211 was to form part. To some extent this was undertaken at the request of Lockheed. By common consent this was a task at which he excelled. But it meant that the division was starved of the directing mind of its managing director at a vital time.

And, as we have seen, as Huddie spent more and more time in the United States of America, problems in design had a ratchet effect as the company strove to meet the promised timetable.

In order to be able to meet the delivery dates stipulated by the contract the production department of DED had laid down stringent 'lead times' early in the project stage of the RB.211 programme. This stringency was known when the contract was signed. In the event progress on the design side was slower than had been envisaged. Despite this the engineering department, under strong pressure from the production department, released drawings to production long before an experimental -22 engine first ran. Releases to production were made at substantially earlier stages than had occurred on previous Rolls-Royce projects. The increasing pressure on the designers led to the release by them of designs that they knew might well prove to be unsatisfactory. Some designs were even released to production without any prior testing whatsoever. This led to a situation in which modifications were being called for in ever increasing numbers, adding again to the burdens of both design and production. Parts were made and scrapped at a rate greater than ever before in Rolls-Royce. This was a direct result of the assumption by Rolls-Royce, at a time when the design was not stabilised and the research not adequately completed, of a contractual obligation to deliver a new engine to a fixed time scale. The pressures upon the designers resulting from these lead times were exacerbated by the demands of Lockheed that drawings should be produced to a rigid programme schedule.

The inspectors came to the conclusion that it was the technical setbacks that caused the financial difficulties.

The technical problems encountered in the course of tackling the engineering and production tasks were also beginning to create financial difficulties. Financial forecasts and statements are only an evaluation in money terms of underlying physical activities. In June 1969 Mr Metcalfe presented his report to the DED board estimating that the RB.211-22 project, over its whole life, would lose £16.3 million, whereas the previous projection had shown a profit of £46.9 million – an adverse swing of £63.2 million. At the same meeting a draft 1970–74 forecast was considered by the DED board. This showed that by 1973 an upswing in research and development expenditure and a reduction in the level of forecast profits would adversely affect the cash position of the division by some £87 million . . . suggesting severe financial problems in the future.

Where the inspectors were highly critical was in their comment that key people did not react quickly enough to the implications of these forecasts and did not keep the main board fully informed. They wrote:

The full gravity of the situation revealed by these documents, however, does not seem to have been conveyed to the main board at its next meeting, when Sir David Huddie's report was merely –

Programme Status
There are still serious mechanical and performance problems with the RB.211 but in our judgement the task remains very difficult but achievable.

The main board was advised also that the draft 1970–74 forecast was 'unacceptable'. Certainly those members of the main board who were also members of the DED board were aware of the full facts, but the main board as a whole does not appear to have been adequately alerted. An alarming number of main board directors (including, according to his own testimony, Sir Denning Pearson himself) were not conscious of the existence of the Metcalfe report until after the receiver was appointed over a year and a half later. One of them ruefully conceded that this created a picture of 'rotten communications'. We are left with the impression that on occasion those who reported to the main board upon the state of the RB 211 failed to do so in terms which sufficiently indicated its gravity. The information systems in operation in Rolls-Royce, and communications generally within the company, were not as they should have been. Typically there was failure to convey information simply, concisely and pungently. This was not conducive to the best use of the company's resources. This may go some way to explain why the magnitude of the difficulties encountered with the RB 211-22 was not adequately appreciated by many main board members until a comparatively late stage, despite the fact that in September 1969 the following passage is to be found in Sir David Huddie's written report to the main board:

Meeting our programme on the RB.211 now represents a formidable task and we are putting the programme on an emergency basis . . . but we cannot yet put on paper a specification which would meet our contract commitments.

The dilemma, in retrospect, was clear. Rolls-Royce had won this contract, vital for its own future and significant in national terms. It felt it had to fulfil it, at whatever cost. But the cost was going to be huge. As we have seen, Metcalfe summed it up succinctly when he reported that 'any steps one could reasonably take to limit or reverse the adverse financial trends would inevitably prejudice the achievement of the technical objectives of the contract specification in the time available'.

And, crucially, it was not only the development costs that had soared – from an original estimate of £89.3 million to £170.3 million by September 1970 (£1.6 billion to £3.1 billion in today's terms) – but also the estimated 1971 production cost of an installed engine had increased from £153,655 to £237,452, *a sum greater than the selling price of the engine*. Furthermore, as we have seen, every new forecast produced in 1970 showed a sharp escalation of cost over the forecast produced only weeks earlier.

The inspectors were sharply critical of the lack of communication at the highest levels of the company.

It is a wry commentary on this depressing story that not until its very final stages did those having the direction of the company recognise that financial disaster loomed ahead. In the chairman's statement of 22 June 1970 shareholders had been told that the board was satisfied that the company had sufficient working capital. This was reiterated at the annual general meeting on 21 July 1970, though estimates by then available forecast a substantial cash shortage . . . Yet again, on the day in November 1970 that Lord Cole was appointed chairman, little more than two months before the receiver was to be appointed, the company had announced in its half-yearly financial statement (appendix 13) that 'satisfactory arrangements' had been made with the banks, and that subject to unforeseen circumstances the additional resources negotiated in that month would meet the company's cash requirements. One would look far for more striking confirmation of the financial consequences of seeking to navigate uncharted seas.

The inspectors' final conclusion is difficult to disagree with thirty years later. They wrote:

At the end of the day the question remains – why did this great company end its days in receivership? Its main board included men who would have adorned any boardroom. The company was not lacking in skilled engineers. It was not lacking in aero-engine experience. It was not lacking in advice from outside advisers and consultants of HMG. It was respected by its competitors. Its products had an enviable reputation. Yet it crashed monumentally. There can be no doubt that the effective cause of the crash was the drain upon the company's finances brought about by the RB.211 project. To be sure, the company was not without other problems; for example, the merger with Bristol Siddeley some years earlier had increased its liquidity troubles. But in the event this was of relative unimportance. The RB.211 venture apart, none of the problems facing Rolls-Royce would have given rise, immediately or remotely, to the burdens which submerged the company in February 1971. The size of the engineering task was seriously underestimated. From this basic fact flowed the appalling consequences recounted in this report. Once the contract was signed no system of financial control could have significantly influenced the course of events. We have been critical of the decision to conclude the contract. Yet although it failed to achieve its goal, in fairness it must also be stated that in relation to the magnitude of the challenge involved the performance of the company on the project thus undertaken was remarkable. The extent to which countless individuals contributed exceptional personal efforts to the task should not go unrecorded. This is all the more notable if one recollects the lack of leadership from which DED suffered for much of this critical time. Sir William Cook when asked about the way in which the project was conducted, felt able to say to us 'it was done professionally'.

Reluctantly, the inspectors felt compelled to lay the final blame on the two most senior executive directors of the company.

If the board as a whole did not fully appreciate the extent of the risks involved we do not in all the circumstances find this surprising. There is truth in the view that Rolls-Royce tended to be a company dominated by engineers. The board, as it appears to us, treated this essentially as a matter upon which it should be guided by the assessment of those experienced aero-engineers from among its members who were most closely concerned with the project. This meant pre-eminently Sir Denning Pearson and Sir David Huddie. Their standing in Rolls-Royce was such that they could and did influence the decisions of their colleagues in this matter. [Jim Keir, later a main board director of Rolls-Royce and in the late 1960s an executive assistant to Huddie, said in August 2000, 'Huddie was a great engineer who inspired nothing less than hero worship. If I have one criticism it is that he did not keep his fellow directors fully informed.']

The inspectors continued:

Their knowledge and experience should have enabled them to recognise the awesome dangers of the venture. But they shared an abiding conviction that without an involvement in this sector of the aero-engine market the future of Rolls-Royce would be but a shadow of its past. We are bound to say that in our view they allowed this conviction to colour their approach. To conclude a contract with a US manufacturer

155

and become for them not merely a consummation devoutly to be wished, but a task in which failure would be unthinkable. This is not the attitude of mind best calculated to produce the most objective analysis of the risks involved. Nor did it. The terms ultimately negotiated meant a commitment for Rolls-Royce which placed at hazard its continued existence. Sir Denning Pearson and Sir David Huddie were men who had devoted their lives to Rolls-Royce. The company owed them much. Nevertheless they advocated the acceptance of this rash commitment. They were equipped to recognise and warn against its imprudence, even if others less expert were not. In omitting to do so they failed properly to discharge the responsibilities of stewardship which rest upon the directors of a public company.

The author was not able to interview Sir Denning Pearson as he died in 1992, but Sir David Huddie told him just before he died in 1999:

I'm not blaming anyone else . . . We had promised a bit more than we could perform. Things were not deplorable. Flight engines were four months late. Four months is four months but it is not deplorable. The deplorable thing was the cost . . . The accountants never got cash-flow into our heads.

The business careers of both Sir Denning and Sir David were finished. In their retirement, and both lived into the 1990s, they will, we hope, have gained some satisfaction that the RB 211 and its derivatives proved to be every bit as successful as was hoped in the 1960s and are the foundation of the company's success today. Indeed, without the vision of Huddie and his appreciation that the future of Rolls-Royce lay in producing engines of the size and calibre to compete in the world market for civil airlines, primarily the USA, Rolls-Royce as we know it today, would not exist. And Pearson understood that and backed him. Pearson's two big mistakes were his failure to spend enough time at Westminster and in the City, organising the necessary funds to see the development of a world-beating engine through its inevitable teething problems, and his failure to appoint a strong leader in engineering following the tragically early death of Adrian Lombard.

CHAPTER SIX

TAILORING CARS FOR CUSTOMERS

THE POST-WAR RANGE
THE COACHBUILDERS
SILVER CLOUDS AND SILVER SHADOWS
COLLABORATION WITH BMC
CRISIS
VALUE OF THE NAME
MERGER WITH VICKERS

THE POST-WAR RANGE

THE RECEIVER, RUPERT NICHOLSON of Peat Marwick Mitchell & Co., almost certainly on the insistence of the Government, decided to split the Motor Car and Aero Engine Divisions into two and a new company, Rolls-Royce Motors, was formed, taking over the assets of the Rolls-Royce Motor Car Division and the Oil Engine Division (which would henceforth be called the Diesel Division). How had the Motor Car Division fared since the war?

The origins of the post-war Rolls-Royce Motor Car Division were laid, both in terms of location (Crewe) and of designs, well before the war. We shall see later in this chapter how Crewe was selected. The designs developed from the decision by Hives in 1937 to create separate Aero and Chassis Divisions. Through that year he gradually separated Aero and Chassis, but left the technical and engineering functions till last because of the personalities involved. The choice for chief engineer of the Chassis Division lay between R.W.H. Bailey and W.A. (Roy) Robotham – 'Rumpty' to some; 'paddle-arse' to others. Hives chose Bailey, and Robotham felt aggrieved. Nevertheless, time was not wasted in feuding and both men got on with the work of developing cars for the future. As Alec Harvey-Bailey, R.W.H. Bailey's son, wrote in *Rolls-Royce, Hives' Turbulent Barons*:

As the rationalised range, with the new engines, started to appear it was evident how well he [Robotham] and By [R.W.H. Bailey] had co-operated.

Robotham worked towards achieving a 'rationalised' range and thanks to him the Rolls-Royce Wraith and Bentley V used common cylinder head assemblies. R.W.H. Bailey favoured the straight-eights as power units for the Phantom III replacement and for more sporting cars, and Robotham agreed with him. The 1940 Motor Show would have had a new straight-eight Rolls-Royce and Bentley, as well as a Bentley V with the new inlet-over-exhaust engine in the six-cylinder range. There would also have been a Wraith replacement and a small three-and-a-half litre Silver Ripple Rolls-Royce.

Unfortunately, the outbreak of World War II in September 1939 meant abandoning all these plans. Bailey was asked to return to the Aero Division by Hives, and Robotham became chief engineer for the Chassis Division. He began to formulate plans for the post-war range. As well as the rationalised range replacements for the Phantom III, Wraith and Bentley V, powered by the B range engines, and sometimes referred to as the Senior Range, in due course he also began work on smaller cars – the Junior Range – with another smaller range of engines, the four-cylinder J40 and six-cylinder J60. He also included ideas for small cars such as the 10 hp Myth and 6 hp Wisp.

Interestingly, in view of what developed later, Robotham also thought of commercial vehicles and a tractor, suggesting that Rolls-Royce take a licence from an established tractor manufacturer.

Development work on cars had continued during the war. Robotham said in his book *Silver Ghosts and Silver Dawn*:

In the autumn of 1944 our situation in the post-war car market was very satisfactory, for we had an up-to-date design for a range of chassis which had had more pre-production testing than any of the Rolls-Royce cars with which I had hitherto been associated.

However, no work had been done on the body side and it was clear that coachbuilt bodies made in the pre-war period would no longer be satis-factory from a quality point of view. Furthermore, with the rise of costs, especially labour costs, they would be prohibitively expensive. How were they to find a modern, mass-produced design? Ivan Evernden had worked on bodies under Royce and produced some drawings for a general purpose body. Robotham asked his friend, Spencer Wilks, Chairman of Rover, for his advice knowing that he had bought annually 20,000 quality bodies for his Rovers before the war. Wilks advised him to have a set of body tools made by Pressed Steel Corporation at Castle Bromwich.

However, Robotham still needed a design for his body and he was fortunate that John Blatchley wanted to return to motor car design, having spent most of the war at the company's aero engine design headquarters at Hucknall, designing sheet metal cowlings to cloak aircraft engines, a job he later described as 'intensely boring'. Blatchley had worked in the 1930s for Gurney Nutting, the prestigious coachbuilder, where he had become design chief. He remembered it well.

We were in the business of virtually tailoring cars for customers. We rarely made two alike. It was a wonderful learning process.

When Blatchley heard about the new car the motor division was planning, he presented himself to Robotham who was delighted to take him on board. Blatchley said later:

They had a working prototype but no one there had a clue about body design. My first job was to try to put a little finesse into it – the prototype had no elegance at all. I was the designer caught between Rolls engineers and the industrialists at Pressed Steel, who were going to make the bodies, and it was too late to change much except insist the door hinges were concealed internally and make the roof pillars look less clumsy. I also had to completely design the interior – seats, dash-board, trim, the lot.
    Working in a disused squash court behind the factory [Blatchley is right about the squash court, but mistaken about the location; it was not behind the factory but at Robotham's house, Park Leys, near Duffield], I had to make a car that would be stamped out by the thousand look like a craftsman-made motor car.

And Blatchley was not only a brilliant stylist; he could, as Robotham said:

work up the engineering drawings of a complicated unit such as an instrument panel: in other words, he did not style something which could not be made.

He also learned from Pressed Steel how to make quarter-scale clay models. From Blatchley came the Bentley Mark VI, announced in 1946, and later its near-identical sister, the Rolls-Royce Silver Dawn – the model that re-established Rolls-Royce car manufacture after the war.
    In the meantime, Robotham had to establish whether Pressed Steel could make the bodies and at what cost. He first talked to them in January 1944 and established that they could make the tools within twelve months and, to his horror, that the tools would cost at least £250,000 (about £12.5 million in today's terms). He also calculated that the cost of each body would be about half that of a Park Ward body (as we have seen, Rolls-Royce had bought Park Ward in 1939). However, the quantities were small and the cost

reduction would not be felt until at least 5,000 bodies to the same design had been made. Hives gave his approval for the expenditure on the necessary tooling. On 23 October 1946, the first post-war car was handed over to the Gaekwar of Baroda, the first post-war customer.

The resumption of car production, after the war, took place in the Crewe factory that had been erected in 1938 as one of Derby's shadow factories. (It could be argued that Crewe was Rolls-Royce Derby's *only* shadow factory, as the factory at Millington near Glasgow was a Government owned *and* managed factory. The boss was a civil servant and his subordinates Rolls-Royce secondees.) As the possibility of another war with Germany increased – Winston Churchill from his isolated position in the countryside of Kent warned in ever-more strident terms of its inevitability – the British Government finally realised how unprepared the country was. Its intelligence had established what a formidable war machine Germany was building, and by 1938 plans were put in hand to match it. With its Merlin engine, Rolls-Royce was a key supplier for the new fighters, the Hawker Hurricane, and the Spitfire being developed by Vickers Supermarine, and the Air Ministry wanted to be sure that not only would the Derby factory be able to produce the engine round the clock, but also that there would be other factories that could do so as well.

The board had initially rejected the idea of Rolls-Royce shadow factories when the Government had first suggested it in 1934. But when Hives became General Manager in 1936, he reversed the policy, and he and Arthur Sidgreaves, who had become Managing Director in 1929, began negotiations with Colonel H.A.P. Disney, Director of Aeronautical Production at the Air Ministry. Two schemes were discussed. The first was for car chassis production to be removed from the Derby factory, leaving it free to concentrate on aero engine production, and Hives pursued this to the point of looking for a site at Burton-on-Trent, a few miles south-west of Derby. The second was for the building of a second factory that could provide capacity for 5,000 Merlins per year, and that would be used in the event of the Derby factory being destroyed.

It was this second scheme, alongside an expansion of the Derby plant and additional subcontracting, that was chosen. However, Hives and Sidgreaves were worried about who would control the factory. In a note to Hives, Sidgreaves said:

If we have to carry all the Government procedure and Government staffs, I do not know how it is going to work. We naturally do not want to hand over a lot of our orders to a Government-run factory.

Hives and Sidgreaves secured what they called 'parental control' by insisting that the Ministry, having paid for the plant to be built and equipped, should lease it to Rolls-Royce.

The site suggested by Hives and Sidgreaves to Colonel Disney was Shrewsbury, but this received objections from the Ministry of Labour. Hives suggested Worcester, Stafford or Crewe. Both the Air Ministry and the Ministry of Labour favoured Crewe, and on 24 May 1938 the decision was made.

From that day, events moved quickly and by November the first trainees were producing engine parts, and the first Merlin engine was completed on 20 May 1939. Throughout the war the Crewe factory concentrated on producing aero engines, and, when the war ended, the board needed to decide where to manufacture both its aero engines and its cars. By 1945, Rolls-Royce controlled seven million square feet of factory space, nine times that of ten years earlier. There were plenty of options and it was decided that the manufacture of motor cars would not return to Derby but would move to Crewe.

It was felt that the relatively modest demand for luxury cars in an austere post-war world could be met from a square footage somewhat less than the whole of the Crewe factory, and part of it was sub-let to other companies. The new general manager (later managing director) was Dr. Llewellyn Smith. He had been deputy works manager of the Hillington factory near Glasgow and he was made chairman of a committee in 1944 to discuss post-war car production. The other members were Robotham and John Morris, Crewe's first works manager. It could be said that Robotham was unlucky not to have been given the job of running the car company after the war. It was, after all, thanks to him that the design team had stayed together and operated so well. However, running a design team is one thing, but running a factory with several thousand employees quite another, and Hives obviously considered Llewellyn Smith the right man for the job. We saw earlier how Hives had preferred R.W.H. Bailey to Robotham in 1937. Clearly his reservations about Robotham were still present in 1945.

Educated at Rochdale High School, Manchester University and Balliol College, Oxford, 'Doc' Smith had worked on both the car and aero engine sides of Rolls-Royce in the 1930s. In 1939 he and Hives were awarded the Manly Memorial Medal for their joint paper on high-output aircraft engines.

The first post-war car, the first complete car built by the company, was not a Rolls-Royce but a Bentley Mark VI. The board was well aware that the Labour Government, elected in 1945, was likely, in the spirit of the times, to tax luxury cars heavily, and therefore felt that the prospects for the Bentley would be better.

Announced in spring 1946, shortage of raw materials and components delayed full production until the autumn and deliveries until the end of 1946. A few Mark Vs had been built before the war and, heavily used during

161

"Excuse me, Miss, but are you the owner of the Rolls?"
"No: but thanks awfully for asking me."

The Labour Government after the war were determined that the social divide of the inter-war years would not be recreated.

the war, much was learned from them. (Two prototypes – the Rolls-Royce Silver Dawn 80 and the Bentley Scalded Cat – had been sent to Canada for safety. They were shipped back, arriving at Liverpool on 30 August 1944. The Silver Dawn 80 was again used by Hives as his personal car, and was eventually taken off the road and scrapped in July 1950 after driving 111,000 miles. The Bentley returned to the Clan Foundry fleet and was withdrawn in August 1951 after driving 143,000 miles.)

162

However, the four-and-a-half litre, six-cylinder engine was new and had been designed to put into operation the ideas on interchangeability of component parts. It had covered thousands of miles in cars and in a potential replacement for the Bren Gun Carrier as part of its development testing. *Autocar* described it as 'one of the world's engineering masterpieces'. In its Bentley version a little silence was sacrificed in the cause of performance.

The engine had overhead inlet and side exhaust valves, an arrangement originally developed by Royce for the engine of his first 10 hp car! The cylinder head was aluminium and the top ends of the cylinder bores were protected by a coating of special corrosion-resistant chromium. The gearbox in unit with the engine was the classic Rolls-Royce silent type, with four speeds and syncromesh on second, third and top, the direct drive. As in the pre-war cars, the gear lever was sited on the right beside the driver's seat. The brakes, hydraulic on the front wheels, mechanical on the rear, had a servo drive from the gearbox, making the car do the work after light pressure on the brake pedal. This was essentially identical to the servo unit introduced with the four-wheel brakes on the 1923 Silver Ghost. Every Rolls-Royce and Rolls-Royce Bentley had the gearbox-mounted mechanical servo until the Silver Shadow and Bentley T were launched in 1965.

The chassis had a massive deep-sided box-section frame, with exceptionally strong cruciform central bracing. It carried a new kind of body mounting – twelve rubber bushes mounted on vertical axes.

The first post-war Rolls-Royce was the Silver Wraith and this was a throwback to pre-war days in that it was offered as a chassis only. It was not until 1949 that the first Rolls-Royce complete with a standard steel body was produced. Called the Silver Dawn, it was introduced initially only for export and was virtually a Bentley Mark VI with a Rolls-Royce radiator, although it had twin fog-lamps and heavier bumpers and an engine the same as that in the Silver Wraith with only a single carburettor. The interior was identical to the Bentley except for the arrangement of the fascia, which had separate gauges grouped around a central speedometer.

The launch of the Bentley Mark VI and the Rolls-Royce Silver Wraith were successfully achieved though there were several doubts expressed about the future for luxury cars in a world ravaged by six years of war. In Britain itself, the climate could hardly have been worse. To the surprise of many, in the General Election of 1945 the British people not only threw out the Conservatives, led by the lionised Winston Churchill, they also voted the Labour Party in with a majority that would brook no opposition to their plans for continuing and extending the State control that had been established to win the war. (Labour actually won 210 seats from the Tories.) And the new administration, with many of its leaders such as Clement Attlee, Ernest Bevin, Herbert Morrison and Hugh Dalton having run major departments in

Churchill's wartime Coalition Government, went off with a bang. On 16 August 1945, the King's Speech foreshadowed the nationalisation of the Bank of England, the coal industry and civil aviation; the establishment of a national health service and increased social security; the repeal of the Trades Disputes Act of 1927, and Government drives to produce more houses and more food.

Dalton wrote: 'There was an exhilaration among us.' However, he and his colleagues soon had to face the reality of Britain's economic position. The damage to the country's position in the world was considerable in World War I, but it paled into insignificance compared with its sacrifices in World War II. Britain had lost one-quarter of its national wealth – £7,000 million. As a deliberate act of policy, the country had sacrificed two-thirds of its export trade, and its economy had been totally distorted to the cause of winning the war. By the end of the war, only 2 per cent of British industry was producing for export, and yet it needed to import two-thirds of its food and the bulk of its raw materials. Its merchant shipping had been reduced by 28 per cent. There was perhaps some consolation in that Britain was better off than most of Europe.

There was only one country to turn to, and that was the United States of America – and Maynard Keynes was despatched to negotiate a $6 billion loan. The terms imposed were stringent, even harsh. *The Economist* condemned them.

Our present needs are the direct consequences of the fact that we fought earliest, that we fought longest and that we fought hardest. In moral terms we are creditors; and for that we shall pay $140 million a year for the rest of the twentieth century. It may be unavoidable, but it is not right.

And as the immediate post-war years passed, conditions did not improve. The Labour Government pressed on with its nationalisation programme but at the same time was forced to introduce bread rationing, an extreme measure that had not been found necessary during either of the world wars. And in the winter of 1947 the weather took a hand. Coal supplies ran out and, as Roy Hattersley put it, in his book *Fifty Years On*:

The House of Commons debated the fuel crisis on 5 February. Shinwell, instructed by Attlee to abandon his usual vacuous optimism and tell the country the hard facts, set out the emergency measures which echoed round Parliament and the country 'like a thunderclap'. Some powerstations would close so that available coal supplies could be concentrated on others. Control orders rationed the supply of domestic electricity to limited hours each day. As an immediate result, two million workers were laid off. Within a fortnight unemployment had risen to 2.5 million.

This winter crisis of bread rationing and fuel shortage was quickly followed by the first of many sterling crises. To add to the woes of the Rolls-Royce Car Division, its products were not seen as being in tune with the more egalitarian times. Purchase tax on cars costing more than £1,000 was doubled to 66.66 per cent. In 1947, a Bentley Mark VI cost over £3,300 (inclusive of purchase tax). A Daimler was half that price, and small popular cars from Austin, Morris and Ford were only a tenth of the price.

The one consolation was that the Government and the country needed every dollar they could get, so at least there was applause for each Rolls-Royce or Bentley that could be exported.

In spite of all these difficulties, the company had not lost its knack of getting good reviews in the press. On 5 April 1946, *Autocar* wrote:

Perhaps the particular charm of the Silver Wraith lies in the effectiveness with which all the undesirable manifestations incidental to the development of power by machinery have been skilfully exorcised. The result is a car which is like a living thing. An imaginative person might easily believe that the Wraith's feelings could be hurt by a carelessly casual or definitely dangerous driver. Offended dignity might cause it of its own volition to move quietly off to more congenial company! Perhaps such imaginings arise from the extraordinary responsiveness of the car.

And we have seen how Rolls-Royce, from its earliest days, had shown a flair for securing the endorsement of well-known and respected motorists and this ability continued in the post-World War II period. Raymond Mays, a racing driver with an enormous fan club in the 1940s, said of the Mark VI saloon:

Make a list of the world's fine automobiles – French, Italian, American, anything – and ask yourself this simple question: Could I, after driving 600 miles or more from home base to race venue, cruising at eighties and nineties for hours at a stretch in the process, could I willingly and zestfully set myself, the same evening to thrash the same car around a completely unfamiliar circuit for further hours at a stretch? Yes, and enjoy it . . . any fit and habitually fast driver could do it, and go to bed without an ache in his body, on any saloon Bentley that had Rolls-Royce craftsmanship behind it.

The Silver Wraith continued in production until 1958 and carried a greater variety of coachwork than any other post-war chassis. The companies Park Ward, H.J. Mulliner, Hooper, James Young and Freestone Webb, and a number of smaller coachbuilders, all made elegant coachwork for this chassis. The majority were limousines or saloons, although Mulliner made some *sedanca de ville* designs and also some coupés. This proved to be almost the swansong for most of the coachbuilders and would seem to be an ideal place to review their contribution to the Rolls-Royce story.

THE COACHBUILDERS

As we saw in Part One of this history, until after World War II Rolls-Royce produced only chassis and the bodywork was produced by outside companies. No history of Rolls-Royce motor cars would be complete without some comment on the leading coachbuilders that did so much to provide the styling and glamour on top of the excellence beneath the bonnet.

Park Ward, taken over by Rolls-Royce in 1939, was founded in 1919 in High Road, Willesden, North London by W.M. Park and C.W. Ward who had been employed by Sizaire Berwick (a company sued by Rolls-Royce for imitating its radiator shape). Park was the tool-room foreman and Ward in charge of body production. The company produced top-class coachwork not only on Rolls-Royce chassis but also on those of Daimler, Mors, Lanchester and Sunbeam. From a small start, they took over the stables of London General Omnibus Company.

Park Ward's initial work for Rolls-Royce was to put coachwork on limousines bought back from the Government and on First World War staff cars. It was not until 1921 that Park Ward put its first body on a new Rolls-Royce, but, by the end of the 1920s, 90 per cent of its work was on Rolls-Royce chassis. In the 1930s, Rolls-Royce felt sufficient confidence in Park Ward to place the research and development work on the three-and-a-half litre Bentley with it and, at the same time, took a considerable financial stake in the firm.

During the 1930s, Park Ward developed an all-steel body framework, enabling it to produce bodies of standardised design in relatively large numbers. This gave the company a dominant position among Rolls-Royce coachbuilders and in 1939 Rolls-Royce bought its entire share capital.

During World War II, Park Ward's precision engineering skills were put to use in the war effort in producing aircraft components such as bomb doors and cowlings for the de Havilland Mosquito, plus cowlings for Hurricane and Spitfire fighters and Lancaster bombers. The company also produced engine mountings and air intakes for all the leading British aircraft builders and supplied ambulance bodies at the remarkable price of £21 each. At the same time, it produced wooden ammunition boxes for the Royal Navy and bamboo-framed canvas kites for the protection of merchant navy shipping against air attack.

After the war, Rolls-Royce's use of bodies from Pressed Steel restricted Park Ward's role, but it still provided coachbuilt bodies for the Silver Wraith and on some Mark VI and Silver Dawn chassis, as well as on Bentley Continentals. In 1952, the chief styling engineer at Crewe, John Blatchley, was given responsibility for the external design of the Park Ward coachwork.

At the end of the 1950s, Rolls-Royce bought another coachbuilder, H.J. Mulliner & Co., and later amalgamated it with Park Ward. This company went back as far as 1769 in Northampton, although H.J. Mulliner & Co. was founded in 1900 when Henry Jervis Mulliner bought the coachbuilding firm, Mulliner London Ltd., from his cousins, Arthur Mulliner of Northampton, and A.G. Mulliner of Liverpool. Born in 1870, H.J. Mulliner had spent two years in Paris learning how coachbuilding could be adapted to the new motor car.

Mulliner became friendly with Charles Rolls. Indeed, Mulliner was a founder member of the Automobile Club of Great Britain and came to know not only Rolls but also Claude Johnson. Mulliner's business premises were located in Brook Street, Mayfair, close to Charles Rolls's first offices for C.S. Rolls & Co., which were also in Brook Street. By 1905, Rolls's main headquarters and showroom were in Conduit Street. As business grew, Mulliner moved to a factory in Bedford Park, Chiswick, in West London.

Mulliner produced reliable if unremarkable coachwork for Rolls-Royce up to World War I and during the 1920s, but it was in the 1930s that the firm produced its most elegant bodies when it developed its 'High Vision' concept. The principal feature of this design was a larger-than-normal glass area, and this exterior work was matched by interior work of the highest standard. As Martin Bennett said in his excellent book, *Rolls-Royce and Bentley, The Crewe Years*:

The name H J Mulliner and Co on a car's sillplate was a guarantee of exterior elegance, unsurpassed interior luxury and finish, together with a high level of constructional integrity.

In World War II, Mulliner also manufactured aircraft components such as engine cowls for Mosquitos and resumed work for Rolls-Royce after the war, producing nine of the seventeen bodies of the Phantom IVs sold to royalty and heads of state – including the one built for Princess Elizabeth and the Duke of Edinburgh. (Mulliner had also built the body on the Daimler bought by Queen Mary in the 1930s.)

In this post-war period, the company was almost the only coachbuilder making *sedanca de ville* coachwork. The company dominated the production of coachwork for the Bentley Continentals, working closely with both Ivan Evernden and John Blatchley, and becoming closely involved with wind-tunnel testing at Hucknall. Following the Rolls-Royce purchase in 1959, Mulliner and Park Ward were fully combined in 1962.

After the amalgamation, Mulliner moved out of its Chiswick factory and in with Park Ward in Willesden. After the introduction of the Silver Shadow and Bentley T series in 1965, Mulliner Park Ward continued to build bodies on Silver Cloud III and Bentley S3 Continental chassis for a few months but began to turn its attention to body engineering rather than traditional coachbuilding. The only Rolls-Royce car with a full chassis being produced by this time was the Phantom V, to be succeeded in 1968 by the Phantom VI, and from 1973 both chassis and body production of the Phantom VI was carried out at Willesden. Further work was done on the Corniche and Camargue in the late 1960s and during the 1970s, but the slumps of the early 1980s and early 1990s finally brought closure of the Willesden operation and a move (if only on paper) to Crewe.

Another supplier of Rolls-Royce bodies was Hooper & Co. (Coachbuilders) Ltd., which could be traced back to 1807. As with other coachbuilders, the arrival of the motor car brought changes, and Hooper concentrated on Daimler and Rolls-Royce. In 1938, Hooper bought another long-established coachbuilder, Barker & Co., and built new premises in Acton, West London, and during World War II the amalgamated firm concentrated on aircraft component work. In 1940, it was bought by Daimler, itself acquired by the Birmingham Small Arms Company (BSA), at the end of the war. After some flamboyant work for the Chairman of BSA, Sir Bernard Docker, and his ostentatious wife, on Daimler cars, Hooper concentrated on Rolls-Royce and Bentley in the 1950s. However, the firm ceased manufacture in 1959, and concentrated on maintenance and repairs of Hooper bodies built in earlier years.

The last independent coachbuilder to put bodies on Rolls-Royce and Bentley chassis was James Young Ltd. Established in 1863, James Young became famous for its lightweight Bromley Brougham, and it was no surprise that its early bodies for motor cars were widely praised for their elegance. They were soon producing bodies for Alfa Romeo, Austro Daimler, Bugatti, Hispano-Suiza, Isotta Fraschini, Lanchester, Mercedes and, of course, Rolls-Royce.

Responsible for a number of design innovations, James Young also contributed to the war effort in both world wars, building lorries, armoured cars and ambulances between 1914 and 1918, and forty-five gallon jettison tanks for Spitfires, shells, inflatable rafts, ammunition boxes and feed necks for Hurricane fighters, Air Ministry tarpaulins, gun covers and mobile canteens between 1939 and 1945. It was at the James Young factory that a Mark V Bentley was completely destroyed in 1941. The very first production postwar Mark VI Bentley was given a saloon coupé design by James Young.

The firm had been acquired by the Rolls-Royce agent Jack Barclay Ltd. in 1937, and was joined after the war by Gurney Nutting, which had produced some sporting designs for Rolls-Royce and Bentley in the 1930s. The firm continued to produce bodies solely for Rolls-Royce and Bentley until the mid-1960s and the arrival of the Silver Shadow and Bentley T series.

Finally, we should mention Freestone & Webb Ltd., the smallest of the post-war survivors in the coachbuilding business. Formed in 1923, the firm operated, like Park Ward, from Willesden, and worked from the beginning both on Rolls-Royce and Bentley chassis when Bentley Motors was still independent. The firm concentrated on one-off bodies rather than the batches of their competitors. Bought by the car distributor H.R. Owen in 1955, it delivered its last cars in 1958.

There were, of course, many other coachbuilders, including Brewster, Cockshoot, Cunard, Arthur Mulliner, Rippon, Salmons, Thrupp & Maberly, Vanden Plas and Windover, which were substantial firms, and Abbott, Arnold, Binder, Victor Broom, Caffyns, Carbodies, Carlton, Cole, Connaught, Croall & Croall, Crosbie & Dunn, Fleetwood, Flewitt, Framay, Fuller, Gill, Grosvenor, Hall Lewis, Claude Hamilton, Hamshaw, Harrison, Lancefield, Lawton, Maddox, Mann Egerton, Maudsley, Mayfair, Maythorn, Melhuish, Morgan, Motor Bodies, Nordbergs, Offord, Van-Den-Plas (not to be confused with Vanden Plas), Watson, Weymann and Wylder, which were somewhat smaller. There was also a considerable number overseas.

The first Rolls-Royce completely built at Crewe was the Silver Dawn (the Silver Wraith had continued to have bodies made by specialist coachbuilders). It made its first appearance at the International World's Fair in Toronto and then in New York in 1949. Chassis, wheelbase and body were the same as the Bentley Mark VI standard steel saloon, while the engine, with its single carburettor, came from the Silver Wraith.

Next came the Bentley R-type. Styled by John Blatchley, from a distance it was hard to see any difference from the Mark VI. However, the boot was extended (giving more room), the rear wheel arch was lower and the rear wing extended. Its extra length gave it more elegance, though some, rather unkindly, called it the 'big boot, big bore Mark VI'. The Continental version, introduced for the export market in 1951, was designed by Ivan Evernden, who said of it:

After World War II, coachbuilders both at home and abroad made some very elegant bodies for the Bentley Mark VI chassis. But none possessed the total qualities of 'Olga', [the prototype Continental] for she evolved with a purpose, not only to look beautiful, but to exhibit those characteristics which appeal to the connoisseur of motoring; a high maximum road speed coupled with a correspondingly high rate of acceleration, together with excellent handling qualities and 'roadability'.

As we have seen, wind-tunnel tests were carried out at the Aero Engine Division's flight establishment at Hucknall to reduce drag and improve lateral stability at high speeds. H.J. Mulliner produced an all-metal body instead of the traditional timber and metal custom-built body. The frame was built of light alloy sections, panelled with aluminium, reducing the overall weight to 33 cwt. The result was the fastest four-seater saloon in the world and probably, at £6,000 for overseas buyers, the most expensive too!

Initially destined only for export markets, it was released on the home market in 1952, as was the Rolls-Royce Silver Dawn. The 1952 Motor Show was the first show since the war when the Rolls-Royce stand was free of 'export only' signs. The motor car design, styling, development and experimental departments had been transferred from Clan Foundry, Belper to Crewe in April 1951.

The company now understood that if it was to revive its fortunes in the potentially lucrative North America market, it would need to take more notice of customer desires. Neither the Bentley Mark VI nor the Silver Wraith had been offered with left-hand drive until 1949. In 1947, four Silver Wraiths and three Mark VIs had been sent on a 20,000 mile exhibition tour of the United States, but all seven had been right-hand drive vehicles. The Silver Dawn, with left-hand drive, was Rolls-Royce's first serious attempt to regain its position in the USA after the war. Rolls-Royce's marketing efforts in the USA were given a boost by a brilliant advertising campaign conceived by the legendary David Ogilvy, a Scotsman who went to New York after the war to set up the agency Ogilvy & Mather. His by-line for Rolls-Royce adverts was:

At 60 miles an hour the loudest noise in this new Rolls-Royce comes from the electric clock.

The advertising trade magazine, *Campaign,* rated it the best campaign for any product in the history of advertising.

*"The loudest noise in **this** Rolls-Royce doesn't come from the electric clock."*
Copyright 1958, The New Yorker Magazine, Inc.

170

The next major development was the introduction of the first fully automatic gearbox made in Britain. In the USA automatic transmissions had made great headway since the war. They were standard on expensive cars and available as an option on almost every car. The engineers at Crewe began work to develop an automatic gearbox and soon found that they were coming closer and closer to the General Motors Hydra-Matic system, which had sixteen years of development work behind it. In 1952, Rolls-Royce began to manufacture the General Motors gearbox under licence.

SILVER CLOUDS AND SILVER SHADOWS

In designing the seminal Silver Cloud, launched in 1955, John Blatchley admitted to being influenced by the cars coming out of Detroit in the 1940s and early 1950s. He said:

It was difficult not to be influenced by American cars, and I was particularly amazed by Cadillacs.

His first efforts at a new car for Rolls-Royce after the war were rejected as 'too modern'.

I was asked to do a sketch of something more traditional, more in keeping with the Rolls image, which I did in about 10 minutes. It was taken into a board meeting and they decided to make it there and then.

Blatchley later stressed the importance of teamwork.

You must have the engineers on your side because there's always the challenge of maintaining the room inside the car for the passengers against the encroachments of the engineers who want larger silencers, a larger petrol tank and speakers for the stereo system.

Blatchley found both Harry Grylls and 'Doc' Smith supportive, as they knew that appearance and spaciousness were essential. He also remembers two comments from Hives: 'Blatchley, never be ashamed of the size of the Rolls-Royce radiator' and 'It's a waste of time testing in wind tunnels. It's not relevant to people's idea of a Rolls-Royce.'

An innovation with the launch of the Rolls-Royce Silver Cloud (list price £4,796 10s 1d including tax – about £120,000 today) was the simultaneous introduction of a Bentley version (£127 cheaper than the Rolls-Royce). The engine for both was a 4,887 cc version of the six-cylinder 10E engine of the Silver Dawn and Bentley Mark VI.

The Rolls-Royce Bulletin of January 1956 was to say:

There was, among the motoring intelligentsia a gasp of surprise, when it was announced that the new Rolls-Royce Silver Cloud and Bentley S Series were, apart from the form of the radiator, in fact exactly the same machine.

Early in 1956, power-assisted steering was offered as an option – though at first for export only. By this time, especially in the USA, it would have been difficult to sell a luxury car without such a facility.

In 1959, the Silver Cloud Series II and Bentley S2 were introduced with a more powerful V8 engine with a 6.2 litre capacity. This increased performance by 30 per cent.

As Martin Bennett put it:

Over the years 1946 to 1959 the six cylinder F-head engine grew in swept volume from 4,257 to 4,887 cc. The final version of that power unit, with 8:1 compression ratio and two inch S.U. carburettors, represented its practical limit of development. Without serious loss of refinement . . . no further significant power output could reasonably be pursued. With solid competition, particularly from across the Atlantic . . . the in-line six had reached the end of the road. The further requirement called for a power unit that was at least as light as the six, requiring no more bonnet space, while being quieter, more refined and potentially much more powerful. This is precisely what was achieved with the light alloy V-8 announced in August 1959.

Three years later some styling changes were noticeable with the launch of the Silver Cloud III and Bentley S3. The bonnet was made to slope downward slightly, improving forward vision, and the front wings were restyled. The twin headlights became four. As with the Series II, continued improvements were made to the interior layout.

Tony Brooks, idol of the Formula One circuit of the 1950s, said that, after driving the Silver Cloud III for no fewer than 2,700 miles in 1964:

The pleasure that I found in driving the Silver Cloud came from a balance of qualities that is unique in my experience. The result is that the car gives you high performance motoring as near effortless as it can be under all the varied road conditions experienced on test . . . Motoring of this kind cannot be anything but expensive, but I would say that in terms of sheer motoring pleasure, safety and durability, the Rolls-Royce is excellent value for £5,500. As a big, luxurious car that can nevertheless be driven in a highly sporting manner, there is nothing quite like it.

The next big advance in technology for Rolls-Royce came with the introduction, in 1965, of the Silver Shadow and its Bentley equivalent, the T series.

The Silver Shadow evolved from two different design projects – one, a Rolls-Royce, known in the development stages as 'Tibet'; and the other, a Bentley, known as 'Burma'. (There was a tradition of giving experimental cars codenames – for example, the Phantom I was the 'Eastern Armoured Car', the 20 hp the 'Goshawk' and Phantom III the 'Spectre'.) Burma was conventionally sprung whereas Tibet had air suspension (conceived by Royce as early as 1910). Because Tibet lost air pressure overnight, the Silver Shadow was launched with the Burma suspension.

The development work on the two models was merged and the design followed the Burma more closely, although many Tibet features were incorporated in the eventual Silver Shadow. The main engineer behind the development of the Silver Shadow was Harry Grylls, chief engineer of the Rolls-Royce Motor Car Division from its formation in 1954 until 1968. Initial outline plans were laid as early as 1954 and road-testing of the Tibet began in 1958. When it was finally introduced after eleven years of development and refinement, it was described by *Motor* in October 1965 as:

. . . almost certainly the best and quite certainly the most sophisticated car in the world.

Certainly no previous Rolls-Royce model, not even the Phantom III, had introduced so many new features. Perhaps its most revolutionary feature as far as Rolls-Royce was concerned was its monocoque, or unitary construction. This method had become essential in view of the demands of the modern motorist as well as the decline of the coachbuilding industry. Furthermore, customers wanted smaller cars but without losing interior space. Monocoque construction allowed these apparently contradictory desires to be achieved.

To prove to dealers and potential customers that these twin demands had been achieved, the Silver Shadow was demonstrated at the 1965 London Motor Show without its doors, a distinct novelty for Rolls-Royce. Research had shown customers switching to Aston Martin because of the difficulty of manoeuvring the large Silver Cloud in traffic. The monocoque construction made independent rear suspension and automatic height control more practicable. The new model also introduced four-wheel disc braking, and, on the Silver Shadow I and Bentley T1, recirculating ball steering with integral power assistance, along with electric operation of the gear change selector, front seat adjustment and window lifts. The luggage boot also gained both from the more 'boxy' shape and from having the spare wheel stowed beneath it rather than inside it. Space was also found for fuel tanks with a third more capacity. The V8 engine was essentially the same as the Silver Cloud II and III, but it was repositioned to make more accessible certain items which were difficult to service.

The *Financial Times* was to say of the Silver Shadow and Bentley T series:

[A]lthough some feared that the 'majesty' of a Rolls-Royce was lacking, the fact remains that customer response has proved better than ever before, and there are still delays in delivery . . . all the enviable features of every Rolls-Royce that has ever been made are here, plus many new ones that have never been dreamed about.

And Paul Frere, a leading motoring journalist and winner of the Le Mans twenty-four-hour race, said, after driving a Silver Shadow for 2,000 miles on both fast *autoroutes* and through Alpine hairpin bends:

I did the trip from Brussels to Monaco in one day . . . It felt strange speeding down the Autoroute at 110 mph with no noise coming from the engine or road and the air conditioning keeping the temperature inside the car perfect. On reaching Monte, one impression of the Silver Shadow was dominant. I was fresh, relaxed and not in the least bit tired – a remarkable tribute to a car after having driven 700 miles. [Mike Evans remembers the Financial Director, William Gill, telling him that the Bentley S2 was the first new car he ever had and the first car in which he could drive all the way to Gleneagles after a Friday morning board meeting in Conduit Street in London's West End and still be fresh enough for a round of golf.]

Ironically, in view of the fact that it was the Bentley Burma that formed the basis for the Silver Shadow and the Bentley T, when the Silver Shadow was launched, sales of the Bentley were barely noticeable compared with the Rolls-Royce. The company concentrated on promoting the Rolls-Royce brand at the expense of Bentley, and sales of the Bentley T series were only 3 per cent of the Silver Shadow.

There was no doubt that the introduction of so many new features simultaneously caused a considerable number of quality and reliability problems, but 37,000 Silver Shadow and Bentley T series cars were sold in their fifteen-year production run, with a record year in 1978, when 3,347 were built. With hindsight, the sublimation of the Bentley during this period (for some years the Bentley alternative was not even mentioned in the Silver Shadow brochure) is difficult to understand.

COLLABORATION WITH BMC

During the period between the introduction of the Silver Cloud and that of the Silver Shadow, the Motor Car Division, constantly under pressure to make ends meet, entered into a collaboration with BMC.

In the rationalisation of the British motor industry after World War II, the biggest merger was that between Morris, based at Cowley, Oxford, and

Austin based at Longbridge, Birmingham. These two came together to form the British Motor Corporation, or BMC as it quickly became known.

Rolls-Royce had looked at the possibilities of merger and acquisition. As we have seen, Bentley was acquired in 1931 and Rolls-Royce, knowing it should produce a smaller and less expensive car, looked at the possibility of acquiring Rover during the 1930s and during the war, but did nothing. After the war, it looked again at Rover and at Jaguar but again made no move.

However, as developing motor car technology became more and more expensive, Rolls-Royce, notwithstanding the financial help the Motor Car Division received from the Aero Engine Division, looked to exploit its technological know-how beyond the low volume of Rolls-Royces and Bentleys.

At the beginning of 1962, it was announced that BMC and Rolls-Royce were 'examining the feasibility of technical collaboration in the field of motor engineering'. As the *Financial Times* reported in February 1962, Rolls-Royce was certainly suffering from the credit restrictions imposed by the Conservative Government and increased competition in world markets. From a peak of fifty cars a week, production had slumped to as low as only twelve and a new £130,000 paint plant had not been fully commissioned as the throughput did not justify it.

After two and a half years, it was announced that a new joint-venture Vanden Plas four-litre Princess R car would be produced with a 175 bhp six-cylinder Rolls-Royce aluminium engine, the FB60. This announcement did not come as any great surprise as the venture had been common knowledge in the motor trade for some time. What did come as a surprise, both to the motoring press and to some of the key executives both at Rolls-Royce and BMC, was the announcement from George Harriman, the BMC Chairman, that production of the Princess R would be 7,000 initially, quickly rising to 12,000 a year. David (now Sir David) Plastow, at that time on the sales side at Rolls-Royce, remembers conferring with his counterpart at BMC and agreeing that the sales potential was 3,000 cars a year, perhaps 3,500. When they conveyed this to Harriman he said:

Nonsense! 12,000.

However, as the *Financial Times* pointed out, the car was pitched at a level of the market where there had been no growth in the past decade. The selling price was £1,994, and a glance at the competition and the production numbers showed how wildly optimistic was the 12,000 estimate. All three Jaguars – the 3.4 S-type at £1,669, the 3.8 S-type at £1,759 and the Mark X at £2,022 – added together were probably being produced at the rate of 12,000–14,000 a year. The Daimler 2.5 V8 saloon at £1,599 was being produced at 5,000 and the Rover three-litre at £1,641 at 8,000.

Frankly, the car just looked like a BMC three-litre Princess with its hips removed, and the Princess was simply a fancy Vanden Plas version of the Austin Westminster/Wolseley 6/110.

After an initial burst of enthusiasm sales never reached anything like the hoped-for levels and BMC lost interest when it bought Jaguar later in the decade, as the original idea was to compete with Jaguar. From Rolls-Royce's point of view, it was at least an engine contract, and Tom Neville, finance director of the Car Division at that time, remembers that BMC did pay for 15,000 engines, even though only 5,000 were used, but it provided little else. Reg Spencer, a Rolls-Royce engineer for over forty years from 1939 until 1983, said:

During this period we had a short love-affair with the old British Motor Corporation in which we offered them engineering help with their new models in return for our cashing-in on their new body-shell which we would style into new Rolls-Royce and Bentley models, thus saving us a million or two on tooling costs. This relationship created some interesting work but it turned out to be a mis-match of philosophies, the only end-result being the new Princess R car which used our 3.9 litre engine but which included hardly any of the other engineering improvements that we had suggested and was not a great success. Meanwhile we went our own way with the Burma design . . .

There were some other joint efforts with BMC. Crewe finished crankshafts on a production basis for the Mini Cooper S and work was done on a sports car designed to rival Jaguar's E Type. It began as an Austin Healey but widened by four inches to accommodate the FB60 engine. This gave 180 bhp compared with the 110 bhp of the Austin three litre engines. Then there was the G range engine which was like the FB60 but with twin overhead cam, triple big bore carburettors and a 6000 rpm capability. It was capable of giving somewhere near 300bhp compared with Jaguar's advertised 265bhp.

All of these exciting developments died when BMC bought Jaguar in 1968.

CRISIS

While the Silver Shadow was proving so popular and winning sales all round a world enjoying growing prosperity, events elsewhere in the company were building up what would bring the greatest trauma ever to hit Rolls-Royce. We have already read about the receivership of Rolls-Royce Ltd. in February 1971 which was essentially an Aero Engine Division problem but, of course, it seriously affected the Motor Car and Oil Engine Divisions.

At the end of 1967, 'Doc' Llewellyn Smith retired as Managing Director at Crewe and was succeeded by Geoff Fawn, who had joined the experimental department of the Aero Engine Division in 1942. After work on the development of gas turbines and nuclear submarine propulsion units and a spell in charge of Aero Engine Division spares, he was appointed Deputy General Manager of the Motor Car Division and made a Motor Car Division board member in 1964, becoming Deputy Managing Director in 1966.

His three years as Managing Director are remembered as ones of tough decisions on costs and the raising of selling prices above the rate of inflation. The control of the Main Board over Crewe was such that the Motor Car Division had to ask for permission to raise prices. As wage increases started to accelerate in the late 1960s, the Rolls-Royce Motor Car Division suffered more than mass-producers such as Ford because of the high labour content in Rolls-Royce cars, and prices were raised 5 per cent in 1969 and another 8 per cent in April 1970. As the *Financial Times* pointed out:

The company has never disclosed the profitability or the costing of its motor-car division to its shareholders or to anyone else. Unlike some famous luxury car manufacturers, Rolls-Royce is certainly profit-making but its profits probably show a fairly low return on the capital employed. The Silver Shadow has been such a success – with waiting lists of 18 months in the domestic market – that it would be scarcely surprising if the company were tempted to improve its margins somewhat.

Dennis Head, who later succeeded Geoff Fawn as Managing Director of the Aero Engine Division in Derby, remembers Fawn saying to him:

Dennis, when I went to Crewe they were losing money and I said to them 'What's the waiting list?' They replied '180 cars'. So I put the price up and three months later I said, 'What's the waiting list, now?' '500.' So I put the price up again and three months later asked again. '1,000.' 'Dennis, we're not selling cars, we're selling f---ing jewellery. The higher the price the more people want them.'

Fawn was rewarded for his efforts at Crewe by an appointment as Managing Director of the Aero Engine Division on 1 January 1971 (as close a definition of a hospital pass as one could hope to get). His successor at Crewe was David Plastow who came with a tradition of the motor industry in his blood, his grandfather having made motorcycles and his father having worked in the motor trade all his life. Plastow himself had joined Rootes from school and then Vauxhall, where he worked for five years before his two-year National Service stint in the Army in the mid-1950s. While in the Army, he met the Rolls-Royce sales manager for the north of England who offered him a job. After a spell on the road, he was appointed factory sales manager in 1960, and in 1963 he was made sales manager for military and industrial products. In 1967, he was appointed marketing director responsible for all the Motor Car Division's products and soon began to make changes.

*"The very latest information on delivery is three years nine months, sir,
and not, as I said in error a minute ago, three years."*

At that point, the US market was taking a third of the division's output and Plastow, feeling this was too high, organised a drive to increase sales in Europe, Australia and South Africa. As a result, sales to these markets were 80 per cent higher in 1969 compared with 1968. At the same time, sales into the British market remained level while those to the USA were lower. The result by 1970 was 50 per cent going to the British market, 25 per cent to the USA and 25 per cent going to other export markets. The total production of Silver Shadows, Bentleys, and Park Ward and Phantom VI limousines, 1,850 in 1969, was being raised to 2,000 in 1970. Plastow also pruned the UK dealer network from sixty-eight to thirty-five to maintain the reputation for quality of service.

However, Plastow soon faced the problem of Rolls-Royce's receivership. As we saw, the receiver, Rupert Nicholson of Peat Marwick, acting almost certainly on the insistence of the Government, separated the Motor Car Division from the Aero Engine Divisions and launched it as a new company, Rolls-Royce Motors, taking over the assets of the Rolls-Royce Motor Car Division and the Oil Engine Division, which would henceforth be called the Diesel Division. At the same time, the Aero Engine Divisions, Industrial and Marine Gas Turbine Division, and Rolls-Royce and Associates (nuclear), were vested in the Government-owned company, Rolls-Royce (1971). Rolls-Royce Motors was to become effective on 24 April 1971, would employ 8,500 people, and would own assets of about £20 million. Its annual turnover was £30 million.

As always in moments of crisis and high drama with millions of pounds and thousands of jobs at stake, there were moments of comedy, at least in retrospect. Tom Neville, the finance director of the Motor Car Division (he maintains he was sent from Derby to Crewe when he questioned the wisdom of capitalising engine development) remembers that the receiver gave him authority to spend cash up to £1,000. Nevertheless, one Friday he ran out of money and could not buy food for the apprentices' hostel for the weekend. His solution to the problem was to ring up a local scrap merchant and invite him to come and take some scrap and to 'bring £1,000 in cash'.

Along with the senior partner at Peat's, Ronnie Leach, Nicholson called on Ian (now Sir Ian) Fraser, the Director-General of the City Panel on Take-overs and Mergers, and asked him to take over the chairmanship of Rolls-Royce Motors. They told him the plan was to fatten up the company for sale and use the proceeds to pay off the creditors of Rolls-Royce Ltd., and perhaps provide some money for shareholders.

Fraser told them he had never been chairman of anything and knew nothing about engineering or the motor industry (after service with the Scots Guards during the war and ten years with Reuters, his career had encompassed banking). According to Fraser, Leach and Nicholson said:

We don't want an engineer. We've got plenty of those.

After time for reflection and discussion with some of the key executives involved, Fraser agreed to take the job on a part-time basis. He had found Plastow 'young and impressive' and Tom Neville, the finance director, 'bright'. Others, he felt, were 'not as good as they should be'. His view was that British management in industry was not on a par with that of France, Germany, Italy and Japan, and indeed management in the metal-bashing industry he described as 'deeply distressing'.

However, he felt he could build on Plastow and Neville, and he set about creating a board, having agreed with the receiver that he must have some outsiders.

He brought on to the board, as non-executive directors, Christopher Aston (a Director of Ready Mixed Concrete), Peter Benson (Managing Director and Deputy Chairman of APV Holdings) and Dr. Hans Wuttke (a partner in M.M. Warburg-Brinckmann, Wirtz & Co., and a director of S.G. Warburg).

The head office was established in the existing Rolls-Royce offices in Conduit Street, and Plastow spent probably 75 per cent of his time in Crewe and 25 per cent in London. Fraser remembers that Plastow visited him about three times a month and that Neville sent him a stream of financial information. Fraser saw himself as a monitor and fixer, and said to Plastow that he saw his job as little more than the man who 'would fire' him if Plastow was not up to the job.

Surprisingly, there was about £4 million in cash in the company and Fraser told the receiver that if he removed it, Fraser would not take the job. The company also held a strong order book. The main problem, as Fraser saw it, was the one that bedevilled the whole of British industry at the time – the necessity to carry high stocks because of the regularity of strikes at many of the key suppliers. As Fraser recalled:

Pressed Steel went on strike twice a year so we had to keep bodies in stock.

Furthermore, the company bought no less than 71,000 parts from 4,000 suppliers, this for an output of 2,000 cars. Life was not going to be easy.

In an interview with the American business magazine *Forbes* in October 1971, in answer to a question about survival as a $100 million company when $1 billion companies like American Motors were suffering, Plastow said:

We can increase production from the present 2,300 cars a year up to around 3,000 without any further capital investment and those 700 cars would be worth $21 million at retail. After that we shall go more slowly. It would be easy to make a cheap car and call it the Bentley but we would suffer in the long run.

He knew that after-sales service would have to be improved.

We made a mistake in the Sixties. We expanded so fast in the US that our service lagged.

And 1972 was a record year when 2,470 cars were produced – an all-time high. Sales were 2,473 cars, of which 2,068 were four-door models and 358 two-door. Forty-seven Phantom VIs were sold. The geographical split was: 1,339 in the UK, 629 in the USA, 262 in Europe and 243 in other regions, and the value of deliveries was £22.4 million at retail prices, an increase of 19 per cent over 1971. At the RAC Club in January 1973, Plastow told his audience he expected output to increase by 10 per cent that year.

VALUE OF THE NAME

By this time, the receiver was close to making a decision as to whether to sell the company to a single party or to float the company on the stock market. James Ensor in the *Financial Times* grasped precisely the ethical problem Nicholson faced. (His words could just as easily have been applied in the spring of 1998 as in the spring of 1973.)

This apparently straight-forward situation is complicated by a neat ethical dilemma.

Mr Nicholson, as the Receiver, is duty bound to try to achieve the highest price for the creditors that circumstances will allow. At the same time, he appreciates that Rolls-Royce Motors is not an ordinary engineering company; to many people in Britain and outside it is a symbol which has sometimes been seen as a standard-bearer for British engineering. Clearly a great part of the assets of Rolls-Royce Motors lies in its name, perhaps the most valuable brand name in the world.

It is at least conceivable that the interests of the two groups involved in Rolls-Royce Motors, the creditors on the one side and the employees on the other, will differ. The creditors are clearly primarily interested in achieving the maximum possible recovery of their funds; though some of them would clearly also like, as a secondary interest, to see Rolls-Royce Motors well established with the right kind of shareholders.

And the potential bidders were lining up:

* Tozer, Kemsley and Millbourn – a mini-conglomerate with interests in financing and the import and export of vehicles. (It later came seriously unstuck in the oil-price induced recession of the mid 1970s.)
* Thomas Tilling – importer of Mercedes, Audi and Volkswagen. (It was taken over by BTR in the early 1980s.)
* A consortium of British engineering companies.
* Foreign bidders.

General Motors ruled itself out, in spite of supplying not only the automatic transmission and power steering for the Silver Shadow and Corniche but also the outlet for most of Rolls-Royce's US sales through its dealer network, and apparently the finance people at British Leyland (successor to BMC) told the Chairman, Lord Stokes, that there was quite enough investment necessary in British Leyland itself, without indulging any of his personal fancies. Leyland was, of course, also a significant supplier through its subsidiary, Pressed Steel.

A very interested bidder was Tiny Rowland of Lonrho (soon to be dubbed by Prime Minister Edward Heath as 'the unacceptable face of capitalism'). He had just bought from Dr. Felix Wankel the rights to the rotary engine, of which Wankel was inventor. Rolls-Royce's Diesel Engine Division in Shrewsbury had been the most active British company in developing the Wankel engine at the Crewe factory. Plastow remembers that Rowland came to see him and the Chairman, Ian Fraser, and talked enthusiastically of the great things they could do together. Fraser was sufficiently impressed to ask Plastow what he thought. Plastow was not too keen, pointing out that, traditionally, car manufacturers built their own engines.

Another interested bidder was Hanson Trust, the conglomerate put together by James (now Lord) Hanson, after he had left his family haulage business in the mid-1960s. Despite Hanson's stated aim of sticking to low-tech, basic industries, he saw an opportunity of adding a prestige name to what Alex Brummer and Roger Cowe described in their biography, *Hanson*, as a 'non-descript bunch of companies he controlled'. (In June 1998, Martin Taylor, the former Deputy Chairman of Hanson plc, and by then a non-executive director of Vickers, confirmed to the author at the Vickers EGM that James Hanson had been attracted by the name.)

According to Brummer and Cowe:

Hanson Trust joined with financial partners to make a consortium bid for the car-maker. The £34 million tender was higher than the competing bids, but the government decided that it would do better to float the company on the stock market.

And indeed, in early May it was announced that all tenders had been rejected (it was rumoured that Lonrho had offered £40 million) and that an offer for sale would proceed that would value the company at £38.4 million, made up of 36 million ordinary shares at 90p each and £6 million of convertible loan stock. Rothschild and Cazenove swung into action and within hours had secured underwriting for the whole issue of what was the biggest equity funding operation in Britain up to that time.

Forecast profits for 1973 were £4.5 million and the price of the shares represented a prospective price earnings ratio of 16:1, quite a glamorous rating, but, as *The Times* put it:

Rolls-Royce is Rolls-Royce and there is no other.

Rolls-Royce (1971) Ltd. granted the Motor Car Company the right to use the name, Rolls-Royce, on existing cars and their derivatives but not on anything else.

Its uniqueness was encapsulated in many ways by the salesmen or 'inspecting engineers' as they were called. One, who served Rolls-Royce for sixty years, was Roger Cra'ster.

In Part One of this history we saw how Charles Rolls and Claude Johnson sold Rolls-Royce cars in the early days using their contacts and building the cars' reputation in endurance tests. We have also seen how competitive the car market has always been. Certainly it was a growth market, but at the same time there were plenty of manufacturers fighting tooth and nail for their share of the growth.

Born in Edinburgh in September 1915, Cra'ster was educated at eight schools, from two of which he ran away. He was clearly nonconformist from an early age. Eventually, to the relief of the school world, he reached eighteen and wondered what to do. His uncle suggested he work at Rolls-Royce as he

Dr. ('Doc') Llewellyn Smith. An Oxford graduate, he had
been deputy works manager of Rolls-Royce's Glasgow shadow
factory during the war, before being appointed to run the
Motor Car Division at Crewe. He supervised the return to full
production in the post-war years.

The Bentley Mark VI was the first car built at Crewe and the
first complete car built by the company. Cautious about the
prospects for luxury cars in a country that had just elected a
Labour Government with a large majority, the Rolls-Royce
board felt the prospects for Bentleys were better than those
for Rolls-Royces.

The Rolls-Royce Silver Dawn was initially built only for export and was virtually a Bentley Mark VI with a Rolls-Royce radiator, but with extra attention paid to quietness.

The Rolls-Royce Silver Wraith. *Autocar* wrote in April 1946: 'Perhaps the particular charm of the Silver Wraith lies in the effectiveness with which all the undesirable manifestations incidental to the development of power by machinery have been skilfully exorcised.' The Wraith was produced as a chassis, the practice before the war, and the coachwork was added by specialists to suit the customer.

Harry Grylls (centre) became chief engineer of the Motor
Car Division when motor car engineering was transferred from
Clan Foundry in Belper to Crewe in the late 1940s. The
styling genius of John Blatchley (on his left) created the
Bentley Mk VI R-type, Silver Cloud and Silver Shadow.
Roney Messervy (on Grylls's right) had worked at Hucknall
during the war, as had Blatchley. He returned to the sales
and marketing side of the Motor Car Division.

The Rolls-Royce Phantom IV which was given to his
daughter, the future Queen Elizabeth II, by King George VI
on her marriage to Philip, Duke of Edinburgh, in 1947.
It became established that the Phantom IV would only be
sold to royalty and heads of state.

When designing the Rolls-Royce Silver Cloud, which
was launched in 1955, John Blatchley confessed to being
influenced by Detroit, saying: 'It was difficult not
to be influenced by American cars, and I was particularly
amazed by Cadillacs.'

The actor, John Mills, with his Bentley S2 which was introduced
with the Rolls-Royce Silver Cloud Series II in 1959 with a more
powerful V8 engine that increased performance by 30 per cent.

A Bentley Continental, with coachwork by H.J. Mulliner,
established new standards for high-speed, high-performance,
continental touring.

The US President, Dwight Eisenhower, with Prime Minister
Harold Macmillan, in a specialist Rolls-Royce at Heathrow
Airport in 1959.

Roger Cra'ster organised many a spectacular presentation
of Rolls-Royce motor cars, especially in the USA. Here, he is
driving a Silver Cloud into the Hotel Olympic in Seattle.

The Rolls-Royce Silver Shadow and its Bentley equivalent, the T series, evolved from two different design projects known as 'Tibet' and 'Burma'. The *Financial Times* would say of them when they were launched in 1965: 'all the enviable features of every Rolls-Royce that has ever been made are here, plus many new ones that have never been dreamed about.'

Sir David Plastow, pointing out the finer features of the Silver
Shadow to Sir Leslie O'Brien, Governor of the Bank of England,
at the Earls Court Motor Show in 1969. Plastow worked
his way up the Motor Car Division of Rolls-Royce and took over
when Geoff Fawn was transferred to Derby in January 1971.
He led Rolls-Royce Motors through the public flotation in 1973
and on to the merger with Vickers in 1980. He became
Managing Director of Vickers and said of the merger: 'To be
realistic there is no conventional synergy in the deal. We
are taking two small engineering companies, and putting them
together to make a sizeable one in international terms.'

The Rolls-Royce Corniche. Launched in the immediate
aftermath of the receivership, it soon built up a two-year
waiting list in spite of its list price of £14,000.
They were soon changing hands at prices up to £18,000
(£270,000 in today's terms).

The Rolls-Royce Camargue. Designed by the Italian coachbuilder Sergio Farina, it was launched into the teeth of the oil-price-induced world recession in 1975. Furthermore, it was priced at £29,250 – twice the price of the Silver Shadow and six times the price of a Jaguar XJ6. (Adrian Lombard's son, Roger, was project engineer on the Corniche and overall project manager on the Camargue.)

The Bentley Mulsanne was designed by Fritz Feller who said:
'Nothing in this life is so dull and miserable as "the average"
or the "mean". Once we throw away the concept of excellence
and perfection, we take away the excitement and
incentive for living.'

had always shown an enthusiasm for cars and Rolls-Royce made the best cars. His parents paid £300 for him to become a premium apprentice and we would have to say he was the perfect example of why Hives thought it was a poor scheme. Cra'ster hated it and thought that 'working in a noisy, smelly workshop with one week's holiday a year was absolute hell', especially as he was paid the princely sum of five shillings and ninepence halfpenny a week (about 29p or £14.50 in today's terms!). Somehow he stuck it for four years and was then sent to see Elliott who suggested he might work on aero engines. Cra'ster said he was not interested in aero engines.

Fortunately, his next port of call was Harry Grylls, running the experimental division of cars. Cra'ster found this 'absolute heaven' but after a time, feeling he needed some sales experience, he used some old skiing friends, the Appleyards, to give him a job in their car distribution business. Within two months, he received a letter from Robotham suggesting the job of inspecting engineer in London. This was very attractive, offering as it did a secretary, a Rolls-Royce complete with chauffeur and a salary of £350 a year. Cra'ster was extremely fortunate to be offered this appointment which had traditionally been given to a number of 'well-bred' men such as Don Hanbury, Dick Abbot-Anderson and Shane Chichester, the last a relative of Winston Churchill. His job was to visit customers to make sure they were happy with their car and, after a slightly sticky start (he was informed by the butler of his first customer – 'I'm sorry sir, she died last night'), he enjoyed it enormously. Unfortunately, within four months war broke out and he was sent home to await instructions. He worked under Robotham during the war and immediately afterwards joined the sales side again, this time covering the north of England, Scotland and both Eire and Northern Ireland. Alas, this time there was no secretary, no Rolls-Royce and no chauffeur.

Again, his first customer provided Cra'ster with some problems. He lived near Chesterfield and owned a 1926 Rolls-Royce. When they tried to start the car, it sounded rather rough and would not start. Cra'ster opened the bonnet and fiddled with the carburettor but still it would not start. He realised he was going to have to look at the manual and feeling this would be embarrassing in front of the customer, said to him:

You seem to be very interested in what I'm doing, sir. Perhaps you'd like to follow it in the book?

Somehow, to his great relief, Cra'ster managed to start the car, but worse was to come. The customer was so impressed that he insisted Cra'ster always come to service his car, saying:

I don't want anyone else touching it.

In 1953 he was sent to the USA to visit dealers and look after Rolls-Royce customers, and his moment of fame arrived in 1957 when he was preparing to set up a show in the Olympic Hotel in Seattle to demonstrate the new Silver Cloud. This was due to happen in the ballroom on the first floor and Cra'ster was presented with a problem when the elevator broke down. He solved it by dismantling the revolving doors at the main entrance to the hotel and then driving his Silver Cloud up the steps into the lobby and on up the next set of steps into the ballroom. The show was not a great success – he sold one car – and Cra'ster was offered a consoling drink by a man who turned out to be the air correspondent for the *Daily Mail*. Over the drink, Cra'ster gave him the full story of how he got the car into the ballroom and, within a few days, it appeared on the front page back in England.

The Chairman, Lord Kindersley, was so impressed he wrote personally to Cra'ster and six weeks later Cra'ster was promoted to export sales manager. From there, he moved on to assuming responsibility for the Queen's cars. A Phantom IV had been given to the future Queen Elizabeth II by her father King George VI on her marriage to Philip, the Duke of Edinburgh, in 1947. It became established that the Phantom IV would only be sold to royalty and heads of state. Owners included the Aga Khan, the Sheik of Kuwait, the King of Iraq and the Shah of Iran who bought three. The Phantom IV was followed in 1959 by the Phantom V, built to special order, and then, in 1969, by the Phantom VI.

When the unusual and expensive Rolls-Royce Camargue was introduced in 1975, Cra'ster knew that the Shah of Iran would want one (and as the price of oil had just quintupled, he could certainly afford one). He drove to Zurich to meet the Shah, who summoned him one evening, saying: 'Let's try it out.'

To which Cra'ster replied: 'You can't try out a new car in the dark, your majesty.'

'Well, let's go and sit in it, then.'

Cra'ster remembers that they tried it out properly the next day, enjoying an escort of police both in front and behind. And his influence with the Shah was used by the Managing Director, David Plastow, when he wanted to talk to the Shah about buying tanks for which Rolls-Royce would make the engines. Cra'ster secured the audience, Plastow secured the order – 1,000 at £75,000 each – but unfortunately, as we shall see, the Shah fled the country before the order could be confirmed and delivered. Clearly, with Cra'ster it was not what he knew but whom he knew.

The company at this point embraced not only cars but also industrial loco-motives, light aero and military tracked vehicle engines, aero engine components and even golf club heads, but cars were the core of the enterprise, providing half of the turnover and profits and employing more than half of the 8,166 workforce. As well as the main factory at Crewe, there were three other factories. There were two factories in London within two miles of each other in Willesden. The coachbuilding activities of Mulliner Park Ward – making bodies for the highly successful Corniche and the Phantom VI – were to be expanded, and the two factories were to be consolidated with the one at Hythe Road. The Diesel Division, employing 2,300 in Shrewsbury, produced diesel engines for automotive, railway, industrial, earth-moving and marine use. The six-cylinder Eagle engine for heavy lorries was offered as a standard or optional engine by all British truck manufacturers using engines in this power range. The Diesel Division was expected to contribute 19 per cent of profits from 17 per cent of turnover.

At the Crewe factory, the company also manufactured military engines such as the multi-fuel K60 for combat vehicles, which could run on petrol, paraffin or diesel fuel. These military activities, apart from anything else in the group, made the Government look askance at any foreign buyer of the company. The 'B' range petrol engines were used in the Alvis Saladin armoured car, Saracen armoured personnel carrier, Stalwart high-mobility load carrier and the Ferret Scout car. The revolutionary Swedish 'S' tank, needing a very compact diesel engine, was using one from Rolls-Royce. We have already seen the work being undertaken on the Wankel rotary engine, and Plastow saw great potential for this engine. Light aircraft engines had been produced since 1963 under licence from Continental (now Teledyne Continental) Motors of the USA.

Finally, the foundry at Crewe produced golf club heads, beer barrel bungs, rifle pieces, parts for Rolls-Royce cars and the Flying Lady mascot. However, most of the foundry's output was devoted to nozzle guide vanes for gas turbine aero engines, mainly undertaken on a subcontract basis for Rolls-Royce (1971).

Rolls-Royce Motors faced the problems that all other luxury car makers had faced. As we have seen, few survived even in the early days and over the years more had gone out of business, adapted their approach or been taken over by larger groups. Famous names of the 1930s – Auburn, Armstrong Siddeley, Hispano-Suiza and Facel-Vega had gone. Others – BMW, Alfa Romeo and Lancia – had survived by abandoning luxury saloons and sports cars in favour of less glamorous, but more profitable, smaller cars. Yet others – Ferrari, Maserati, Aston Martin and Lamborghini – had made alliances with larger stronger groups who were making money out of more prosaic products.

The Rolls-Royce Motor Car Division at Crewe itself had struggled to make money in the 1960s, relying on selling jet engine components to its parent company and marketing light aircraft engines to balance the books. The situation was not helped in the early 1960s when the Conservative

185

Chancellor of the Exchequer, Selwyn Lloyd, disallowed tax relief on expensive company cars. Fortuitously, the trauma of the receivership coincided with strong demand for the Silver Shadow (the initial quality-related teething troubles had been overcome) and especially for the Corniche, and meant that the car company started its new independent life with a waiting time of over a year on the Silver Shadow and over two years on the more expensive Corniche. And, as we have seen, profitability and return on capital at 10.9 per cent were quite respectable. This was a better return than was being achieved by the mighty General Motors and way ahead of Europe's most profitable car companies, Daimler-Benz and Volvo.

And Rolls-Royce was in a strong competitive position in the luxury car market. In the UK, where it made half its sales, there was no direct competitor at all. The nearest was Jaguar but even its most expensive model, the Vanden Plas XJ12, cost £4,000 compared with £10,400 for the Silver Shadow. In the USA, its prices were way above Cadillac and Mercedes-Benz, and as these two raised output they left the top end of the market to Rolls-Royce. The equally expensive Ferraris, Maseratis and Aston Martins catered for a different market. Plastow rejected the idea of greatly increased production and direct competition with Mercedes and Jaguar.

We are utterly committed to a narrow range of business and we are able to devote the considerable engineering talents in the company entirely to this.

If the £10,400 Silver Shadow seemed expensive, the coachbuilt Corniche, introduced two weeks after the receivership in February 1971 with a list price at £14,000, was in such demand it was changing hands between £17,000 and £18,000. Plastow was taking steps to raise output – from 350 to 600 a year.

In this apparently serene situation, three problems lay ahead; two foreseen, the other totally unforeseen. The first foreseen problem was that of the inevitable change of model at some stage. Experience showed that this would be when the company was at its most vulnerable – Rolls-Royce had really put itself under great strain in the 1960s when its new model meant a new suspension, new braking system and a new type of monocoque body shell.

The second foreseen, but nevertheless difficult problem, was coping with the increasingly stringent safety and pollution requirements being demanded in the USA. These requirements posed special problems for the smaller manufacturers with their limited resources. Plastow was to give great credit to the designer John Hollings for seeing the company through the increasingly stringent conditions placed on all motor car manufacturers.

The third, and unforeseen, problem hit not only Rolls-Royce but the whole world at the end of 1973.

The rate of inflation had started to accelerate in the 1960s as successive Governments – Conservative and Labour – pursued a policy of maintaining full employment at all costs. This was the avowed, and in many ways worthy, aim of all politicians after World War II. No one wanted a return of the high unemployment levels of the 1930s and each successive Government implemented Keynesian policies of keeping demand high using, if necessary, and especially at election times, public money to prime the economic pump. The result, especially in a Britain with some archaic industrial management and workforce practices, was the creation of greater demand than supply – the classic cause of inflation. The general index of retail prices, which had risen only 0.5 per cent in 1959 and 1.1 per cent in 1960, was accelerating at 4.7 per cent in 1968 and 5.4 per cent in 1969.

We have already seen how appalling the economic environment became in the 1970s. Initially, Rolls-Royce Motors coped with this dire situation rather well and, in early 1975, reported profits of £5.4 million on sales of £58 million for 1974 – making it almost certainly Europe's most profitable car maker. But the company was not unscathed. Sales in the USA had held up well and, not surprisingly, increased in the Middle East. However, in the home market the waiting list for Corniches had virtually disappeared, as had the premium price for low-mileage, second-hand Corniches. This suggested a blocking of the distribution pipeline if dealers were stuck with second-hand models. The company eschewed the classic response to such a situation – launch cheaper versions – and introduced the world's most expensive car, the Camargue.

The Italian coachbuilder, Sergio Farina, had designed a Bentley coupé in 1968. Built around the Silver Shadow chassis, it resembled the famous Bentley Continentals. Only one was produced and, although it sported some rather unattractive features, most noticeably the rectangular headlights, the overall effect was quite dashing and elegant. The purchaser was James (later Lord) Hanson, who had just left his family haulage business to build one of Britain's largest companies, initially Hanson Trust, subsequently Hanson plc.

Rolls-Royce was sufficiently impressed to invite Farina and his company, Pininfarina, to design a flagship car, a luxurious model that would surpass even the elegance of the standard Silver Shadow and the Mulliner Park Ward designs. The result, known as the Camargue, was launched in January 1975. A less propitious economic environment could scarcely be imagined. When it was conceived, the world economy was booming and everything in the garden looked rosy. However, as we have seen, the world financial situation deteriorated sharply in late 1973, and into this world of straitened

187

circumstances and expensive petrol, the large and distinctly thirsty Camargue was launched. Some 530, including one Bentley, were built.

And the company priced the Camargue at £29,250 – 35 per cent more than even the Phantom VI, twice the price of its own Silver Shadow and six times the price of a Jaguar XJ6.

Both the names, Corniche and Camargue, were a tribute to Sir Henry Royce and his connection with the South of France. The Corniche is the strip of land stretching along the Côte d'Azur. The name had been registered by the company as long ago as 1939. Camargue came from the Ile de la Camargue, in the delta of the Rhône, also on the Mediterranean coast. Over half of the Camargue was reclaimed from the lagoons, and the main industries are cattle and sheep-raising, and wheat, wine and rice production. Its chief sport is bull-fighting but, in contrast to Spain, the bull is neither injured nor killed.

Someone remarked that launching the Camargue at this time was a bit like the band playing on while the *Titanic* sank, but Plastow disagreed, saying that the development costs had been relatively small as it shared the mechanical units of the Corniche, and tooling and jigging costs had been kept very low. With its high price, the Camargue offered a very attractive return on its investment costs. And sales? Plastow said:

The first year's production was sold unseen and unpriced, a high proportion in what might loosely be described as OPEC markets.

Following the successful flotation in 1973, Rolls-Royce Motors' profits rose steadily, if unspectacularly, through the mid-1970s, in spite of the oil-price induced recession of 1974/5. By May 1977, the company felt strong enough to make an £8.4 million bid for the truck maker, Foden. In retrospect, and perhaps also at the time, it was a slightly bizarre move, the only logic apparently that both companies were based in Cheshire.

Foden was in a weak position following an ambitious expansion programme – put in train in the strong demand market of the early 1970s – which ran into the mid-1970s crisis of recession, and in 1975 the company lost over £1 million. The company was effectively rescued by the Government and then bailed out by a group of City Institutions, which gradually nursed it back to health. Nevertheless, in the first four months of 1977, it sold only 365 vehicles compared with 376 in the same period the year before. Volvo, the largest importer, was selling twice as many trucks in Britain.

Rolls-Royce pointed to the logic of combining the engineering expertise needed for producing the world's highest quality motor cars, with that needed for producing some of the best trucks on the market. Certainly both companies had a tradition of labour-intensive manufacturing.

The company's offer was turned down by a majority of Foden's shareholders; Rolls-Royce did not increase its terms and the offer quietly lapsed.

The replacement for the Silver Shadow, the Silver Spirit, was launched in 1980. David Plastow was to say later that he felt the board meeting in 1976, at which they debated whether to introduce this model, was the most crucial of his career. At the time, following the fuel crisis of 1973/4, voices were raised for caution and a postponement of any new model. However, courage won the day. The Silver Spirit was based on the Silver Shadow II base unit, using the same engine and drive train but with a revised rear suspension and Mineral Oil hydraulics, first used on the Corniche and the Camargue. The styling was more angular than the Shadow, and the Spirit appeared much larger although it was only three inches longer and two inches wider. The main difference giving this appearance of extra size was the increase of 30 per cent in the glass area. At the same time, the Silver Spur replaced the Silver Wraith II and the Bentley Mulsanne replaced the T2 series. Modifications were also made to the Flying Lady mascot, the Spirit of Ecstasy. Safety laws were affecting bonnet mascots or 'hood ornaments', as they were called in the USA. These were accommodated by the creation of a retractable Spirit of Ecstasy, which withdrew into the radiator shell if struck.

The stylist with the most influence on the new cars was Fritz Feller, an Austrian-born engineer blessed with diverse talents. He had spent his first years at Rolls-Royce working as an engineer and his work on the Wankel engine in response to the Government request for high specific output engines for future tanks was later described by David Plastow as the work of a genius. In 1969, he was asked to become head of Car Styling. He was an accomplished violinist and painter but had no direct experience in this field. Nevertheless, the Silver Spirit and the Bentley Mulsanne bear testimony to the success of this appointment. Feller himself said:

Nothing in this life is so dull and miserable as 'the average' or 'the mean'. Once we throw away the concept of excellence and perfection, we take away the excitement and incentive for living.

And of his styling for Rolls-Royce he said:

Our next car must be the natural successor to our last car. We must aim to maintain product identity, not merely by using such obvious means as our interlocked ROLLS-ROYCE symbol, our mascot or our distinctive front grille . . . Even without noticing these obvious features, anyone looking at the cars we have developed over the years should know that they are looking at a Rolls-Royce. Our designers must be bound by our traditions.

And there was plenty at stake in replacing the Silver Shadow series, which had been Rolls-Royce's most successful series with sales of 32,300 cars in the fifteen years since it had been introduced – 17,000 of them exported.

MERGER WITH VICKERS

On 25 June 1980, it was announced that Rolls-Royce Motors and Vickers were to merge. The announcement was not greeted with rapture. Indeed, many saw it as a merger of weakness rather than strength. Ian Fraser, Chairman of Rolls-Royce, later admitted:

Vickers had no products and no management and we were short of the cash needed to develop the business.

Vickers, Sons & Company Ltd. had been incorporated on 17 April 1867 and gradually established itself in the difficult trading conditions of the subsequent twenty years. In 1887, it entered the armaments industry when the board invested in plant for the construction of guns and armour. It was a seminal moment. As J.D. Scott put it, in *Vickers, a History*:

In the vital years from 1897 to 1914 this new and powerful combination of Vickers, Son & Maxim Limited [Vickers had bought Maxim] was to be one of the chief bulwarks of British naval strength . . . It was, quite obviously and quite deliberately, challenging Armstrong Whitworth for the leadership of the British armament industry.

Britain was rearming furiously in the first decade of the twentieth century as the naval threat from Germany grew, and while Armstrong built the gun mountings for the *Dreadnought* battleships, Vickers built the engines as well as some of the ships. At the same time, Vickers made sure it was one of the first companies involved in the new weapon of war, the aeroplane. The first Vickers monoplane was built in 1911. At the Olympia Aero Show in 1913, the company exhibited the prototype of a machine that was used extensively in the early years of World War I. This was the FB1 (FB stood for fighting biplane), a two-seater pusher biplane with a Vickers machine-gun mounted in the nose and operated by the observer. During the war, Vickers designed about thirty different aircraft. The most famous, the Vimy, was the result of a request from the Air Board to design a twin-engined night bomber. Originally fitted with two 200 hp Hispano-Suiza engines, the Vimy could lift a third of a ton more than larger rivals when Rolls-Royce Eagle engines were fitted. The Air Board ordered 350, expecting to bomb Berlin with them and win the war. However, before they could be used, the war ended and both the Vimy and the Rolls-Royce Eagle engines became world-famous in the more peaceful feat of flying Alcock and Brown in the first non-stop flight across the Atlantic. (For the full story of this remarkable feat, refer to the first volume of this history.)

The near-slump conditions of the 1920s brought crisis to the steel, shipbuilding and aircraft industries and forced Vickers into a merger with Armstrong Whitworth. In 1927, a new company, Vickers-Armstrong Limited was founded, and, as rearmament became again a national priority, it prospered in the manufacture of a new generation of armaments, ships (including submarines), aircraft and tanks. As H.M. Postan said in his book, *British War Production*:

Organization for tank design in the War Office was rudimentary in the extreme, and but for the solitary and pioneering efforts of the designers at Vickers-Armstrongs the country would have possessed no facilities for the design and development of armoured vehicles.

In the air, the Supermarine company – which had been bought by Vickers in 1928 after the formation of a separate Vickers subsidiary, Vickers Aviation Ltd. – developed and produced in great numbers two of the most important combat aircraft of World War II, the Spitfire and the Wellington. (Again, see the first volume of the history for further details.) Besides these two aircraft, Vickers also built many other military and civil aircraft between the wars, including commercial versions of the Vimy, as well as the Vulcan, Viastra, Vanguard, Vellore and Vellox airlines. Military aircraft included the Victoria, Vernon and Valentia transports, Vildebeest single-engined torpedo bomber and its Vincent general-purpose derivative, the Virginia bomber of 1936, the first Royal Air Force aircraft to incorporate the geodetic construction technique designed by Barnes Wallis, which was later used in the Wellington bomber as well.

The Wellington twin-engined bomber first flew in 1936, entering RAF service in 1938. It could carry almost as great a tonnage of bombs as the four-engined Boeing B-17 Fortress over a similar range, and became the main bomber of the RAF in the early years of World War II. It was superseded by the four-engine bombers, notably the Avro Lancaster. Nevertheless, 11,641 Wellingtons were built by Vickers during the war, as well as nearly 20,000 Spitfires and Seafires. However, Vickers' biggest commitment was still in shipbuilding. In the early part of 1940, Vickers had in hand, at Barrow and Newcastle, naval orders worth £60 million (about £3 billion in today's terms).

At the end of World War II, Vickers was forced to look for new products to replace the weapons of war. As we saw earlier in the chapter, one of the areas in which it invested heavily was crawler tractors. Unfortunately, competition from the USA proved too severe and the venture was abandoned. The truth was that Caterpillar's tractors, using fixed suspension, were more efficient. Vickers used tank suspension which was unnecessarily complex. For a time, the aircraft-building side of the business was extremely successful as George Edwards (later Sir George Edwards, OM) pushed through the Viking and then the Vickers Viscount, which, as we saw in Chapter Two, became Britain's best-selling passenger aircraft. Thereafter, Vickers' and Britain's fortunes as aircraft manufacturers declined sharply, with limited sales for the Vickers Vanguard, the Vickers VC10 and the BAC 111.

During the 1970s, Vickers was emasculated by the Labour Government which nationalised both its shipbuilding and aircraft interest, which at that time represented half its sales and two-thirds of its profits. The shortage of top management at Vickers became only too apparent when Dr. Bill Willetts, recruited from Plessey Telecommunications in March 1980, resigned within a month for 'personal reasons'.

Rolls-Royce Motors' financial problems stemmed from difficulties in 1979 on a number of fronts. Demand for diesel engines had fallen and an expected order for 1,000 tanks to go to Iran had been cancelled following the fall of the Shah. There had also been considerable industrial unrest among the engineering unions. At the same time, investment in a new Rolls-Royce model – to be launched in the autumn of 1980 – was proving as expensive as all new models had proved to be in the past. By the end of 1979, borrowings had reached nearly 75 per cent of shareholders' funds and seemed destined to reach 100 per cent by the end of 1980.

There was brave talk to the press and to City analysts of 'a move forward into the 1980s', but David Plastow, destined to be Managing Director of the new Vickers group, said:

To be realistic there is no conventional synergy in the deal. We are taking two small engineering companies, and putting them together to make a sizeable one in international terms. In the end we will insulate ourselves from the business cycles and troughs.

Much of the press comment was critical and this remark from Melvyn Marckus was typical:

The press conference called on Wednesday to unveil Vickers' £38 million takeover bid for Rolls-Royce Motors was memorably unimpressive. In the claustrophobic atmosphere of Morgan Grenfell's parlours Sir Peter Matthews, chairman of Vickers, and Ian Fraser, head of Rolls-Royce, struggled laboriously to convince their audience that the formation of Rolls-Royce Vickers was a good idea. They failed.

In spite of this criticism, the deal went through, and Vickers and Rolls-Royce faced the 1980s together. The recession of the early years of the decade, again set off by a hike in the price of oil, proved just as difficult for Vickers as it did for other British manufacturing companies (Rolls-Royce Motors' sales fell by 25 per cent in 1982).

We have already seen how Britain and much of the rest of the world suffered in this recession, but by 1983 the worst was over and an economic upswing began which lasted for the rest of the decade, with most businesses prospering.

And Vickers, with Rolls-Royce Motors in the van, prospered too. In early 1985, sharply higher profits were reported – up from £19.5 million to £30.8 million, helped by £14.1 million contributed from Rolls-Royce. The year 1985 was even better, with pre-tax profits rising a further 48 per cent to £45.1 million. The Rolls-Royce figures were helped by increased sales of Bentley cars.

The introduction of the Bentley Mulsanne and especially the Bentley Mulsanne Turbo in 1982 began the revival of the Bentley name, which had been allowed to wither almost to nothing in the 1970s. In the first few years after the war, Bentleys accounted for almost 65 per cent of Crewe's production, but by 1980 this had dwindled to about 5 per cent. Rather than shelving the name, a move apparently seriously considered, the company decided to revive it and aim for a 60:40 split, Rolls-Royce to Bentley. This target was met in the home market by 1987, helped by the launch in 1984 of the Bentley Eight – priced at just under £50,000 to attract purchasers who had always thought Rolls-Royces and Bentleys were too expensive for them. Back came the mesh grille to remind people of the early glory days of the Bentley Boys.

Rolls-Royce and Bentley motor cars continued to make money for Vickers for the rest of the 1980s. More exciting times came along in the 1990s, but the detailed story will have to wait until the third volume of this history.

CHAPTER SEVEN

WE ARE ENGINE MAKERS

A NEW FAMILY OF ROLLS-ROYCE ENGINES
INTO THE TRUCK MARKET
THE 'THRILL A MINUTE' MILITARY BUSINESS
'THINK ABOUT NUCLEAR ENGINES'
ROCKETS AS WELL
THE SOUBRIQUET OF HMS ROLLS-ROYCE
ELECTRICITY, GAS, OIL

A NEW FAMILY OF ROLLS-ROYCE ENGINES

IT IS GENERALLY ACCEPTED THAT it was W.A. Robotham's idea that Rolls-Royce should manufacture diesel engines. In Part One of this history, we saw how Robotham had kept together the Rolls-Royce car specialists at Clan Foundry in Belper, and how he had served as chief engineer of Tank Design at the Ministry of Supply. He felt obliged to resign in the autumn of 1943 when his boss, Oliver Lucas, did so after a disagreement with Duncan Sandys, Parliamentary Secretary to the Ministry of Supply. However, he had not wasted his time at the Ministry from Rolls-Royce's point of view, for he returned with the news that Rolls-Royce's rationalised range of petrol engines would be used to power a broad spread of army vehicles. This would become a very useful source of turnover for the Car Division for many years after the war. The 'B' range comprised four-, six- and eight-cylinder in-line engines named B40, B60 and B80. The range shared many common components, being based on a common bore and stroke, plus an overhead-inlet side-exhaust-valve layout. They owed their origin to the pre-war decision to develop 'rationalised' four-, six- and eight-cylinder engines for the new range of motor cars, whose development was slowed down by the outbreak of war. The 'B' range became the standard power unit for all the British Army's wheeled combat vehicles. Full-scale production of the engines began at the Crewe factory in 1947.

When the decision was made to manufacture complete cars at Crewe, Robotham, with every justification, must have hoped that he would be chosen to head the newly revived Motor Car Division – but it was not to be. Dr. Llewellyn Smith, whom he had originally engaged and who had assisted with the management of the Glasgow factory throughout the war, was appointed a main board director and placed in charge of the Motor Car Division. Robotham wrote in his book, *Silver Ghosts and Silver Dawn*,

I have to admit that this was a disappointment to me.

According to Alec Harvey-Bailey, writing in his *Rolls-Royce, Hives' Turbulent Barons*:

He seriously considered resigning and concentrating on his farm.

Fortunately, Robotham did not resign but concentrated on an idea he had developed on a recent visit to the USA: the design of a range of diesel engines, using light alloys wherever possible. After all, who knew more about light alloy engines than Rolls-Royce? He had been impressed both by the size of their trucks and the lengths to which they were going to reduce weight – high-tensile steel chassis, aluminium cabs, no front-wheel brakes, spare wheels or tool kits. As a result the 'C' range was born, again, as with the 'B' range, in four-, six- and eight-cylinder in-line engines using the principles of rationalisation. Initially, twenty light-alloy six-cylinder development engines were built, some going to Shell Research at Thornton for evaluation. Six were ordered by Bernard Sunley who was selling US Euclid earth movers through his company, Blackwood Hodge, but he never received them because another candidate appeared (i.e. Vickers) and took priority.

According to Robotham, Hives had appeared to take little interest in this diesel development, but suddenly:

Our diesel engine project acquired an almost embarrassing glamour.

The reason was that Vickers, an important customer of Rolls-Royce throughout the war and currently developing the Vickers Viscount, had decided to develop a range of crawler tractors. Vickers had already built a number of Shervic crawler tractors from cannibalised American Sherman battle tanks to send out to the British Government's West Africa Ground Nuts scheme. The Vickers plans were on a scale that persuaded Jack Olding to give up his lucrative Caterpillar dealership to concentrate on setting up the sales and service organisation throughout the world. (He would later have reason to regret this decision.) And, in time, the Oil Engine Division also had reason

195

to regret its support of the Vickers tractor venture, as it led to Robotham turning down Bernard Sunley's request for a worldwide franchise for Rolls-Royce diesel engines. This rejection led Sunley to persuade the US company Cummins to set up the manufacturing of their NH-type engines in Scotland, thus establishing in the UK a strong and powerful competitor. [In 1954, senior executives of Cummins suggested the Oil Engine Division give up manufacturing their own engines and make Cummins engines under licence.]

In the short term, the Vickers project seemed a godsend. Initially they proposed to manufacture the mid-size tractor of the range, which would require an engine of some 180 bhp, about the output of the engine Rolls-Royce had been developing. And it requested (apparently there was no written order) 1,220 engines. Vickers were not prepared to pay a premium for light alloy engines because they wanted weight for improved traction, and so the engines were redesigned in cast iron. The option of returning to light alloy was never considered again.

On 6 July 1951, the *Derby Evening Telegraph* reported:

A new Rolls-Royce oil engine [Robotham insisted on calling it an 'oil' rather than a 'diesel' engine], designed and produced in the company's new factory in Victory Road, Derby, will power a new crawler tractor with which Britain will challenge America's monopoly in the heavy tractor and earth-moving field. The engine is a six cylinder 12.17 litres supercharged example of the new 'C' range, the first batch of which is destined to power the Vickers-Armstrongs VR180 crawler tractor.

And Hives stated strongly Rolls-Royce's commitment to the project:

This is the start of a new family of Rolls-Royce engines. We are treating this new Oil Engine Division very seriously. It is not to be looked on as an appendage to the other divisions of the company.

However, there were plenty of problems to come. Robotham wrote later:

There were clouds on the horizon, and customer complaints began to multiply. Fortunately for us the engine did not get much criticism, but this was mainly because the troubles with the rest of the machine prevented very many hours of continuous operation being achieved. I began to wonder how long Vickers could continue flying heavy replacement parts to the remote corners of the earth, a most expensive operation. The infant Rolls-Royce diesel division was certainly beginning its life in a somewhat precarious manner, and there seemed to be little doubt that we were heading for serious trouble.

By autumn 1953, when Rolls-Royce was delivering twenty engines a week to

Vickers, Robotham was told that production of the tractors had been halted and would not be resumed until all the faults had been ironed out. Other customers were sought, but the numbers were hardly sufficient to make the business viable and almost every customer needed their engines tailored to suit their particular application. Vickers eventually started their production but in 1958 decided to close the business after selling over 1,500 tractors world-wide. This left the Oil Engine Division in a very difficult position. Vickers had lost most of the sales and service organisation set up by Jack Olding and was forced to compete with Caterpillar, Cummins, GM and Deutz – all about ten times its size – without a tried and established application for other markets. However, there were a number of embryo applications in the oil-field, earth-moving, marine, generator/compressor sets and railway fields, but demand was relatively small.

Considerable effort was made to find new customers. In 1955, the Division participated in the British Trade Fair in Baghdad in the heart of the Middle East oilfields. And in 1957 it staged an exhibition on the new Sinfin B site factory at which some 27 different applications of the engines were shown, including Scammell oilfield trucks, ERF lorries, Rotinoff cattle trains, generator sets, compressors, marine equipment, coal winding gear, Woodfield Hoist drilling rig, Euclid dumper trucks, the Vickers tractor, plus railcars and shunting locomotives at the nearby British Rail workshops.

However, Robotham's answer was to pursue merger possibilities and he talked to a number of companies including Lister and AEC (eventually to be taken over by Leyland). Both were keen apparently. But, according to Robotham:

By this time the company's involvement in the civil aviation aircraft engine market had assumed such proportions that prudence caused the majority of the Board to decide against further expansion and diversification in the diesel field lest the company become involved in over-trading.

To add to his difficulties, the Aero Division needed the space the Oil Engine Division occupied on the Sinfin B site in Derby. Other sites – from Jarrow in the north-east to South Wales – were being considered when the Sentinel company in Shrewsbury was put on the market. Founded in Glasgow in 1905, Sentinel had moved to Shrewsbury in 1914 and built itself up to be the most prolific steam wagon manufacturer in the UK. By 1950 it had taken a licence for Ganz diesel engines for its truck manufacture, although still making large steam wagons for South America. It was also manufacturing machine tools, aircraft tow tractors, generating sets, steam shunting locomotives and many other small products. Sentinel was half-owned by British Metal Industries and British Oxygen – both of whom were probably relieved when Rolls-Royce took it off their hands in 1957.

197

At least the Division now had its own space and it meant that the practice of sending crankcases, crankshafts and cylinder heads from Sinfin to Rolls-Royce's Glasgow factory – to be machined on the old Merlin aero-engine tooling – could be stopped. However, the acquisition brought plenty of new headaches. As Robotham put it:

They were losing money on activities about which I knew absolutely nothing . . . they really had no viable product whatever.

This move to Shrewsbury heralded a turn in the Division's fortunes. International Harvester (IH) of the USA were at the time considering increasing their Doncaster production to include a British version of their TD.20 crawler tractor. A special, naturally aspirated C6 was developed and successfully engineered into the tractor design, which proved to be a very successful marriage with some thousands of engines delivered over the ensuing years. One of the biggest projects was for irrigation canals in the lower Nile as part of the Aswan Dam project. Unfortunately IH would pay only £760 an engine whereas Vickers had been paying £1,350 – although the TD.20 used a normally aspirated specification whereas the Vickers engine had been supercharged. Nevertheless, as Peter Vinson, originally personal assistant to Robotham in the early 1950s and ultimately Sales and Marketing Director of the Oil Engine Division, put it:

That brought the colour to our cheeks but we were desperate to replace the Vickers volume . . . It forced us to get to grips with cost reduction.

The ultimate decision as to whether to take on such a large, and potentially loss-making, contract lay with the Rolls-Royce Finance Director in Derby, William Gill. The Division itself, conscious of the need for extra throughput in the factory, recommended acceptance, and indeed this was what happened. As Vinson had said, it meant the need to get to grips with cost reduction.

Even with the IH contract, the Division desperately needed further orders. Fortunately, the early link with Shell spread to other oil companies and was another plank on which the division built up demand. Using the wide power range of the rationalised engine design, potential customers could draw benefit from using similar parts in marine, electric generation, compressors and other oilfield equipment. Soon there were 100 marine engines in work boats servicing Shell's offshore platforms on Lake Maracaibo in Venezuela. The Kuwait Oil Company had engines in Scammell oilfield trucks, generators sets, drill rigs and other applications. Oil companies in Nigeria, Borneo, and Trinidad became customers, but all these companies later found it more economical to subcontract this work to specialist contractors, most of whom were American with their own equipment. This substantially reduced the potential for Rolls-Royce engines.

From the outset, engines for electric generator sets were an important market which throughout its life became the most consistent sector for the division. Peter Vinson remembered that:

On one of my early trips to the Middle East, I met Dick Bird, MD of Petbow, who was there selling generator sets to the oil companies. The result of this meeting led to a long association with Petbow who became a valued customer for our engines throughout the life of the division. Petbows gave us our first major order after February 71 for £3m. Their first Rolls-Royce powered generator sets ran in 1954 and their first export order was for three sets for BP in Tanzania which were still in service some 23 years later. [Robotham had joined Petbow Ltd. after he resigned from Rolls-Royce in 1963, following a disagreement with the Board. He believed Rolls-Royce should abandon civil aviation altogether and leave it to the Americans. Hives was clearly right to question Robotham's judgement in the 1930s and after the war.]

The business built up to the point where every UK generator-set manufacturer was offering Rolls-Royce engines, and French and South African dealers were also manufacturing sets. One application was the provision of the total electrical supply over five years for the 52 mile Orange/Fish river diversion tunnel where 48 turbocharged eight-cylinder Petbow sets were run together in parallel.

The promise offered by the generator set market led to a decision in the later 1950s to complement the engine range with a new industrial 'D' family with twice the swept cylinder volume. Designed under the leadership of John Read, the 32 litre DV8, the only version of the range developed, saw application in many generator sets up to 600 kW (electrical) output over the next twenty-five years. It was also in demand in railway applications, serving as the power unit for a number of shunting locomotives and was used on the famous South African Blue train for air-conditioning services.

Major sales and service operations were developed with Rolls-Royce of Canada and Australia.

Ken Wright, Managing Director of the Australian operation, was one of the post-war Rolls-Royce characters. While looking after marine engines on Lake Maracaibo, he spent time in the local gaol following an alleged minor motoring offence. [Presumably, he had refused to pay off the policeman who arrested him.] Later he became service manager at Shrewsbury before moving to Rolls-Royce of Australia. He was known throughout the world for his skills as a service engineer and also as an amateur magician.

The division was also asked by its dealer in Mexico for a licence to manufacture the six-cylinder engines there. The financial calculations indicated there was no case to go ahead. The dealer, Sr Antonio Aspra, appealed directly to Sir Denning Pearson (who insisted on calling him Sr Aspro). The appeal was allowed on the basis that the company, as distinct from the Division, was for strategic reasons, keen to have a substantial presence in Mexico. A joint-venture agreement was negotiated and a factory established at Tlaxaca near Puebla.

The operation got off to a difficult start with a bank as senior partner. Eventually, Shrewsbury managed to appoint their own man to run the operation. Bob Hickman was appointed Managing Director and remained in that post until the Mexican factory, and its authority to manufacture engines in Mexico, was sold to Volvo.

The Sentinel products current at the time of the take-over had caused considerable anxiety within the Division's management. Decisions were made to cease work on many of the products over a period of time. The aircraft tow tractors were the subject of an MoD contract. Despite the substantial losses that were being incurred, completion was essential. Fortunately, it proved possible to renegotiate the price to recuperate some of the losses.

However, it was critical that the machining and general engineering capacity at Sentinel was filled with profitable products as soon as possible. The machine tool capability was kept intact for a few years. Work was undertaken for Molins for the partial manufacture and assembly of cigarette-making machines. In addition, the Division made a bid for an Aero Division contract to undertake the manufacture of sheet-metal combustion chambers. A contract was obtained from Derby on a competitive and arm's length basis and that work continues to this day.

The locomotive business was a particular headache since there was an outstanding contract to supply steam receiver shunting locomotives to Dorman Long, locomotives that were far from satisfactory. Harold Whyman was given the job of either negotiating out of the contractual liabilities or engineering a solution, and from this dilemma came a design which was to make a substantial contribution to the Division's profit over the next ten years. By this time, the Motor Car Division was manufacturing three-stage industrial torque converters and so work was undertaken to bring back the faulty locomotives and fit them with diesel engines coupled to torque converter transmissions from Crewe. The overall design was completely modernised, and ergonomically arranged controls were fitted into a central cab. The locomotive was so successful that the twin-axle range was complemented by three- and four-axle models which took over 50 per cent of the market. The Sentinel 0-6-0, redesigned by Frank Mamblin, won a Design Council Award in 1968.

To improve the marketing of the locomotive business, the Division acquired a controlling interest in their largest distributor, Thomas Hill (Rotherham)

Ltd., and for some years, the Division became the largest producer of shunting locomotives in the UK.

Before leaving Derby, the Oil Engine Division had held discussions about entry into the wider railway market with the Motor Car Division. Ivan Evernden in the Motor Car Division had for some time been looking at the possibilities of broadening the product range there and had recommended taking a licence for the Twin Disc torque converter. The availability of this transmission could broaden the potential for the Rolls-Royce diesel engines to dump trucks, shunting locomotives and railcars. Discussions were held with the Midland Region of British Railways (BR) to pursue opportunities and two projects were launched. First, two railcar sets of an existing design were to be re-equipped with horizontal six-cylinder naturally aspirated engines while still making use of the existing four-speed transmission and control system. The second, more ambitious, project was for Rolls-Royce to supply a complete set of equipment, from the driver's handle down to the axle-mounted final drive transmission.

Following the successful completion of the tests on the prototype, Colonel Fell was invited to form a new Railway Traction Department (RTD), initially housed at Ascot Drive, Derby. After Fell's retirement, it was moved to Shrewsbury in 1960 and focused on the selling, application and service responsibilities. The fortunes of the Oil Engine Division were at a low ebb at this time but the 1,000 railcar power packs operating on British Rail, and in Australia, Malaysia, Mexico, Portugal, Jamaica and Norway, brought long-term profitable spares demand for up to thirty years – a stabilising influence.

The diesel hydraulic power packs for locomotives marketed by the RTD were in demand by other locomotive manufacturers. A feature of the licence agreement with Twin Disc was that Rolls-Royce could not sell torque converters directly to customers using other makes of engine. This gave some advantage to Rolls-Royce since the matching of the two units could be accomplished more easily in-house than for competitors and as a result Rolls-Royce gained the major share of the market.

In the late 1950s and early 1960s, Colonel Fell saw an opportunity to develop the market for his locomotive transmission system, his patents having been given to the company. He had argued that a multi-engined locomotive arrangement, equipped with automotive-type high-speed engines, could result in reduced costs and fuel usage, compared with the then-proposed diesel electric types. The prototype Fell locomotive had successfully undergone extensive trials on BR using four Paxman engines and it was decided that further prototype locomotives would be built with Rolls-Royce engines and torque converters with new differential transmissions and controls designed and manufactured at Crewe. For the hump shunting and trip working duties, two 600 hp twin-engined locomotives were built by the Yorkshire

Engine Company (later to be taken over by the Diesel Engine Division), and a mixed-traffic 1,500 hp four-engined version in conjunction with the Clayton Equipment Co. The locomotives were successfully trialled but BR's interest evaporated as they struggled to cope with the many problems arising from the new locomotive types they had purchased, particularly the German-designed diesel hydraulic locomotives. As a consequence, Crewe ceased further work on transmissions and the torque converter manufacture was transferred to Shrewsbury.

The huge losses experienced by the Division in the early years at Shrewsbury could not be tolerated and it was clear that the business would not survive unless it was quickly moved into profit. In 1960/1 Bill Harris put forward a plan to Norman Parry, Robotham's tank design assistant from earlier years and by this time general manager, to reduce costs and to move the Division into profitability. None of the Rolls-Royce divisions had experienced redundancy in modern times. The plan required redundancy on a substantial scale – among middle management in particular. To his credit, Parry backed the plan with enthusiasm but Robotham was less enthusiastic. After much discussion, he agreed to allow Parry to put the plan into operation but with the warning that the responsibility lay with Parry. Within twelve months, the Division was in profit – largely due to the drastic cutting of costs, increases in the sale of the Division's natural products and significant reductions in the loss-making products inherited from Sentinel.

INTO THE TRUCK MARKET

As we have seen, Robotham left Rolls-Royce in January 1963. 'Doc' Llewellyn Smith became Chairman of the Shrewsbury operation and Norman Parry, Managing Director. Parry was faced with the need to increase engine manufacturing levels, since output had been static at about 2,000 units per year for some time. His courage had made the Division profitable but profit growth would only come from an increased order level. It had been argued that the Division would not compete head-on with mass producers of truck engines, but recently maximum vehicle weight legislation had changed, with the demand now in the range of the six-cylinder engine. The Division had been discouraged by the failure ten years previously of the trials of a four-cylinder engine in ERF trucks in South Africa and the six-cylinder engine with International Harvester. Gardner, one of the established competitors, was more frugal on fuel and weight, but offered outputs below the optimum level now required.

Peter Vinson, then Director of Sales and Marketing, led the campaign to take a more serious look at this market, of which he said:

It is vitally necessary to increase our number and this is a sector we need to pursue if we are to survive.

Discussion with Scammell and Atkinson, who had been fitting C6 engines into special oilfield and other vehicles, gave encouragement to be more active in this market, as both warmed to the idea of a British six-cylinder in-line 200 bhp engine. John Read, Director of Engineering, felt that many of the problems encountered in Canada had been mastered and that:

We now had an engine comparable with the competition.

Read was shortly to retire in December 1965. When working on the wartime tank designs, Parry had spent much time with Rover, who were developing the Meteor engine, and had met and been impressed by Ron Whiteside, who was now managing the Diesel Department at Rolls-Royce Canada. It was Whiteside who had carried out the installations in IH trucks. Parry therefore invited Whiteside to come to Shrewsbury, take over the Engineering Department from John Read and in particular develop a specialised version of the 'C' range, specifically tailored for the truck market.

Launched at the Motor Show in 1966, the new truck engine with a power range of 205 to 300 bhp was promoted as the Eagle range, an established Rolls-Royce name, the use of which was designed to create an image of quality and durability. Whiteside improved the piston ring specification significantly by re-sourcing from the USA and improved the fuel economy and performance generally. He also styled external features, a novel consideration in those days. Eagle engines were the first to be certified to the new British Standard AU 141 for emissions. The Eagle Mk III 220 won the 1975 Design Council Award for Industry.

Over the next twenty years, the demand for the Eagle steadily grew until it shared the UK market equally with Cummins and Gardner. Early sales were helped by a large order from British Road Services [the ancestor of National Freight in its nationalised days] through Seddon Vehicles and Guy Motors. A further boost was given by its selection by Scammell as the exclusive choice for their new Crusader truck, and later for the new fleets of logistic vehicles for the MoD. Peter Vinson regarded the success of the Eagle truck engines as the vindication of Robotham's original vision.

Following these developments, the Oil Engine Division was approached firstly by Ford and later by the Jaguar and Guy Motors Divisions of British Motor Holdings (BMH) for a vee engine of 170/180 bhp to fit into the new 24 to 28 ton truck ranges. A programme was initiated to investigate the design of a suitable engine based on the bore and stroke of the current LV car engine, to enable existing machinery at the Motor Car Division to be utilised, should manufacture proceed. Ten or so development LD diesel engines were produced with small variations in swept volume, and trial installations in Guy trucks were successfully completed. The Ford opportunity was lost to Perkins but the BMH project appeared to be going well, with planning for series manufacture under way, when BMH merged with Leyland Motors. Sir Donald (later Lord) Stokes, the Chief Executive, was alleged to have said:

We don't need help with diesel engines, thank you.

In 1969, the MoD set up a competition for a new range of engines to power a new family of tracked combat vehicles, principally a future main battle tank (MBT80) and a smaller armoured personnel carrier known as MICV. The current vehicles of these types were powered by an MoD-inspired design of six-cylinder opposed-piston two-stroke engines of some complexity but which were designed to run on an extremely wide range of fuels. Jointly developed with FVRDE, they comprised the L60 for battle tanks, the K60 for armoured personnel carriers and the H30, a small auxiliary power unit, manufactured respectively by Leyland, the Rolls-Royce Motor Car Division and Coventry Climax. In the hands of the army, the L60 was unreliable and although many of the problems related to the complexity of the engine design, the too frequent army maintenance drills were a contributory factor. The specialised nature of the L60 meant that it had no other applications, making it an orphan in development terms with no contribution possible to its development from other applications. Combat vehicles achieve very low running hours and can therefore be more highly rated. In fact, specific outputs are quite often at a level in excess of twice that from a similar engine in commercial use. Rolls-Royce has long been able to obtain reliable high outputs from its engines, making it uniquely qualified to meet Defence Department requirements. The MoD determined that future purchases should be based on a commercial design and have a warranty arrangement that put much more responsibility on the supplier.

As a prelude to entering the competition, Whiteside issued a challenge to his research team to double the output of an Eagle to 600 hp. The power was to be achieved without using abnormally sized charge coolers/radiators, and within a 10 per cent tolerance of commercially acceptable values for specific fuel consumption, smoke, exhaust temperature and maximum cylinder pressures. The targets were met, enabling the Division to offer higher specific outputs with some reduction in life – but well within the needs of a tracked vehicle application. They won the competition against two other entrants, believed to be Leyland and Ricardo, and produced outline designs and concepts for a range of V8, V12 and V16 versions of the 'C' range, to be known as CV. Although new engine designs were now required, the long experience of a similar cylinder size was to prove of immense importance in shortening the development programme. The opportunity was also taken to design the engines to metric standards, thus reducing the opportunities for use of common parts.

Apart from the military applications, these projected new engines were also destined to fill out the generator-set range, since at that time engines were available to power generator sets only up to 250 kW (electrical). With the new CV range, engines for outputs of 300 to 600 kW(e) were now going to be possible. Therefore there could be two markets for these engines, which would justify the development and tooling investment and result in offering the MoD an engine developed from a commercial pedigree rather than an orphan like the L60. There followed a part-funded contract to build nine experimental CV8 engines rated at 750 bhp destined for the MICV project.

These negotiations took place at the time of the receivership on 4 February 1971, which caused some understandable delay to their completion. Fortunately for the Oil Engine and Motor Car Divisions, the receiver put them together under the banner of Rolls-Royce Motors. Initially, there were those at the Diesel Division who were less than enthusiastic about merging with the Motor Car Division, but Freddie McWhister, the receiver's most senior manager, held out the carrot of an eventual public flotation and the doubters were persuaded. And as we have seen, Rolls-Royce Motors was successfully floated on the London Stock Exchange in May 1973.

A new Rolls-Royce Motors board was formed, being in part an amalgam of the two Divisional boards. Shortly afterwards, Norman Parry retired as Managing Director at Shrewsbury and was succeeded by Tom Barlow. Bill Harris became Group Commercial Director and Chairman of the Diesel Division. A very welcome change was that of the division's name to Diesel Division.

However, the receivership did bring some redundancies at Shrewsbury. The change from steam to diesel industrial shunting locomotives had by this time been largely completed and the production line at Shrewsbury was shut down, as was the locomotive design office. Other sections were also closed or slimmed down.

The Engineering Department focused on the design of the 90-degree CV8, at the same time looking at CV8 and CV12 power pack ideas for the various combat vehicle concepts being considered at the Military Vehicle Engineering Establishment (MVEE).

A combat vehicle requires compactness in power packs, gunnery equipment, crew compartments, etc. and in consequence there is ongoing conflict between the various needs. By careful design with in-line cylinder centres, the specific weights and volumes for both the CV8 and CV12 were only marginally greater than equivalent specialised military engines being made in Germany and the USA.

THE 'THRILL A MINUTE' MILITARY BUSINESS

Whereas the railway, generator set and compressor markets had been the mainstay of the 1960s and the truck market was now replacing the railway sector, in the late 1970s and into the 1980s, it was the military market that took up the running.

Internationally, the largest potential customer was the Shah of Iran. In trouble with his British Chieftain tanks and L60 engines supplied by Leyland, the Shah was convinced by David Caldwell, Director of the MVEE, that Rolls-Royce could provide a suitable alternative. The Shah, whose garage was full of Rolls-Royce motor cars, said 'Yes, super!' and ordered 1,500 tanks from the Royal Ordnance Factory at Leeds, to be fitted with complete power packs to be supplied by the Diesel Division. This was great news for the Division but, as David Plastow, put it:

With that thrill a minute quality characteristic of military business, our plans were upturned in 1974 by an urgent requirement for a 1200 bhp engine for a new main battle tank order for the Imperial Iranian Ground Forces. To say that we were poised ready for such a requirement would be an overstatement. The CV12, which we knew we would have to use to satisfy the Iranian project, was at that stage literally a paper engine.

The Division was faced with a challenge not only to design and develop this new engine but to build a manufacturing facility in which to make the engines and assemble and test complete power packs incorporating the David Brown transmission and a yet-to-be-designed cooling group. The tank was to be a revised design of the Chieftain with Chobham armour. Here was a great opportunity for the Division to secure a Government-backed order for complete power packs, with subsequent spares business including funding for the design and development of that power pack. There followed some rather blunt discussions as to how they could deliver these power packs in two and a half years, which required the new facility to be built in eighteen months, including working test facilities for the development phase of the power pack work. Brian Leverton, having been in charge of engineering for just three months, remembers:

I was sent for by Plastow and asked whether we could sign off a production specification in two and a half years. I countered with three years but I was persuaded to settle for two and three quarter years.

On 29 October 1974, Ron Whiteside received a telegram from the Assistant Under Secretary for Sales at the Ministry of Defence, authorising him to 'enter into commitments up to a total of £1 million for items which are essential to the deal'. (This was reminiscent of Beaverbrook's telegram to Hives during the war which read: 'The British Government has given you an open credit of one million pounds', and which we learned about in Part One of this history.)

The CV8 programme was still in its infancy and, to establish a level of confidence in the MoD, one of the experimental CV8s currently under development at 650 hp was successfully run at its anticipated development limit of 800 hp on full power for 100 hours, equivalent to the required 1,200 hp from the projected CV12. It was therefore with some dismay that during the subsequent visit of the Duke of Kent, and just before he was due to see it running, a test CV8 experienced a catastrophic failure, a circumstance which Gerry Collin, deputy chief engineer, fielded with some skill. The tank project duly got under way, based around the 60-degree CV12 which had much the same swept volume and shape as the Meteor tank engine of earlier years worked on by Whiteside. The tank project was code-named 4030 in MoD circles.

A Military Engine Division headed by Peter Vinson as Managing Director, was formed early in 1975 to manage the project, and a complete new factory was begun alongside the existing factory although separated from it by the Shrewsbury to Crewe railway line. The necessary bridge to join the two areas and the new factory were completed on schedule. Meanwhile, the Engineering Department was using the experimental CV8 engines and in-line C6 engines to simulate various performance and design aspects in advance of the arrival of the prototype CV12s scheduled for April 1976. There were some novel features on the CV range, for example the oil backed slip-fit dry liner for ease of overhaul and weight saving, a departure from the 'C' range practice of wet liners. To control maximum cylinder pressures to reasonable limits, the compression ratio was dropped at the expense of creating a more difficult starting ability. This was addressed by the design by CAV of a sophisticated fuel burner in the inlet system to provide supplementary heat to facilitate starting in the prescribed low temperatures. The analogue control system, also designed by CAV, sensed a number of critical parameters and provided fuel-pump governing control for an engine with such high rates of acceleration that no mechanical means then existed to provide safe control. Such systems were amongst the earliest to employ electronics in an environment which was somewhat hostile, even without taking into account the special needs of installation in a battle tank.

The impressive efforts of the team, progressed by Gerry Collin, resulted in the prototype CV12 duly running one week ahead of schedule and the project rating of 1,200 hp being achieved after the first hour. Development proceeded without too many problems although the use of helical gears, whilst justified for quietness reasons, provided some excitement – being coupled to the high-delivery fuel pump with a somewhat lumpy demand. The shape of the power pack compartment was finally agreed, but the transmission heat rejection was underestimated, resulting in the need to make a late change to the cooling equipment.

Sir Alfred McAlpine Ltd. erected the 100,000-square-foot factory on time, an impressive achievement, and machine tools arrived from many points of the world (regrettably too few from the UK).

Power pack deliveries had begun on schedule to Leeds in mid 1977 but the Shah was overthrown in 1979 and the order from the MoD inevitably cancelled. Suddenly, there was an empty factory except for a stack of tank engines and power packs awaiting disposal. The need for the release of CV-engine build specifications for generator-set application was urgent, with development having taken second place to the tank contract. Furthermore, the Eagle engine development had been neglected and there were demands for more fuel-efficient versions with greater driveability.

Although the Iranian tank project had apparently collapsed, the Division continued design work for the Army's projected 'Main Battle Tank for the 80s' (MBT80). Also, there was a continuing demand for Eagle engines and their derivatives from the MoD. Engines for the Combat Engineer Tractor were now in production, and there were low- and medium-mobility logistic vehicle demands for gun tractors and heavy plant transport.

The MICV project for an armoured personnel carrier had been shelved in favour of a smaller MCV80 version to be developed by GKN at Telford and later to be seen as Warrior in various peace-keeping missions such as Bosnia and Kosovo. The Division was asked to supply a complete power pack, including the transmission to be developed and manufactured at Shrewsbury from a General Motors-Allison unit. Towards the end of the 1970s, the King of Jordan placed an order for 250 series 4030 tanks which were to be known as Khalid, and the MBT80 project was gradually merging into a revised version of 4030 which would become Challenger. An automotive version of the CV12 at 650 bhp was also selected for the new Army tank transporter. The MoD had realised that by using the same basic engine design as the main battle tank and its transporter, there would be some commonality of spares and maintenance tanks.

By the late 1970s, the division was working at almost full capacity with both automotive and generator-set engine demand buoyant, particularly since commercial variants of the CVs were now available for marine applications and generation in the 300 to 600 kW(e) range. The new version of the Eagle – to meet more stringent environmental regulation and fuel economy demands – was now under development.

The year 1980 brought another change of ownership, in that Rolls-Royce

Motors was absorbed into Vickers. The new Chief Executive was David Plastow and the Board also included some other Rolls-Royce Motors directors including the Diesel Division's Chairman, Bill Harris. There were few immediate changes, although the factory was one of the first in the Vickers group to abandon separate canteen facilities for directors and management. The new communal cafeteria arrangements were generally welcomed but did provide some anxious moments for some managers who saw their staff eating alongside directors! It was during this period that Prince Charles visited the factory – to obtain first-hand experience of discussions between management and unions – and sampled these new eating arrangements. His PA wrote after the visit:

His Royal Highness was most impressed with all he saw and more so with those he met and certainly left with the feeling that Rolls Royce Shrewsbury have the right idea on management union understanding and co-operation.

Significant orders for Challenger power packs were now being received and substantial development work was also in hand for the MCV80 Warrior vehicle being developed by GKN. However, the demand for commercial engines was now dropping. Like aero engines, a diesel engine is a proprietary component, it is not an end-product. Long-term marketing success is dependent on the success of the end-product. However, the entry of the volume vehicle manufacturers into the top end of the heavy-truck market resulted in the progressive demise of the independent UK truck makers. This diminished that market for the Division. As Peter Vinson put it:

When we first started looking at the heavy truck market there were no less than 26 British firms in the business. By 1984 there were only five in the over 20 tonne sector that were not manufacturing their own engines.

This realisation had been one of the reasons behind the company's failed bid for the Foden heavy-vehicle manufacturer in 1977.

In spite of the successes in varying industries, Peter Vinson looked back on the 35 years of the Division's history as a period of constant struggle against the background of constantly disappearing customers in a market of much larger competitors.

In consequence, in 1983 when the engine output had declined to 3,000, it was decided that the Division should be sold to a bigger engine company who would have the resources and image to enable Shrewsbury products to be sold into new markets. The CV range of applicants, particularly the profitable military business, and the new version of Eagle were complementary to the Perkins Engines range, and so in 1984 the Division was bought by Perkins, a subsidiary of Varity (formerly Massey Ferguson) for £20 million. Perkins continued to sell the very successful engines developed under Rolls-Royce and raised output levels again.

We saw earlier that Hives had expressed some reservations about the viability of a diesel engine division. He had worried that there were too many remote, small-scale operators around the world and consequently too little volume to support a worldwide infrastructure of sales and service.

Nevertheless, Eagle truck engines are still in demand and CV12 power packs are being sold for Challenger. It can be argued that it is the best main battle-tank power pack in the world. The CV8 and CV12 are established in the generator-set market, complementing the C6. Ron Whiteside's vision had provided the base for the CV engines to be world beaters.

'THINK ABOUT NUCLEAR ENGINES'

At 11.00 a.m., Washington time, on 5 August 1945, the President of the United States, Harry S. Truman, announced:

Sixteen hours ago an American airplane dropped one bomb on Hiroshima . . . It is an atomic bomb. It is a harnessing of the basic power of the universe . . . We are now prepared to obliterate more rapidly and completely every productive enterprise the Japanese have above ground in any city. We shall destroy their docks, their factories, and their communications. Let there be no mistake; we shall completely destroy Japan's power to make war . . . If they do not now accept our terms they may expect a rain of ruin from the air, the like of which has never been seen on this earth . . .

The nuclear age was born, albeit in the worst of circumstances. Britain's Government, by this time a Labour one, had to decide whether the country should become a nuclear power. It was not a foregone conclusion. The Labour Party had been full of disarmers before the war, but those in power now, whatever their previous convictions, decided that Britain, still a world power, could not stand back from the development of nuclear power.

Nuclear power did not *have* to be used destructively. Harnessed constructively it could replace coal, which was difficult to extract from the earth and, when burnt, polluted the atmosphere with noxious sulphur dioxide.

'Atoms for peace' was the slogan of the hour, and people dreamt of limitless and cleanly produced cheap energy. The UK quickly realised that it would have to plough its own furrow when the McMahon Act, limiting the free exchange of technical information, was passed in the United States of America. The UK started its own intensive research programme based at Harwell, near Oxford. In 1947, the first graphite-moderated research reactor

GLEEP (Graphite Low Energy Experimental Pile) went critical. Work began at Sellafield for the production of fuel, and the decision was made that gas-cooling would be the system used for the British power plant. By the end of 1950, the Windscale reactor was critical and the peaceful energy programme was well advanced.

However, the Korean War broke out in that year, prompting calls for more fissile material. A fast-breeder programme was set up and, in October 1952, the first UK atomic bomb was detonated. Nevertheless, the civil programme continued. In 1954, the Atomic Energy Authority (AEA) was established and, in October 1956, Calder Hall was commissioned.

While all these developments were taking place, the Admiralty was watching to see if there was a practical possibility of nuclear propulsion for warships. The attraction for submarines was enormous because nuclear energy would be a non-air breathing means of propulsion, which would allow long-range underwater operations. The Admiralty already knew that the US Navy was developing a nuclear submarine. The keel of the USS *Nautilus* was laid in June 1952 and she was commissioned in September 1954.

Fully aware that the Royal Navy would not want to be left behind, Vickers and Rolls-Royce wanted to make sure they would be involved in supplying the ships and the engines. Ernest Hives said to Alex (later Sir Alex) Smith, when he invited him to Derby from his position at the Production Division of the AEA at Risley, near Warrington:

Some day soon, there are going to be nuclear engines maybe for ships and submarines, maybe even for aircraft. We are engine makers, so if there are going to be nuclear engines I want to make sure they are Rolls-Royce engines.

Smith protested that anyone wanting nuclear engines would have to go to the AEA, which knew an awful lot more about nuclear reactors than Rolls-Royce. Hives replied:

Maybe, maybe. But we know an awful lot more about engines. We know a lot more about what it takes to make them reliable. And we will learn about nuclear reactors faster than they will learn about engines. Besides, the Atomic Authority is dealing with great big reactors. They'd be no use for engines. If there are going to be nuclear engines, then the reactors are going to have to be small.

Hives had heard that Smith specialised in small reactors, and persuaded him to join Rolls-Royce to work at Old Hall, a country house at Littleover, on the outskirts of Derby, where Dr. Griffith was already working. Just as he had instructed Griffith at the beginning of World War II to 'go on thinking', Hives now instructed Smith:

I just want you to think. Think about nuclear engines for a start. Just go on thinking.

An advanced project team had been set up under Adrian Lombard's directions, working in the stable block of Duffield Bank House. The team, led by John Hollings, effectively head designer, included Bill Gilligan who concentrated on heat transfer, and Norman Battle. And it was not just submarines that Hives was considering. He said to Hollings one day:

Can atomic energy be used to drive an aeroplane engine?

Hollings replied:

An engine needs heat. Atomic energy provides heat so, in principle, the answer is yes.

As a result of this conversation, Hollings and Gilligan were sent on the three-month reactor course at Harwell and returned as Bachelors of Reactor Engineering at the end of 1954.

By this time the team was too big for the stable block at Duffield Bank House and moved to the Old Hall, Littleover, where laboratories and a rig shop were built in the grounds.

As Smith pointed out in his book, *Lock up the swings on Sundays*:

Trying to design something when you have the knowledge that somebody else has already done it is a very different proposition from probing around trying to assess the feasibility of an idea. There is no question of whether or not it is possible. You know that it can be done. It is knowledge which both helps a designer, and yet raises apprehension in him that he may be found wanting, that he may be unable to find out how it can be done.

The efforts that we were making in Rolls-Royce, at the Old Hall, in 1956 to envisage a nuclear reactor system which could fit into a submarine and be able to propel it were made in the knowledge that the Americans had already done it. They had designed, built, launched, commissioned and operated the USS *Nautilus*, the first ever nuclear-powered submarine which, by all accounts, was a huge success. But we had no inkling whatsoever of how they had done it.

The Royal Navy, as we have seen, had indeed become interested in acquiring nuclear submarines and had set up a small unit inside the AEA at Harwell. In the early 1950s, the US Navy sent two representatives to talk to the Royal Navy with the idea of setting up a joint US–UK programme. However, Admiral Hyman Rickover, who had almost single-handedly driven the US nuclear-submarine project forward, would not even grant the British officers a hearing.

Coincidentally, Smith and a Rolls-Royce colleague were sent to Westinghouse in Pittsburgh (we have already seen in Chapter One that Rolls-Royce had been collaborating with Westinghouse on developing jet engines) in the hope that they would pick up some information from engineers and scientists who had worked on the *Nautilus* but who had by this time transferred to work on nuclear reactors for industrial power production at Westinghouse. They were no more successful than the British naval officers. As Smith said later:

The American engineers and scientists were excellent hosts but, in accordance with the obligations put on them by the American government, they said nothing, absolutely nothing, not so much as a syllable, about the design of the reactor in the *Nautilus*. [The USA Atomic Energy Act of 1954 severely limited the assistance that could be given to other nations.]

However, in 1958 a Bilateral Agreement was signed between the UK and US Governments whereby, in exchange for the US receiving details of the British gas-cooled power reactors, the US would supply one complete S5W (W standing for Westinghouse) reactor and propulsion plant.

Pressure was brought to bear on Rickover, obviously from the US Government, and he came to Britain to visit the Royal Navy, the Atomic Energy Authority, Rolls-Royce, Vickers and others. Before he visited Rolls-Royce, he made it clear that he wanted to meet the chief engineer. By this time Adrian Lombard was chief engineer and he had not been involved in any of the nuclear experimentation that had been going on at Old Hall. He did his best to become as conversant as possible with nuclear energy in the short time available.

When Rickover arrived, it was obvious that he had come to Britain because he had been told to rather than because he wanted to and he spent the morning being rude about British engineering standards. According to Smith, the situation was only saved by Rickover's chief technologist, Dr. Harry Mandl, who every so often would say:

Aw, come on, Admiral, for Chris'sake! It's not as bad as all that. Remember you're in Rolls-Royce. They've got a pretty good track record in designing things.

Hives appeared for lunch and once Rickover appreciated that 'Lord' Hives, the Chairman, was not some banker or lawyer but, as Hives put it, 'just a bloody mechanic', he relaxed. Listening to Hives talk of Royce and the standards of excellence he had set, Rickover became convinced that if he had to co-operate with the Brits then Rolls-Royce was the company with which he wanted to work. He concluded the meeting by saying that he would return to the US and recommend that a party of engineers and scientists should be sent over from Britain, but he was insistent that Lombard be one of them.

Vice-Admiral Sir Robert Hill, the first nuclear submariner to be President of the Institute of Marine Engineers, posed the question: 'Why did Rickover, given his fierce individuality and nationalism, go along with the idea of the USA making the priceless gift of a complete submarine propulsion plant, with full supporting information, enabling the Royal Navy to become a highly competent nuclear navy at a fraction of the cost of independent development of the technology?' When asked about it, Rickover wrote a letter to the commanding officer of the *Nautilus*, in which he said:

I did this because of my feeling of urgency about the international situation, my admiration for the British and particularly my great liking for Mountbatten. [Rickover had struck up a close rapport with Mountbatten as well as with Hives.]

We can only wonder if Rickover would have been impressed by Winston Churchill who, during his second spell as Prime Minister in the early 1950s, subjected his ministers to aerial bombardment while discoursing about the horrors of nuclear warfare. Peter Hennessy wrote in his excellent book, *The Prime Minister, The Office and its Holders Since 1945*:

Often the minister at the receiving end of a furious minute would be summoned to the bedside of the PM the following morning to explain himself.

There, unless the Cabinet or a Cabinet committee he chaired was due to meet, Churchill would lie until shortly before lunch, an unlit cigar in his mouth, his bed covered in papers, a 'Garden Girl' beside it to take dictation. At his feet would be Rufus the poodle, whose breath was likened to a flame-thrower by one of his private secretaries. On his head sat Toby, the constantly twittering budgerigar. Toby, for some reason, was particularly excited by the presence of Rab Butler, the Chancellor of the Exchequer. If Rab was briefing Churchill on the latest strains on the economy, Toby would fly round the room, occasionally opening his bowels on Rab's head. According to one of the private secretaries, Anthony Montague Browne, Toby found the Chancellor's bald head an irresistible target as well as a perch. On one occasion Butler was seen to mop his head 'with a spotless silk handkerchief' and was heard to sigh resignedly, 'The things I do for England . . .'

Toby plainly fascinated the Prime Minister's colleagues. In January 1955, Harold Macmillan, Churchill's last Minister of Defence, was summoned to the bedside to discuss the horrors of thermonuclear war. That night he recorded the scene in a way that defies parody as Toby somehow managed to upstage even the hydrogen bomb. Toby began the meeting sitting on Churchill's head, swooping occasionally to take sips from the whisky and soda beside the Prime Minister's bed:

On 6 July 1951, the *Derby Evening Telegraph* wrote:
'A new Rolls-Royce oil engine, designed and produced in the
company's new factory in Victory Road, Derby, will power
a new crawler tractor with which Britain will challenge
America's monopoly in the heavy tractor and earth-moving
field.' These photographs show a Rolls-Royce C6 SFL oil
engine and a Vickers VR.180 tractor fitted with the engine,
under test in extreme conditions of heat and sand in
Tripolitania in June 1952.

In 1958, Rolls-Royce took over the Sentinel company in Shrewsbury. One of its main products was locomotives and this photograph shows a Sentinel locomotive operating at the side of the Manchester Ship Canal.

The Rolls-Royce Oil Engine Division sold power packs to
British Rail and to railways all over the world. This photograph
shows a Rolls-Royce-powered train in Jamaica.

Sisu trucks in Finland, all fitted with Rolls-Royce
diesel engines.

Scammell became an important customer and encouraged
the Diesel Engine Division to develop a specialised version
of its C range specifically for the truck market.

The Rolls-Royce Diesel Engine Division worked closely with
the British Army, supplying engines for many of their
vehicles such as the Warrior (TOP) and Challenger (BOTTOM).

In 1974 an order came from the Government to supply 1,500
power packs to the Royal Ordnance Factory for battle tanks for
the Imperial Iranian Ground Forces.

The redoubtable Admiral Rickover, who was driving forward
the development of the US nuclear submarine project, was
sceptical of all things foreign but was won over by Lord Hives.
Rickover expected Hives to be some lawyer or banker but was
delighted to find that he was a 'bloody mechanic'.

TOP: The keel of the USS *Nautilus* was laid in June 1952 and she was commissioned in September 1954. Aware that the Royal Navy would not want to be left behind in the nuclear race, Vickers and Rolls-Royce wanted to make sure they would supply the ships and the engines.

BOTTOM: HMS *Dreadnought*, the first Royal Navy nuclear submarine, was a composite design. From the reactor aft, the equipment would come from the USA, while forrard it would be British.

Dounreay in northern Scotland where a prototype nuclear submarine was built. Up to this point there had been plenty of theory. Now the project had to work in practice.

A successful Blue Streak launch from the Woomera range in Australia. Blue Streak was powered by two Rolls-Royce RTZ-2 engines, each giving about 170,000 lb thrust. (For comparison, the RB 211-524 was giving about 60,000 lb thrust.)

The Hovercraft SRN4. Powered by the Bristol Proteus
engine, it entered service with British Rail on the
Dover–Boulogne–Calais crossing on 1 August 1968 and
continued in service into the 1990s.

Rolls-Royce achieved great success in the gas-pumping field.
In 1964, TransCanada PipeLine built gas-pumping
installations to carry gas from Alberta in the west to the
industrial cities in the east. The generating sets
were supplied by Rolls-Royce in conjunction with
Cooper Bessemer, with the Rolls-Royce Avon coupled
to a Cooper Bessemer two-stage, free power turbine.

The opening of the through-deck cruiser shore test facility at
Ansty in 1973 by Lord Carrington, the Secretary of
State for Defence. Seated at the table is Sir Kenneth Keith,
Chairman of Rolls-Royce, behind him Ralph Robins,
Managing Director, Rolls-Royce Industrial and Marine
Division at the time, and to his left, Chief of Naval Staff,
Sir Edward Ashmore.

An industrial RB 211, showing the environmentally friendly
low-NOx DLE (dry low emissions) combustion system.

This Oseberg platform in the North Sea is powered by five
Rolls-Royce industrial RB 211 engines.

'Really, he is [a] unique dear man with all his qualities and faults . . . The bird flew about the room; perched on my shoulder and pecked (or kissed my neck) . . . while all the sonorous "Gibbonesque" sentences were rolling out of the maestro's mouth on the most terrible and destructive engine of mass warfare yet known to mankind. The bird says a few words in a husky voice like an American actress . . .'

Apparently, Rickover was also greatly impressed by the fact that when his electric razor broke down while he was visiting Derby, Rolls-Royce engineers mended it for him while he was having discussions with Hives and his colleagues. He insisted that a direct contract between Westinghouse and Rolls-Royce was to be implemented without the direct involvement of the US Navy, the Admiralty or the AEA. As a result, Rolls-Royce acted as the appointed UK agent in negotiations, and Rickover also insisted that the fuel element and core manufacture to the US design was not to be subcontracted or sub-licensed by Rolls-Royce to any other agency. As a result, Rolls-Royce built at its own expense a core manufacturing facility in Derby, an arrangement that required the surrender, effected at Ministerial level, by AEA of some rights and dues under the Act of 1954 concerning nuclear fuel manufacture in the UK.

However, there were limits to Rickover's co-operation. UK Treasury influence forced an attempt to obtain from the USA the design of their longer-life core (S5W Core 2), but when the Controller of the Royal Navy visited Rickover in September 1962, Rickover refused further help saying that enough assistance had already been provided and the UK should stand on its own feet.

The delegation of thirty members, including five from Rolls-Royce, duly went to the USA and learnt a great deal. Shortly after their return, a licence agreement was signed between Westinghouse and Rolls-Royce whereby a nuclear-powered propulsion system was to be purchased from the USA.

This all changed the pace of progress. The first Royal Navy nuclear submarine, *Dreadnought*, was now going to be a composite design. From the reactor aft the equipment would come from the USA, while forrard it would be British. Project teams were set up and in the contractor team were Vickers as the shipbuilder, Vickers Nuclear Engineering to provide the steam machinery, Foster Wheeler (the British subsidiary of the US parent) would supply the steam generators, while Rolls-Royce would supply the reactor core and the control equipment.

There were lengthy discussions as to who should take the lead, but in the end, and perhaps on the basis that the technology was more akin to aerospace than shipbuilding, a joint company was formed, Rolls-Royce and Associates, in which Rolls-Royce had 52 per cent of the shares and Vickers and Foster Wheeler 24 per cent each. Later, the engineering company,

Babcock and Wilcox, joined to provide pressure vessels. As well as buying equipment from Westinghouse, other equipment, including the hull, was bought from the Electric Boat Division of General Dynamics Corporation of Groton, Connecticut.

Ray Whitfield, who became Managing Director of Rolls-Royce and Associates in 1965, remembered that the consortium, having quoted the Royal Navy a fixed price, were so efficient that the ship was built for much less. The decision was taken to return some of the money.

By this time, 1958, the project had outgrown the Old Hall at Littleover and most of the team had moved to the Sinfin B site behind the old Oil Engine Division building. They were joined by the Vickers and Foster Wheeler personnel from the disbanded Naval section at Harwell. The new team became known as the Contractors Derby Team and one of the naval officers associated with them said:

You could distinguish which company they worked for. Foster Wheeler were the City gents, with bowler hats, dark suits and rolled umbrellas; Vickers were the hairy-arsed shipyard engineers while Rolls-Royce were those with cardigans and ball-point pens.

Rolls-Royce and Associates was not itself a manufacturing company. It was a design, development, procurement and later, when the plant was at sea, product support organisation. Its partners, including Rolls-Royce, were the principal manufacturers. Rolls-Royce produced the reactor core itself and the structural material that made up the core assembly. It also made the control-rod drive assemblies as well as certain small-bore valves. Rolls-Royce won the argument with the Atomic Energy Authority and was allowed to make the fuel elements. It received enriched uranium from which it made the fuel alloys.

Yet again the project at Derby outgrew its premises at Sinfin B and a site was acquired on the Derby ring road at Raynesway, large enough for an office block, a rig shop, the zero-energy experimental reactor 'Neptune' and, on a separate but adjacent site, the factory where the nuclear cores themselves would be manufactured.

And the development work continued. Alex Smith joined the project from Old Hall, and the Stefan Bauer team came back as the Advanced Project Group. The principal task by this time was to produce a core that would give more power and have a longer life. This core was designated Core B. The increased life was achieved by building into the fuel elements 'poisons' that absorbed neutrons but which would burn up during the lifetime of the core. In this way, more uranium could be built into the core at the outset.

Meanwhile, at Dounreay in northern Scotland (in fact you can hardly go much further north on the mainland), a land-based prototype nuclear submarine was being built. Construction, by Vickers Shipbuilders, had begun in 1957 at Boston Camp, opposite the AEA's fast-reactor Dounreay atomic site. This had meant that the roads through the Scottish Highlands had been improved but, as half a submarine set in a circular shielding tank with all the associated machinery was now being constructed, further road improvements were needed. Water chemistry labs, decontamination facilities, a fuel storage pond and a dynamometer to absorb the full power of the submarine all had to be transported to Dounreay. Until this point, there had been plenty of theory. Now the project had to work in practice.

By August 1961, the Dounreay Submarine Prototype (DSMP) was virtually complete and Rolls-Royce took over the running of the plant (it was owned by the Ministry of Defence) to train the team chosen to put through trials. In 1963 DSMP, also known as the Admiralty Reactor Test Establishment (ARTE), was commissioned. However, some of the materials were found inadequate, necessitating a rebuild. The Admiralty took the view that the plant had achieved its task by detecting weaknesses under test and provided the opportunity to modify the equipment before it was brought into service. The plant went critical for the first time on 7 January 1965 and was capable of driving a *Valiant*-class submarine at underwater speeds in excess of twenty-five knots. Indeed, in 1967, HMS *Valiant*, shortly after commissioning, sailed the 12,000 miles from Singapore submerged in twenty-eight days. It was a notable demonstration of how nuclear technology had transformed submarines from slow underwater vessels able to operate at a few knots submerged for up to a day, to warships capable of over twenty knots with the ability to stay underwater for months, operating unseen and undetected.

On 2 April 1965, the Chairman of Vickers-Armstrong handed over the DSMP to the Navy Defence Minister, Christopher Mayhew, who immediately handed the plant over to Sir Denning Pearson, Chairman of Rolls-Royce and Associates, which had been awarded the contract to run the plant under Royal Navy control.

The first reactor plant of the British nuclear submarine programme was designated PWR1 (Pressurised Water Reactor 1). Like all other PWRs, it used uranium as the fuel, which underwent fission to produce heat in a process not needing oxygen. The heat was used to generate steam, which drove the turbo-generators providing the submarine's electrical power and the main turbines providing propulsion power. An abundance of energy meant that the submarine could also make all its own water and oxygen to condition the air for the crew.

The first Core A for the DSMP prototype was manufactured by Rolls-Royce in Derby and was subjected to an accelerated lifetime's run to prove plant and systems before they went to sea. It was depleted by October 1967 and the reactor core was removed between January and March 1968, while at the same time the propulsion machinery was overhauled. Core A was then replaced by Core B. It was the first refuelling of a nuclear submarine reactor in Britain. Core B operated until 1972 when it was replaced by the third and final core for PWR1, Core Z. Cores B and Z were fitted into the *Swiftsure* and *Trafalgar*-class submarines respectively.

By 1976, an uprated PWR was needed to power the Royal Navy's future submarine fleet, and a brand new reactor plant was designed and built – to be known as the Shore Test Facility (STF) – for the PWR2 programme. This time, instead of building the reactor in Caithness, the complete nuclear plant was constructed by Vickers Shipbuilding and Engineering Limited (VSEL) at Barrow and transported the 500 miles to Dounreay by sea. This offered enormous savings in construction time and cost, and allowed parallel construction of the buildings with their support systems and the reactor module.

STF was commissioned on 25 August 1987, ahead of schedule and within budget, and provided a wonderful example of effective project management by Rolls-Royce and Associates, who oversaw all procurement, construction and commissioning, loaded the core and took it to full power. Core G, the first PWR2 reactor core, went critical on 25 August, the day it was commissioned. We shall have to wait until Part Three of this history for further developments of Rolls-Royce's involvement in the nuclear industry.

ROCKETS AS WELL

While Rolls-Royce was involving itself as deeply as possible in nuclear development, it also made sure it did not miss out on any activity in the field of rocketry. We have already seen how the German V2 rockets had threatened London in the latter part of the Second World War.

There is no doubt that post-war rocket development in Britain, the USA and the Soviet Union owed much to pioneering work carried out in Germany before and during the war. As David Williams explained in *A View of Ansty 1935–1982*:

The exceedingly rapid advances made by the Germans in rocket technology both before and during . . . the Second World War, were without parallel elsewhere. By comparison, progress in Britain was negligible. Serious work had started in Germany in the years 1929–1930 by a few groups of private inventors and by 1933 this had attracted the attention of the Army Weapons Group. The major research and development station at Peenemünde was set up in 1937. Here activity was concentrated mainly on bi-propellent rocket engines using liquid oxygen for the combustion of the fuel. The development of large ballistic missiles was carried out, the V2 emerging as the only one to see operational service. Walther, working at Kiel, concentrated its attention on the use of hydrogen peroxide and also on this fuel in combination with suitable catalysts. Rockets propelled by the latter combination were extensively used by the Wehrmacht during the War. By mid 1944, the Allies had become acutely aware of German rocket development and especially of the Messerschmidt Me163B interceptor and of the V2 ballistic missile. Urgent steps were taken as soon as hostilities ended in Europe to interrogate German specialists and much of their know-how became available in the UK.

Peter Stokes in *From Gipsy to Gem* agrees:

German efforts in the war years laid the foundation for all subsequent rocket activity. . . . British aircraft propulsion work through the war years was limited to the application of solid cordite-type propellants for assisted take-off, primarily for Fleet Air Arm operations.

By the early 1950s, the British Government was planning the replacement of the V-bomber force with long-range ballistic missiles as the delivery system for the country's nuclear deterrent. Its code name was 'Blue Streak'.

The USA was ahead of any progress in this field made in Britain and technical assistance agreements were made with US companies. De Havilland worked with General Dynamics's Astronautics Division, and Rolls-Royce, who were entrusted with the propulsion, with the Rocketdyne Division of the North American Aviation Corporation. Before the mid-1950s, Rolls-Royce's prior experience on rocketry was confined to a series of paper studies of rocket propulsion systems. The R.Z.2 engines, which Rolls-Royce built, owed much to the Rocketdyne S.3 family which were used in Thor and Jupiter, and, in modified form, as the twin boosters for Atlas.

The R.Z.2 was a single-chamber engine, burning kerosene and oxygen liquid propellants with turbopump feed. The thrust of each engine initially was 137,000 lb at sea level, growing to 168,000 lb at cut-off at an altitude of 250,000 feet. Later development brought the sea-level thrust up to 150,000 lb. The R.Z.2 was 22 lb lighter and some 10 per cent more efficient than the original Rocketdyne S.3 engines.

In 1957, practical tests with large rocket engines began and British-built gas generators and turbopumps were erected at the Ministry of Aviation's Rocket Propulsion Establishment (RPE) at Westcott. The testing was carried out by a joint Rolls-Royce–RPE team and in 1958 complete engines were fired. These were of the early R.Z.1 engine, virtually a copy of the Rocketdyne S.3, but in March 1959 the first R.Z.2 was run at Westcott and, shortly afterwards, component testing began at Spadeadam in the Cumberland fells close to the Scottish border, chosen for its remoteness (compatibility with security), superior subsoil structure and ample supply of cooling water from the adjacent River Irthing. An R.Z.2 was fired there in August of that year by Rolls-Royce personnel.

The Spadeadam Rocket Establishment was built by the Ministry of Aviation at a cost of £20 million (about £500 million in today's terms) and was managed by Rolls-Royce on behalf of the Ministry. As well as carrying out their own rocket engine testing, Rolls-Royce provided plant services to a rocket vehicle test area and, between 1961 and 1963, de Havilland conducted many static firings of the complete rocket.

While facilities were being prepared at Spadeadam for static firing by de Havilland of the complete Blue Streak missile, other facilities were needed to flight-test it with a dummy warhead. The obvious place was the Woomera live testing range of the Australian Weapons Research Establishment which was already being used for live flight-testing of all the other British military airborne weapons at that time. It was already testing a number of rockets in which Bristol and Rolls-Royce were involved – Bloodhound, Thunderbird, Blue Steel and Black Knight. The Blue Streak test facility was sited 30 miles north-west of Woomera, on a bluff overlooking the salt lake, Lake Hart.

Regrettably, Blue Streak as a military weapon was cancelled in April 1960 but, after strenuous efforts, the rocket was re-employed as the first stage of a three-stage European Launcher Development Organisation (ELDO) satellite launcher. The Australian Government, having invested £9 million (£225 million in today's terms) in the project at Woomera, was keen for the project to continue. Since Blue Streak, mounted with upper stages and satellite, would be no less than 13 metres longer than the Medium Range Ballistic Missile, adjustments had to be made to the launching towers at Lake Hart.

ELDO – with Britain providing the first-stage rocket, France the second-stage, Germany the third, Belgium and Holland the guidance mechanisms and Italy the payload – struggled on through the 1960s and into the early 1970s. Thirteen completely successful Blue Streak firings were made at Woomera, the first on 6 June 1964.

THE SOUBRIQUET OF HMS ROLLS-ROYCE

On 27 December 1952, *The Economist* printed an article under the heading 'Rolls-Royce goes to sea'. The article said:

For some time past Rolls-Royce has been seeking to develop its interests outside the aircraft industry. The two gas turbines shortly to be installed in the Admiralty's ex-gun boat, *Grey Goose*, are the first marine engines that the company has built. [This was not true. Rolls-Royce had built nearly 100 Merlin marine engines and by 1952 was also building the Griffon Marine Mk 101.] They follow on the introduction of a new range of Rolls-Royce petrol engines and the appearance of the first of a similar range of diesel engines. The gas turbine units are still regarded as prototypes, but the initial shore trials have been highly successful. Most of the industrial gas turbines under development are the work of companies with long experience of the design of industrial power units; Rolls-Royce brings to the design of these big gas turbines a unique experience of jet engine design.

The principal contribution of a gas turbine is the great power that it gives for small weight. The *Grey Goose* was previously powered by two 4,000 horse-power steam turbines. The two Rolls-Royce gas turbines that are now to replace them are rated at 6,000 horse-power each, but the new installations save about 50 per cent in total machinery weight, as well as valuable engine-room space. Aero-engine manufacturers probably know more about weight-saving than any other engineers. These two features, lower weight and fewer spares, explain the very active interest that the Admiralty has taken in gas turbine development for use in small craft. The United States Navy has also wakened suddenly and sharply to the possibilities, and is to purchase two Rolls-Royce gas turbines and two built by Metropolitan Vickers, the company's chief competitor in this field. Metropolitan Vickers, in whose drawing office the Armstrong-Siddeley Sapphire jet engine was first designed, has already provided the Admiralty with a gas turbine for installation in a motor gunboat, and two improved versions are now due for delivery.

As far back as 1942, the British Admiralty had begun to take an interest in the use of gas turbines for the propulsion of Royal Navy warships. The Admiralty discussed with Metropolitan Vickers the possible use of the aircraft F1 jet engine. During the war, the Navy received a number of *Captain*-class, twin-screw, steam, turbo-electric frigates under lease-lend from the USA and one of them, the *Hotham,* was retained so that the prime mover of one alternator could be replaced with a gas turbine. English Electric designed and built the 6,500 hp EL 60A, two-shaft, regenerative, open-cycle gas turbine to traditional steam turbine criteria. However, it was too bulky and heavy and, although land-tested, was never installed in the *Hotham* and was scrapped in 1951.

Meanwhile, the first Metropolitan Vickers Gatric marine propulsion gas turbine went to sea in 1947, having replaced the centre-line engine of the Navy's triple-screw MGB 2009. Also in the late 1940s, Rolls-Royce developed a sophisticated and complex marine gas turbine, the RM60, which was designed to provide a more economic and flatter fuel curve across the power range. It was this power plant that was tested in *Grey Goose*.

The two main advantages of gas turbine engines in ships were the speed with which maximum power could be achieved and the low level of maintenance required. The whole apparatus also took up less space, while liquid fuel was easier to store than coal had been. The rapid build-up from cold to maximum power meant there was no need for a ship to keep steam at short notice and the reduction in maintenance work reduced the man-hours required by up to 75 per cent.

Following the success of the Bristol Proteus in the fast patrol boats – it went on to be used in hovercraft and hydrofoils – the Bristol Olympus was also adapted for ship propulsion. In partnership with Brown Boveri, Bristol Siddeley won a contract from the German Ministry of Defence in 1962. Bristol were to produce a marinised gas generator and Brown Boveri a two-stage, long-life, marine power turbine. Although the engine passed its acceptance tests and was delivered to the German Navy's technical establishment at Kiel, the Navy abandoned its plans to build a new class of gas-turbine frigate in favour of destroyers bought from the USA.

Three years later, the Finnish Navy commissioned the corvette *Turunmaa*, the first ship to operate an Olympus at sea. *Turunmaa*, and its sister ship *Karjalla*, used an Olympus of 22,000 shp on the centre shaft, supported by a 1,200 hp Mercedes-Benz diesel on each wing in a CODAG arrangement (Combination of Diesel and Gas).

At the same time, Bristol Siddeley began work on its Admiralty project to provide a marine version of the Proteus gas turbine to be fitted into the Royal Navy's new fast patrol boats. The *Brave* class, fitted with the Proteus, put to sea in 1958.

The following year, the shipbuilder Vosper, which had built the *Braves*, began designing a private-venture, fast patrol boat to be called *Ferocity*. The *Ferocity* was smaller than the *Braves* and would use only two Proteus engines as opposed to their three. Over the next fourteen years, twenty-three ships were built by Vosper and powered by Proteus engines. Although the Royal Navy lost interest in patrol boats, foreign navies, including those of Brunei, Denmark, Germany, Greece, Libya, Malaysia and Sweden, all used Proteus-powered boats.

Bristol's Olympus engine was also adapted for marine use. In March 1959, Bristol published preliminary performance data for a marine unit derived from the Olympus 6, and in 1963 the recently formed Power Division of Bristol Siddeley Engines at Ansty designed a purpose-built power turbine. This was in response to the Ministry of Defence requirement of combined steam and gas turbine machinery in the Type 82 Guided Missile Destroyer, with one Olympus and one steam turbine on each shaft.

As David Williams makes clear in his excellent book, *A View of Ansty 1935–1982*, the fortunes of the marine engines were subject to Government decisions on defence.

The first shore run of the new Olympus TM1A was carried out at Ansty in August 1966. By this time defence policy had cancelled further ships of Type 82 but the conversion of HMS *Exmouth* to all gas turbine propulsion with an Olympus TM1A/

Proteus machinery fit had already started. Further defence policy changes also conditioned by budget restraints, now made it necessary to design the Type 42 which was a much smaller ship than the Type 82 but had, nevertheless, to fit the air defence systems of the latter. There was therefore not enough space for both steam and gas machinery. Though there were good enough arguments for the choice of diesel machinery for cruise there were difficulties again with space. Thus the Olympus/Tyne fit came about and became a classic propulsion system fitted both to Type 42 and to the smaller Type 21. The common propulsion system for both ship types was based on a YARD Ltd design using the 21MW Olympus TM3B and 3MW Tyne RM1A with SMM (Stone Manganese Marine) controllable pitch propellers on each of the two shafts and HSDE (Hawker Siddeley Dynamics Engineering) machinery control. Both ship types were planned for the same timescale. In the event Type 21 (HMS *Amazon*) was the first to go to sea on 23 July 1974.

In 1974, the Royal Navy made the decision to change to all-gas-turbine propulsion for both main and cruise engines. This has been described by some as a radical change as significant as the change from sail to steam, from coal to oil and from reciprocating engines to steam turbine. By the time of the Falklands War (1982), of the thirty-two Royal Navy ships involved in the hostile seas of the South Atlantic, nineteen were powered solely by Ansty-built gas turbines, all but one with Olympus–Tyne combinations. The exception was the aircraft carrier HMS *Invincible*, which was powered by four Olympus engines. The biggest ship in the world to have all-gas-turbine propulsion, *Invincible* was powered by four Olympus TM3B modules mounted in a COGAG arrangement, giving a total power output of more than 100,000 shp. When she was delivered to the Navy in 1980, her chief engineer, Lieutenant Commander Peter Clarke, said of the Olympus engines:

They are magnificent. You just select a button and there's all that power – instantly. . . there's no need for anything to get dirty – and no one to make it so. *Ark Royal*'s watch below consisted of 88 men. We have 11. [*Ark Royal* was the Navy's big fleet carrier and had run on steam turbines.]

As Alan Baxter pointed out in his *Olympus – the first forty years*,

The almost exclusively Rolls-Royce powered *Invincible*, taking into account its complement of Sea Harriers as well as its helicopters, Sea Dart missiles and its main propulsion equipment, soon acquired the soubriquet of HMS *Rolls-Royce*.

The *Invincible* broke all previous Naval continuous carrier operations records by staying at sea for no fewer than 166 days. In the whole fleet, only two gas generators had to be removed in the whole campaign. The Chief Naval Engineer, Admiral Horlick, said:

223

The overall performance was a real tribute to the designers, manufacturers, operators and maintainers.

During the 1970s and 1980s, Rolls-Royce from Ansty also supplied the Royal Navy with an adapted Spey. The gas generator was a derivative of the joint Rolls-Royce–Detroit Diesel Allison TF41 which, at the time, was the most powerful of the Spey family of engines.

While Rolls-Royce became the sole supplier to the Royal Navy, any attempts to secure orders from the US Navy were thwarted by General Electric's stranglehold. Elsewhere the two competed – and in Japan, Rolls-Royce was successful. Forming an association with the Japanese company, Kawasaki, in 1971, Rolls-Royce shipped a complete Olympus module to Kawasaki's Kobe factory the following year. This engine was for demonstration use but in 1977 the first orders for ship sets arrived. The *Hatsuyuki*-class destroyer was powered by the twin Olympus–Tyne configuration, and the *Ishi Kari* class, the first new Japanese frigate design since the mid-1960s, used a single Olympus in conjunction with a diesel engine which was used for cruising. Gradually, more of the engine manufacture was undertaken in Japan, until the local manufacture content reached 75 per cent, with the remaining 25 per cent supplied in kit form from the UK.

ELECTRICITY, GAS, OIL

Following the Sandys White Paper in 1957 which had effectively signalled a cut in Government defence expenditure, Bristol Siddeley looked at diversification possibilities. One of them was the production of Maybach diesel engines under licence, while another was the design and production of gas turbine engines for use in both marine and industrial environments.

The necessity of cutting costs at Bristol had led to an attempt to save electricity by the use of diesel engines supported by a marine Proteus engine in a peak-lopping role. Hearing of this the South Western Electricity Board asked Bristol to look at modifying a Proteus so that it could act as a generator for peak and emergency loads, especially for remote communities where existing connections with the National Grid were poor. As a result, in 1959 a plant, powered by a Proteus, was installed at Princetown in Devon. It was the world's first remotely controlled gas-turbine generating station.

Encouraged by this success, the recently formed Power Division of Bristol Siddeley Engines at Ansty began to look at the possibilities for a much broader use of aero derivative engines and later that year a proposal was put to the Central Electricity Generating Board (CEGB) that it should consider using larger blocks of power, burning diesel rather than kerosene in an adapted 200 Series Olympus engine. In September 1960, the CEGB ordered a prototype set which was installed at Hams Hall near Birmingham in July 1962 and commissioned two months later.

Not surprisingly there were some teething problems, especially in the gas generator and automatic controls of the set. Nevertheless, the advantages to the CEGB were soon evident, especially as the demands on the national system increased rapidly. Over the following fifteen years, gas turbine plant was installed on 29 sites in England and Wales. The first overseas installation for an Olympus-powered generating set was at St Helier in Jersey which, with a capacity of 50 megawatts, was the largest diesel generating station in Europe. By September 1970, the CEGB had installed a total of 137 Rolls-Royce industrial gas turbines: 88 Avons, 42 Olympus and 7 Proteus.

Rolls-Royce at Derby had quickly appreciated the potential and adapted the Avon for power generation. In the 1960s, Rolls-Royce favoured *supplying* the Avon as a gas generator to a main contractor such as English Electric or AEI, while Bristol preferred to *act* as a main contractor. After the merger of Rolls-Royce and Bristol, there was much debate as to which was the better approach. As the 1970s progressed, punctuated by the oil crisis of 1973/4 and the quintupling of the price of oil, it became clear that larger sets generating more power were necessary. Furthermore, competition with GE and others became severe. One approach was to band existing Olympus engines together to provide more power and this was tried with some success. The alternative was to develop new engines. As David Williams put it, in *A View of Ansty*:

In principle, however, the case for the large engine remained. By 1976 strong competitive pressures within the industry had resulted in price per kW being halved in real terms over the decade. The economies of scale had created a trend to larger sets and engines of higher power. Since RB223 had not survived, the next best big engine available to Ansty was the Olympus 593, the Concorde power plant. As the ultimate development of the Olympus family of aero engines it was the sort of engine we knew a great deal about. It was the world's largest aero core capable of 75% more power than the preceding industrial Olympus. Very importantly, it had been designed and developed to cruise at supersonic speed at pressure levels within the cycle much nearer to industrial conditions than was the case for any other aero derivative. The essential modifications to the aero engine involved relatively low risk.

A definitive design was prepared with a new three-stage power turbine. The objective was a single ended set of 48MW and a double ended set of 96MW. It seemed to be perfectly obvious that the Company should go ahead with this engine, which, by this time, had a 30 year pedigree. Its credentials were impeccable; calculated return on resources was acceptable though a major cost reduction plan for the gas generator was necessary. The programme started in 1974 with a target date for entry to market of 1981.

Not long after the Industrial 593 programme started, exceptional increases in material costs and a cutback in the aero production programme led to an unacceptable level of price quoted by Bristol for delivered gas generators.

Calculated returns on resources employed in the launch fell below the threshold of acceptability. To achieve a price competitive with published bids for sets from other manufacturers in 1975 meant a considerable reduction. Pending acceptance by HMG of the principle of applying differential overheads, the programme was frozen. In due course activity was resumed but this turned out to be only a temporary stay of execution for in 1979, the Olympus 593, at a time of recession and in the face of irresistible price competition, notably from North American products, sank for good.

In spite of these difficulties, successes were still achieved with existing engines. Two SK 30 Olympus generating sets were commissioned at Burntisland on the east coast of Scotland in 1979 and were floated out 500 miles to Conoco's Murchison Field. Four SK 30 M Olympus sets were also used on the Brae A rig. In Saudi Arabia, two 20-megawatt Olympus-powered generating sets were installed to provide the power for the Military Hospital in Riyadh and were commissioned in 1979, just in time for the Queen's visit in an Olympus-powered Concorde.

A market where Rolls-Royce achieved great success was that of gas pumping. In 1964, TransCanada PipeLines built gas pumping installations supplied by Rolls-Royce in conjunction with the US company, Cooper Bessemer (later Cooper Industries, recently acquired by Rolls-Royce). The engine in the generating set was a Rolls-Royce Avon which was coupled to a Cooper Bessemer two-stage free power turbine which drove the pipeline gas compressor. Ten compressor stations were installed, carrying natural gas from Alberta in the west of Canada to the industrial cities in the east. The plant was required to run for virtually the whole year in remote spots and reliability was therefore of paramount importance. The Avon had proved its reliability in the air. Nevertheless, when first installed in Canada, the average running time between overhauls was only 1,500 hours. Some modifications improved this to an impressive average of 21,000 hours by 1971. Such was the proven success of the Avon in the market – not only of gas pumping but also of gas re-injection, chemical processing and oilfield pressurisation – that by 1982, 571 Avons had been sold or were on order in these markets. They had been sold in 33 markets, including 56 engines to the Soviet Union.

The worldwide recession of the early 1980s affected the Industrial and Marine Division just as badly as the rest of Rolls-Royce. Both the power generation and mechanical drive markets declined sharply from the peak of 1980. Over-ambitious planning of new power stations in the 1960s and 1970s had resulted in saturation of the electrical generation sector. At the same time, there was a slow down in the growth of world energy demand which reduced the demand for electrical generation and oil and gas pumping sets. Furthermore, there was a shortage of finance for new projects in developing countries. To cap it all, the naval market declined as well.

Competition, mainly from the USA, in the 10–60 megawatt range was severe in all three market sectors: marine, mechanical drive and power generation. As well as General Electric and Westinghouse in the USA, the main competitors were Brown Boveri in Switzerland and ASRA-Stal. However, as with the Aero Engine Divisions, the Industrial and Marine Divisions were developing plans for growth in the 1990s and we shall have to wait until Part Three of this history to see how they developed.

CHAPTER EIGHT

WINNING BACK CONFIDENCE

LOCKHEED IN TROUBLE AS WELL
K SQUARED
'THE MINERS CANNOT STOP THE COUNTRY'
THE COMMERCIAL SURVIVAL OF THE COMPANY
'SO LONG AS I AM NOT BUGGERED ABOUT'

LOCKHEED IN TROUBLE AS WELL

THE COLLAPSE OF ROLLS-ROYCE was obviously a serious blow to Lockheed, whose own financial state was parlous. On 28 May 1970, *The Wall Street Journal* had written, under the headline 'Ailing Lockheed urged by Pentagon to seek Bank Aid, Merger as well as Federal Help':

WASHINGTON – The Pentagon told congress that it hasn't yet found a solution to Lockheed Aircraft Corp.'s tough financial problems.

But the Deputy Defense Secretary David Packard stressed that it will require private as well as Federal assistance to enable the defense contractor to get back on its feet and continue major military projects.

Testifying before the Senate Armed Services Committee, the Pentagon's No. 2 man declared, 'It isn't my intent today to present to you a resolution for the Lockheed financial predicament as much as I would like to do so.' But, he added 'I recognise the necessity to segregate [Lockheed's] commercial from Government business, and I am convinced that no resolution can be achieved short of bankruptcy and reorganisation under Chapter 10 of the Bankruptcy Act, without financial support from the private sector.'

Therefore, Mr. Packard said he had urged Lockheed's management 'to seek financial assistance from the banks and even through merger in order to preserve their capability'.

228

'Even with the most favourable resolution of the Government contract problems that I could consider at this time,' Mr. Packard said, the company would find itself short of cash unless it also got private help.

When Dan Haughton was given the dreadful news about his engine supplier in early February 1971, he faced a monumental crisis of his own. He knew he needed another $350 million in outside capital from banks and customers to bring the TriStar into full production, whether with the RB 211 or another engine. Lockheed's banks were unwilling to advance more money without some form of Government guarantee and that would require Congressional approval. This approval would not be easy to attain from a Congress that had become wary of Government support for faltering private enterprises. *Fortune* magazine wrote in June 1971:

Nine customers and 24 banks had to be kept in the game, if the game were to continue, and all – each with its own needs, interests and responsibilities – had to be convinced that the final deal represented the best possible outcome for them and their stock holders. And all had to reach that decision at the same time, though each preferred to wait until the others – and the US and British governments as well – had acted . . . Secretary of the Treasury John Connally, the Texas Democrat who monitored the negotiations for the Nixon Administration, put it best. 'Dan,' he told Haughton, 'your trouble is you're chasing one possum at a time up the tree. What you've got to do is get all those possums up the tree at the same time.'

The sums were enormous. At the time of Rolls-Royce's collapse, more than $1.7 billion had been committed to the project – $990 million by Lockheed, $350 million by its subcontractors, and $400 million by Rolls-Royce and its subcontractors. Lockheed's investment came partly from its own resources but mostly from its customers and banks. Some 178 TriStars had been ordered by six airlines, two investment groups and a British holding company at a cost of about $15 million each – a total of $2.67 billion – and these customers had made advance payments of more than $200 million. Twenty-four banks, led by Bankers Trust and Bank of America, had loaned Lockheed $350 million, and the company had put up as collateral a considerable part of the property it owned, as well as the shares of two substantial subsidiaries – Lockheed Shipbuilding and Construction, and Lockheed Electronics.

In the previous March, Haughton had told the Deputy Secretary of Defence, David Packard, that Lockheed faced what he described as a 'critical financial problem' if the Pentagon did not settle promptly the contract disputes on four of the company's defence projects or, at the least, provide interim financing while the disputes dragged on. More than $750 million was at stake in claims involving the C5A cargo plane, the Army's missile and a group of Navy ships. Just before Haughton flew to London in early February 1971, he reluctantly accepted a settlement that brought Lockheed's losses on the disputed defence contracts to $480 million. Fortunately, all but $190 million had already been written off. Nevertheless, it gave the company an after-tax loss of $80 million for 1970, against a previously reported profit of $10 million. It also reduced its net worth from $331 million to $240 million. But it drew a line under the losses and Lockheed had survived. Haughton was able to arrange $250 million in additional financing – $100 million from the airlines buying the aircraft. However, when Rolls-Royce went bankrupt, all these arrangements fell apart.

Haughton was now deep into negotiations with not only Rolls-Royce but also, more especially, the British Government. The initial reaction from the Government was that Rolls-Royce's capability to supply the armed forces of the UK and its allies must be kept intact, but that the RB 211 would have to be abandoned. As we have seen, Frederick Corfield, Minister of Aviation Supply, argued in the House of Commons that the engine was not essential to Britain's remaining in 'the big-engine league'. (It is difficult to see what other engine was going to keep her there.)

In spite of this negative approach, Lord Carrington, the Minister of Defence, commissioned yet another cost and technical study, and Sir Stanley Hooker, Fred Morley and Arthur Rubbra (described as the 'three ferrets' by Carrington) concluded that the Rolls-Royce board had been too pessimistic in assessing the faults and future of the RB 211. As a result of its report, the Government concluded that another £120 million would be enough to complete the development of the engine.

In early March, Carrington offered Haughton a deal whereby the price of the three-engine set would rise from $840,000 to $1.18 million and that the British Government would put up £60 million of the development cost if Lockheed would put up the other £60 million. Haughton was aghast. How could he add Rolls-Royce's problems to his own? And anyway, the British estimates of the money needed were double those of his own experts. He rejected the offer. Negotiations continued, with the British Government, reasonably enough, requesting guarantees. It was becoming clear that US Government intervention was essential.

While all this was going on, Haughton had to keep his airline customers happy and, fortunately for him, they were not troubled by the delays. The Chairman of TWA, Charles C. Tillinghast Jr., knowing his own cash-strapped position, said:

I would say a delay of a year would have as many advantages as disadvantages, maybe more.

For TWA it was more important that the TriStar survive than that it be powered by the RB 211, and during February and March 1971 a re-run of the original competition to supply the engine for the TriStar took place. Once again the RB 211 won. Its technical difficulties seemed to have been overcome. On the Derby test beds, the engine was coming up towards 42,000 lb thrust, and fuel consumption was within 2 per cent of specification. We have seen how Rolls-Royce originally offered Lockheed the RB 211-06 with a thrust of 33,260 lb on a 90-degrees-Fahrenheit day. By early 1968, the promise of 40,600 lb thrust on an 84-degrees-Fahrenheit day had been given though the engineers expected to achieve this only in the second year of service. On 4 February 1971, the very day of the receivership, Ernest Eltis had sent a telex saying 38,500 lb thrust had been achieved. Apart from thrust considerations, neither General Electric nor Pratt & Whitney would match the British financing.

Pratt & Whitney accepted that it had lost, but General Electric mounted a vigorous campaign, saying that Government support for Lockheed should only be forthcoming if the company bought engines made in the United States of America. Again, the US Government rejected this argument. If it was prepared to support Lockheed, it felt that the company should then be free to buy the engines it felt were most suitable.

Of Haughton's customers for the TriStar, Delta Airlines came closest to defecting. Delta was profitable (it made $44.5 million in 1970) and would have little trouble in financing a switch to Douglas's DC-10. It was also concerned that two of its closest competitors, American and United, were expected to start flying the DC-10 in September 1971 (Douglas was six weeks ahead on its test schedule). On 18 March 1971, Delta signed a letter of intent to buy five DC-10s for delivery in 1972 and 1973. It did not cancel its orders for twenty-four TriStars, and Delta Chairman Charles Dolson told Haughton that if he put the TriStar deal back together, Delta would stick with its orders. Nevertheless, it put pressure on the British Government and Haughton to reach a satisfactory conclusion to their negotiations as quickly as possible.

In March a British delegation, under the leadership of Lord Carrington, went to the USA and, after days of intense negotiation, met Dan Haughton at the British Embassy in Washington. At noon on 25 March, they reached an agreement whereby the new company, Rolls-Royce (1971) Ltd., would produce 555 engines. Lockheed agreed to an increase in price of $180,000 an engine.

This meant, with extra airframe costs due to delays, an increase of $643,000 per aircraft – still less than the cost of a change of engine. Nevertheless, Haughton had to win approval from the airlines and to reschedule their advance payments.

Carrington also wanted assurances that 'there will be a requirement for their engines upon delivery' – in other words, that TriStar production would continue. This condition sent Haughton off to the Secretary of the Treasury, John Connally, to arrange Government backing for the additional $250 million of bank loans he would need. On 6 May, Connally had a meeting with President Nixon at the White House and emerged to say that the Administration would send the loan-guarantee legislation to Congress. He said:

The health of our aircraft industry is essential to the nation's commerce, employment, technological development and protection.

Did this mean that the troubles of Rolls-Royce and Lockheed were past? It certainly did not. The engines still had to be improved and manufactured, the airframes had also to be manufactured and the whole package delivered to the airlines on the newly agreed delivery dates.

In comparison with all the trauma of the previous four years, the next few were relatively calm – although the fact was that TriStar and the DC-10 were competing directly in a market segment in which the industry estimated that the total sales would be about 750 aircraft. (As 249 TriStars and 446 DC-10s were sold, that was a remarkably accurate estimate.) Roy Anderson, Chairman of Lockheed from 1977 to 1985, told the author in 1998:

We sold 250 TriStars and there are still 200 flying but we realised we were losing money – we calculated we lost $2.5 billion – and in 1983, just as we were about to sign an order with Indian Airlines, I had to tell them we were halting production.

Anderson also recalled how close the vote in Congress had been and that Dan Haughton had said to the Lockheed lawyer, 'Bring out the bankruptcy file.' He referred to the competition from Douglas, who had, of course, taken advantage of the problems at Rolls-Royce by referring to the TriStar as the 'airplane without an engine'. However, in his view Rolls-Royce had 'busted their ass' and produced 'a fine, long-lived engine'.

Looking at it from Douglas's point of view, Jackson McGowen, who became President of the Douglas division in the late 1970s, told the author in 1998 that the rivalry between Lockheed and McDonnell Douglas benefited neither, only Boeing (which acquired McDonnell Douglas in 1997). He said:

I met Dave Huddie at the Bel Air Country Club and told him that if Lockheed would quit the 1011 we would take over the customers and would fit the RB.211. I then talked to Lockheed and offered $200 million to take over the TriStar. I also went to see Ronald Reagan, then the Governor of California, when Lockheed were trying to secure government backing for their loans, and told him that if the government backed Lockheed both Lockheed and Douglas would lose and Boeing would be the winner. Reagan told President Nixon that he would not support him in backing the loan. I talked to our congressman who then talked to 'Mac' [the boss of McDonnell Douglas] who said he 'couldn't take a position.' I cried.

Back in Derby, Rolls-Royce pressed on to solve the outstanding problems on the engine. Flight-testing took place throughout 1971 and up to April 1972 when the engine was certified, and within days it was in airline service with Eastern Air Lines and TWA. Of its initial operation, Phil Ruffles said later:

The reliability of the engine in the early years was worse than that of the preceding generation of engines with problems in virtually all modules, due largely to the fact that the development programme, of necessity, had concentrated on performance. Reliability was steadily improved as modifications were implemented, a process requiring the close co-operation of airlines and aided by the modular construction of the engine. Arguably the most serious service problem, although quickly solved, was the uncontained failure of two fan disks in late 1972 and early 1973 which fortunately caused no critical damage. The cause was the sensitivity of the new titanium disk alloy to dwell time at maximum load during the normal cyclic loading experienced with each flight, the susceptibility of the material being dependent on its manufacturing history. This was a new phenomenon to the industry. Prior to changing the material, all disks had to be inspected, a difficult process on the enclosed centre engine. In some cases this was repeated every 50 flight cycles, causing a great deal of disruption and additional costs. Nevertheless, the replacement material was certified within 6 months and the retrofit of the whole fleet was completed just over a year after the original incident.

Following the development difficulties, the life achieved by the combustion chamber and HP turbine blades was very short, typically only 800 hours, and urgent action was taken to improve the design. Additional cooling was introduced in the combustor along with two additional film cooling rows near to the blade trailing edge. Although this improved the situation, giving hot section lives of around 2000 hours, the performance was still not satisfactory. The combustor life was further improved by replacing the sheet metal heat shields with cast ones, while still more film cooling, this time at the leading edge, increased the turbine life to over 3000 hours. In addition to cooling problems, blades were suffering severe erosion caused by carbon shed from the burners. This was solved by a new burner design which prevented carbon build-up by air-washing those surfaces prone to it. However, the turbine blade problem was eventually solved when the forged Nimonic blade was replaced by a cast, directionally solidified blade with an advanced multipass cooling configuration. Cooling air is passed up and down within the blade before ultimate discharge through film cooling holes. This blade entered service in 1979 and had demonstrated outstanding integrity, achieving service lives in excess of 20000 hours and later versions are judged by operators as the best turbine blades in airline service today.

Looking back from the year 2000, because the fan disc problem was overcome, it does not seem too serious. But it was very serious at the time. Jimmy Wood remembered that he went to see Dan Haughton in Burbank who said to him:

There are engineers here who think we should be grounding all the aircraft. [A step that would have threatened the future of the engine again.] I've arranged for all our senior and research engineers to come to the boardroom. You tell them what steps you are taking.

Wood told them that there were two manufacturers of the fan discs, one in the USA, the other (IMI) in Britain. The US manufacturer was the more experienced and Wood had arranged with them to step up production so that the discs could be replaced on all aircraft.

Haughton asked Wood to wait outside while they discussed the situation. After five minutes, he was summoned back into the room to be told by Haughton:

We agree with you. We will not recommend a grounding.

We shall follow the progress of the RB 211 and its derivatives later, but now we need to consider how Rolls-Royce as a company recovered from the shock of its bankruptcy.

K SQUARED

As we have seen, after the successful negotiations with Lockheed in early 1971, the British Government formed a new company called Rolls-Royce (1971) Ltd. It also restructured the board, appointing Sir Arnold (now Lord) Weinstock who was at the height of his powers, having put together GEC, AEI and English Electric, and Gordon (later Lord) Richardson, the Chairman of the stockbroker J. Henry Schroder Wagg & Co., and later to become Governor of the Bank of England. Sir William Cook, former scientific adviser to the Ministry of Defence, and Sir St John Elstub, Chairman of Imperial Metal Industries, were also invited to join the board. Cook and Elstub, along with Professor Douglas Holder (who held the chair of Engineering Science at

Oxford), had already advised the Heath Government on the technical and financial viability of the RB 211 following the bankruptcy. According to Alex Brummer and Roger Cowe in their biography *Weinstock*, he was:

dragooned into joining the nationalised company's board by his old financier, Sir Kenneth Keith, the man who had won him AEI. He stayed for slightly longer than his two-year term . . . But he did not enjoy the experience and this remains the only outside directorship Weinstock has ever accepted.

Stanley Hooker was also appointed to the board and he recalled later, in his book *Not much of an Engineer*:

Thus I found myself, four years after my retirement, occupying the very chair promised to me by Hs 25 years previously! But elation was the last thing I felt. Under me the great team of engineers, and indeed the whole vast Derby works, was completely demoralised. Many were looking for someone else to blame. I called a meeting of the entire engineering staff and explained the exact situation. I then appealed to their loyalty to the good name of Rolls-Royce to get the RB.211 quickly certificated and delivered to Lockheed. I promised 100 per cent support to their efforts and asked any doubters to leave at once . . .

Rather remarkably, within days of first being called up to Derby a few weeks before, I had been able to put my finger on several crucial faults and to have them rectified very quickly indeed. But one cannot truly know a piece of machinery as complex as the RB.211 without living with it from the start. So I did not attempt to run the show but left the day-to-day programme to Ernest Eltis, whom I had displaced as Technical Director, but who gave me the most loyal support, and to his assistant (and an old colleague of mine from Barnoldswick) Johnnie Bush. They proved an inspiration to their teams, and worked closely with Chief Designer Freddie Morley and his assistant John Coplin.

I bore the ultimate responsibility, however, and soon decided to invite my old tutors, and perhaps the greatest of Rolls-Royce engineers, to join me to form a kind of Chief of Staff committee. I asked Cyril Lovesey and Arthur Rubbra, both well over 70, to come back into the thick of it. They responded with alacrity, and I cannot describe the comfort in seeing Rubbra poring over the drawings, and to discuss the forthcoming programme with Lovesey. Their vast experience was immediately put to use, and I pay tribute to the way they gave this to the new company, as to the old – and, in Lovesey's case, he gave all his remaining years.

Hooker recalled that Geoffrey Wilde took over design and development of the troublesome turbine blades, Harry Pearson returned from retirement to take over performance analysis and Peter Colston, 'within the hour of each of our decisions, prepared formal engineering requests and delivered them to two miracle-workers, Eric Scarfe (Production) and Trevor Salt (Experimental Manufacture).'

While these engineers were ironing out the wrinkles on the RB 211-22, it was becoming clear that greater thrust was soon going to be necessary if the engine was to compete successfully with Pratt & Whitney's JT-9D and General Electric's CF6-50. A second-generation RB 211 was going to be necessary and Hooker set 50,000 lb thrust as the target, concentrating on the fan which provides about three-quarters of the thrust. Here is Hooker again:

One of the basic design objectives was to make the second-generation RB.211 fit the TriStar, including the centre (tail) installation, and if possible we wanted the new engine to be installationally interchangeable. Roy Hetherington, the fan and compressor expert, succeeded in designing a new fan passing 20 per cent greater airflow within the same diameter. We also redesigned the IP [intermediate pressure] compressor to pass more air through the core, but here we ran up against a seemingly insurmountable snag. There was a massive steel aircraft mounting ring in the way, and this was regarded as sacrosanct.

I sent for Freddie Morley, who was up to his neck in design mods for the current engine, and, pulling his leg, said, 'Freddie, I hear that you will not allow us to change the mounting ring, so that we can redesign the IP compressor and raise the thrust to 50,000 lb.' I knew perfectly well that he had never heard of the proposal, and he almost burst a blood-vessel. His response was 'Nobody has told me a f------ thing about the f------ ring!!' [Fred Morley only ever used one adjective at work; he was known as 'four-letter Morley' or as 'Lord F'ingham'.] He departed in high dudgeon, to reappear a few days later with a splendid drawing of the redesigned mounting, fully interchangeable with the existing one but giving us the room we needed.

Thus we completed the design of the superb RB.211-524 series. It was to be another two years before, under Sir Kenneth Keith [as Chairman], we were at last permitted to go ahead with full development.

While these developments were taking place, Stewart Miller – who, as we shall see, played a very significant role in further development of the RB 211 and in the development of a Rolls-Royce collaboration engine of the 1980s, the V-2500 – wrote a paper, *The Value of Advanced Development*, which addressed the problem of designing and developing new engines, while at the same time avoiding the financial catastrophe that had just overtaken the company. His opening sentences stated the position clearly:

The recent history of aero engine development in the U.K., particularly that of the RB 211, tells us that there is something wrong with our approach to the development of advanced engines. As technology levels and the size of individual programmes have increased, the technical and financial risks involved in launching a major project have become very significant. The actual development programmes have been significantly longer and more expensive than planned production and there have been substantial losses in the early phases of production. It is doubtful if another major project will be launched unless some means can be found to lessen the risk.

Miller's idea for the future was that Rolls-Royce should build step-by-step on its experience and avoid being pressurised by customers to make promises which would possibly be very difficult to fulfil. As he said:

Although a broad programme of research in advanced technology has been pursued for many years, it has become evident that the step between the basic research and the commitment of a full development programme at a high technology level and against a tight timescale, both demanded by the customer, is too great.

He wanted to bridge the gap between promise and fulfilment in a 'methodical manner' and this would mean creating a category of work – to be known as 'Advanced Development' – within the total research and development effort. This would mean a full-scale engine, core and/or rig work to demonstrate the feasibility of a particular technology and to obtain advanced design information **before** a commitment to a full development programme was made. To Miller, the emphasis should be on testing in the engine environment.

He felt it was most important, if costs and timescales were to be predicted accurately, that recent development experience should be studied:

Any full development programme has as its objective the achievement of an engine specification within defined timescale and costs and the success or otherwise of the programme can therefore be measured by comparison of the actual time and cost taken to meet the specification against the estimates made before starting.

He noted that in the USA, advanced development had been approached in a more structured way. The proportion of funds allocated to activities prior to main development was 35 per cent and the proportion prior to commitment to production 58 per cent. Miller felt these were 'truly remarkable figures', especially when compared with the 10 per cent for work prior to full development for the UK.

Miller concluded by saying that a *systematic* programme of Advanced Development offered the best means of ensuring success in future development programmes on advanced engines.

Many said that, until the arrival of Keith, Rolls-Royce felt leaderless. They looked to Dan Haughton of Lockheed to provide the leadership and vision for both companies.

Sir Kenneth (now Lord) Keith was invited by the Conservative Government under Edward Heath to take over the chairmanship of Rolls-Royce, and did so on 1 October 1972. He was to say later that his predecessor, Lord Cole, wanted Ian Morrow to succeed him, but neither the Government nor the Rolls-Royce board was in favour. John Davies – Secretary of State for Trade and Industry, and a former managing director of Shell Mex and BP, as well as Director General of the CBI – knew Keith from their work together on the National Economic Development Council (NEDC or 'Neddy'), and told Heath that Keith was the right man. Keith was persuaded to take the job. He was to argue later that he took it reluctantly.

Keith, with Rolls-Royce's confidence at rock-bottom, was a shrewd choice. Undoubtedly an extrovert, his personality and experience were admirably suited to raise morale within the company. He was to become known as 'K squared'. During World War II, he had served in the Welsh Guards, reaching the rank of Lieutenant Colonel. After distinguished service in North Africa, Italy, France and Germany, where he was mentioned in despatches, he was awarded the Croix de Guerre with Silver Star. After the war, he moved into merchant banking and also served on the boards of a number of large public companies – including Legal Star, Beecham, and Times newspapers. By 1972 he was Chairman of the merchant bank Hill Samuel & Co. and also Deputy Chairman of British European Airways. His contacts throughout business and Government were legion and, as we shall see, he was to use them to good effect.

Keith was not going to find things easy. Apart from the Lockheed TriStar, the RB 211 was struggling to find business elsewhere. A personality clash between Lombard and Sutter had meant it was not initially bought by Boeing for the 747.

When Rolls-Royce initially approached Boeing in the 1960s to offer its new engine, the RB 211, for its great new aeroplane, the Boeing 747 – 'Jumbo jet' to the public; 'Widebody' to those in the aircraft industry in the USA – it was disappointed to find that Joe Sutter was the chief design engineer who had gained control of the new aircraft on which Boeing had bet its company. Adrian Lombard had already come across Sutter when he had visited Boeing to try to sell the Conway for the 707. He had upset Sutter by going above him to Boeing's President, Ed Wells, when Sutter refused to give Lombard Boeing's technical manuals on the nacelles that housed the engines on the wings.

To Rolls-Royce, the nacelle was a very important part of the engine because it determined how easily it could be opened up for maintenance and how the engine behaved in flight. The nacelle produced drag, so its shape was very important. Ideally, Rolls-Royce liked to supply the nacelle itself, having made sure it fitted well with its engine. On the Conway, Lombard insisted that Rolls-Royce provide the nacelle, but then realised that Boeing knew more about hanging a nacelle on the pylon of a wing than Rolls-Royce. He asked a Boeing engineer to give him Boeing's book on nacelles. When the engineer consulted Sutter, Sutter responded:

Hell with that! That cost Boeing a lot of money. We're not handing that over.

Lombard complained to Wells who was only too aware of Rolls-Royce's connections with BOAC, which tended to request Rolls-Royce engines on its aircraft. Eventually, Boeing sold the information to Rolls-Royce, but Lombard and Sutter harboured resentment over the incident. As Clive Irving put it, in his book, *Wide-Body, The Triumph of the 747*:

Lombard never forgot it was Sutter who had thwarted him. Sutter, in turn, was always leery of Lombard and his works, although he was as respectful as anyone of the British company's pedigree. Who could not be? There was always a touch of class about anything that Rolls-Royce turned out; the aura of that resplendently gleaming radiator on the limousine influenced feelings about the company as a whole, even though you didn't build jet engines the way you did cars. 'Hand-built' was one hell of a sales pitch, and it was nice to think that the aero engines would receive the same lapidary care. But Sutter wasn't a sucker for auras. In his book, when it came to jets, the guys at Pratt & Whitney and General Electric were as classy as the Brits.

As we know, if Boeing had bet the company on the 747, Rolls-Royce had bet the company on the RB 211 and the Rolls-Royce sales people had been talking to the power plant engineers in Sutter's team from the day that details of the engine became available. As soon as Sutter became involved, his former resentment seemed to tip him towards scepticism. He said he could not see the point of three shafts instead of two, and as for the revolutionary new Hyfil blades, he apparently dismissed them as 'oatmeal'.

In spite of this opposition, Lombard persisted, inviting Boeing engineers to Derby to show them how three shafts produced a smoother, better-balanced engine. However, in a re-run of the 707 nacelles, Lombard would not hand over details on the Hyfil blades. The reaction of Sutter was predictable:

Hell! We're not gonna put an engine in an airplane when we don't know what's in it.

Again, Lombard complained to Wells and Wells agreed to visit Derby with a team working on the 747, but not including Sutter. On that basis, Lombard agreed to give him a private briefing on Hyfil. At the same time, Wells knew that Sutter must be convinced and persuaded him to meet Lombard in New York. They met for breakfast, and when the author interviewed him over thirty years later, Sutter still remembered the breakfast in every detail.

We met at 7.30 am in Lombard's suite in the Ritz Towers on Park Avenue and a breakfast was laid out sufficient for twenty people even though there was just the two of us. We went through the details of the engine and finally came to the Hyfil blades. I asked Lombard how they would stand up to bird impact. He told me forcibly that Hyfil could stand any type of bird impact and still keep spinning. He went into his bedroom and came out brandishing a Hyfil blade like a Samurai sword. The next thing, with a cry of 'Watch this' he brought the blade down on the coffee table and smashed the table to pieces.

Sutter said later that he was not impressed by this dramatic show of Hyfil's strength and, of course, as we know, Rolls-Royce was forced to abandon it. And, for the moment, indeed for a number of years, it was forced to abandon its attempts to convince Boeing to put the RB 211 on the 747. Clive Irving summed it up:

For those who had noticed – and Lombard certainly had – there was also an instructive lesson in the RB.211 episode about how the Boeing Airplane Company worked. It was no good trying to seduce either the senior officers of Boeing nor the foot soldiers if the chief engineer opposed you. Sutter was not even a vice president. [He told the author that, although he was in charge of 4,500 engineers, he was only paid $29,000 (£12,000) a year at that time.] But once he made up his mind that the Rolls-Royce engine had serious defects, he was implacable.

In spite of his contacts, Keith knew from experience that dealing with Government was not as straightforward as dealing with businessmen, and as the Government was his master at Rolls-Royce, he was determined at an early stage to establish his remit exactly and to make sure that he would have full Government support in carrying it out.

Keith saw seven key factors for the successful recovery of the company. It was essential both for national reasons and the commercial viability of the business that Rolls-Royce maintained its position as a competent designer and maker of *military* engines. It was also essential to ensure a continuing market for the big *civil* engine, the RB 211, not only in the Lockheed TriStar but also in other aircraft. Additionally, Keith and his board knew that if Rolls-Royce was to stay in the big engine market, a new engine in the 30,000 lb thrust range would have to be developed. However, Rolls-Royce should continue to develop and exploit its existing engines wherever they proved a viable alternative to expensive high-technology replacements.

At the same time, the company should consolidate and expand its share of the small engine market, especially in the helicopter and light aircraft markets, and should give more attention to the adaptation of its engines to the industrial and marine markets and exploit Rolls-Royce's manufacturing and technological expertise in any area where it was relevant.

The present situation of this engine is that Rolls-Royce's chances of selling many more RB 211 engines are poor, unless Lockheed sell more TriStar L1011-1 aircraft. This they are not likely to do unless they launch and promote the -2 version of the TriStar.

Dan Haughton, as you know, is anxious to launch the -2 version, but cannot do this without airline launching customers. He requires a minimum of two such customers and, to ensure confidence, BA ought to be the first. Given a BA commitment, he is confident that Air Canada will provide the second essential order, and I should certainly be prepared on that basis to join with David Nicolson and Dan Haughton in a visit to Canada with that objective.

Lockheed and Rolls-Royce are completely confident that the -2 TriStar has considerable sales prospects once the aircraft can be launched and promoted.

Keith did not find it easy to find the right management structure. One of his early actions was to remove Geoff Fawn from his post as Managing Director of the Derby Engine Division. This was a controversial decision at the time, and remains so to this day, with some saying that Keith had no alternative and others saying that Fawn had already done a good job when moved from Crewe to Derby just before the receivership and was continuing to do so. Not even Geoff Fawn's best friends would try to argue that he was the easiest colleague to work with and there is little doubt that he found those without engineering experience not to his liking. For his part, Keith knew he had been put into the position of Chairman to provide strong leadership and that meant not tolerating what he saw as insubordination. Air Chief Marshal Sir Peter Fletcher, Controller of Aircraft at the Ministry of Defence, wrote to Keith on 5 April 1973:

You will appreciate that I have some of my people resident at your factories and that my head quarters staff are in constant contact with yours at all levels in the technical, contractual and financial fields. In addition, in over-simplified terms, I 'manage' for the Government virtually every civil and military engine under development or in production at RR(71). There is also the fact that I have to control, and in most cases carry out, the detailed negotiations in collaborative projects. This detailed negotiating responsibility has emerged during the last two or three years from the way in which the collaborative management structures have evolved; in earlier times people at my level have been concerned mainly with policy. So, for what it is worth, my staff and I can take a view of your Company and what it does from a variety of angles that cannot be matched anywhere else.

The first problem is the replacement for Fawn. I am afraid I must start by saying that there is nobody in the Company who could run Derby nearly as well as Fawn: my staff, the head engineers at Strike Command and I are unanimous on this. We have always recognised that he is an obstinate 'rough diamond'. But it was he who brought about the great improvement in Service support at Derby and it was he who did more than anyone else to ensure that the RB.211 has just kept abreast of its problems, the fan disc problem in particular. Stan Hooker might have reservations on the value of Fawn's role in the 211 story, but then he and Fawn have never really 'clicked'. Having said all this I must agree that from what I know of the circumstances and his behaviour I do not see how you had any alternative but to remove him. But his departure has left a morale problem [some would say that it improved morale on the basis that Fawn was unnecessarily confrontational and belligerent] and a hole that cannot be adequately filled.

It is easy to have sympathy with both Fawn and Keith. Fawn, the dedicated engineer, would, along with plenty of other engineers in Rolls-Royce, have been fed up by early 1973 with what he would have seen as constant and largely unnecessary interference by people who, in his opinion, had no knowledge of aero engines or how to develop, manufacture, sell and service them. (I am sure he would have expressed it more strongly that that.) Lord Cole, Keith's predecessor, had made a career with the detergent and shampoo group Unilever (he had probably been chosen by the Government because he was *not* an engineer), and had spent most of his time in West Africa. (Jimmy Wood, effectively Fawn's deputy, remarked that his only claim to fame was to have been involved in the Labour Government's failed groundnut scheme in West Africa in the late 1940s. Wood also said that Keith had told him – when he relieved Fawn of his post – that Wood was lucky he was not being fired as well.)

From Keith's point of view we have to understand that he saw it as essential that he establish his authority in a company that to outsiders had run itself badly and needed to face the facts of commercial life. According to Don Pepper, Commercial Director at the time, Keith could not even get Fawn to agree on a date when they could meet at Derby, and when Fawn flew out to the Far East when Keith had expressly instructed him to remain in Derby to supervise the ironing out of problems on the RB 211, it was the last straw. Apparently, Weinstock described Rolls-Royce as 'ungovernable'. Keith knew he had to change that.

To find the right structure, and also to find the people, was going to be difficult. Keith knew that he would constantly have to deal with Government and he therefore appointed Sir William Nield, a former Permanent Secretary in the Cabinet Office, as his Deputy Chairman. This is what Sir Peter Fletcher had to say about that appointment:

The Deputy Chairman could help on the interface with your 'owners' (DTI), deal with external relations on the civil side and run a central secretariat. I believe Bill Nield can fill this slot, but I must advise you that many people I have spoken to in aerospace circles at home and abroad interpret the appointment of a non-aerospace official to this position as simply reflecting the Government's wish to have an ex-government man in a government-owned company. [This was slightly unfair to Sir William Nield who had served in the RAF and Royal Canadian Air Force throughout the war. He had also distinguished himself at the Ministry of Food after the war. Perhaps not so important except that it helped him relate to people – he rolled his own fags and could drink most men under the table.]

Fletcher was very anxious that Rolls-Royce appoint another deputy chairman and one with engineering, and preferably aero engine experience on the military side. He was in no doubt as to Rolls-Royce's pivotal position, writing to Keith:

I strongly recommend that you inject somewhere into the top management someone clearly identifiable as relevant to your military business. At this point I must stress that I am not putting emphasis on the military side just because I am a serving officer or because I know the military side best. As it happens the nature of my work is such that I have to know as much about the civil programmes as I do about the military programmes (it is easier to read across from military to civil than vice versa – after all the RAF's transport fleet is numerically larger than BOAC and BEA taken together) and I need no educating into recognition of the great importance of the civil business. It would be very damaging to both the Company's interest and the national interest if there were a withering civil programme, not to mention damage to your defence interest because the more successful the Company is overall the greater is the insurance that our needs will be adequately met and at an acceptable cost. But in the final analysis there are some harsh realities to be faced and they again reflect a difference between the civil and military sides of your Company's work. If you ran into major difficulties on the civil side, for example if the fan disc problem got worse instead of better, it would be financially serious for you, the Government, Lockheeds and some airlines; but if the worst came to the worst there are other aircraft and other engines that could do the same airline tasks. That is not the position on the military side. The MRCA [Tornado] will provide the RAF with its strike, reconnaissance and air defence capabilities until the 1990s or beyond. If things went wrong there is no aircraft that could be purchased to replace it. There is no engine that could replace the RB199. Although the Governments and the air forces of the three countries are fully behind the programme, there is a fragility about it because of the rather unpredictable situation in Germany and, to lesser extent, in Italy. There would not need to be a catastrophe on the engine; if a major development problem, and there will be some, were badly handled at a difficult time the MRCA programme could be in jeopardy. In fact the whole of the RAF's front-line strength is dependent in absolute terms on the development, production, overhauls and precision of spares within RR(71). If things went wrong there would be no alternative options: there would be a reduction in our defence capability, and that of our allies, that could only be restored by RR(71). For this reason, and it is really the central issue, the Company's top management should reflect at least as much knowledge and experience of all that is involved on the military side as it has on the civil side, and this at present it does not do.

Fletcher had agreed that Keith had been forced into sacking Fawn. However, finding a good replacement would not be easy.

I am afraid I must start by saying that there is nobody in the Company who could run Derby nearly as well . . ., my staff, the head engineers at Strike Command and I are unanimous on this.

On the engine front, Keith's inheritance was not without hope, but there were many difficulties to be addressed.

On the civil side, the main hope for the future was the RB 211 and there were still problems with the fan disc. The Olympus powering the Concorde would have a rosy future if the Concorde's future was rosy. By 1973, few thought it was. There seemed to be an opening for an engine in the 30,000 lb thrust area although General Electric, possibly with SNECMA and Pratt & Whitney, was already in the field. A collaboration would seem to be the ideal way forward towards a quieter replacement for the Spey, Conway, JT-3 and JT-8.

On the military side, Rolls-Royce's major development programme was the RB 199 for the Tornado. This was to be supplied not only to the Royal Air Force but also to the German and Italian air forces – and on the basis of the number of aircraft planned, it looked as though 2,200 engines would be required. Export potential, both for the existing engine and its derivatives, would seem to be considerable.

Another military engine was the Adour where the Jaguar programme would call for about 1,400 engines. Again, export interest was high. The Pegasus was being bought for both the UK and US Hawker Harriers, and an advanced Harrier would require an upgraded Pegasus.

On the small engine front, about 1,100 BS360 engines would be required for the Lynx helicopter, while about 600 Astazou II engines would be needed for the Gazelle and Jetstream.

Regarding relationships with collaborators and customers, Keith was clearly going to need all his legendary charm. In the letter referred to earlier, Sir Peter Fletcher was scathing about Rolls-Royce's recent performances, saying:

The 67th Meeting of the BOD with the firms in December 1972 was devoted to a discussion of the RB199 and it was hoped that Turbo Union would take it as an opportunity to pour oil on troubled waters; but far from it. In the event the RR(71) team spoke in a regrettably arrogant manner, did not address themselves to the main problems, said their time could have been better spent at Bristol and clearly exacerbated the situation; in fact making the Germans very angry and destroying much confidence in the Company and the engine. This meeting is fully documented at NAMMA and in reports to CAS and to me.

Following the meeting the Germans and Italians demanded a live demonstration of the engine. The demonstration took place at Bristol on 21st February 1973 and was fortunately highly successful. It has served, at least for the time being, to satisfy the Germans that the engine is technically sound. But all this was an unfortunate and unnecessary situation and the Germans are more sensitive than they would otherwise have been on the engine situation in general and will be even more difficult to satisfy when there are further problems.

Fletcher also said:

Leaving aside the RB.211, about which the French had developed an almost pathological hostility, customers and industry in France had a number of legitimate complaints that were imperiously brushed aside and the French were invited to realise – as one official put it to me – that it was a privilege to do business with Rolls-Royce. When the Cole–Morrow regime came in the French were not very impressed because the new management did not appear to have much connection with aerospace. For the first year I repeatedly lectured them using the theme that they must forget the past and that all would be well under the new management, not a task that I relished because I did not really believe what I was saying. Now there is another new management. They have been very quiet about this: they are waiting to see what happens.

He concluded:

The lesson from all this, I suggest, is that RR(71)'s tactics must be carefully tailored to the differing situations in the various customer countries and its strategy, in addition to being based on good performance, must be to 'win friends and influence people' since there is much rebuilding of confidence to be done.

As it happened, Keith was an ideal man to 'win friends and influence people'. Whereas Sir Denning Pearson, although having spent a great deal of time in London, based himself in Derby, Keith remained firmly rooted in London in an office in St James's Square, close to Whitehall and not far from the City. He used his contacts for all he was worth, whether they were prime

245

ministers, presidents, dictators, kings or Arab princes. He was soon writing to the Prime Minister Edward Heath, saying in a letter dated 2 February 1973:

I hope you will not mind my writing to you on a subject which is causing me considerable concern.

I and others are worried that emotive statements might be made by people in very responsible positions, criticising certain United States and other airlines for their decision not to buy Concorde, and attributing to the Americans somewhat unworthy motives.

The chairmen of Pan American and T.W.A. happen to be friends of mine, and, as you are aware, I was a member of the British Airways Board during discussions with your ministers about the purchase of Concorde by B.O.A.C.

From what I know, I am convinced that the American airlines made their decision on purely economic grounds. The goodwill of these airlines – and, of course, other overseas operators who have decided not to buy Concorde – remains of the utmost importance to Rolls-Royce. Our commercial shirt is on the RB.211 and possible derivatives of that engine. It would be a great pity if things were said by Ministers, Members of Parliament or others in positions of great influence which might tend to antagonise unnecessarily our good friends and potential customers – particularly in the United States.

The predicament in which we find ourselves is not going to be improved by blaming potential customers who have not placed orders.

Heath replied that he had just returned from the USA and said he had 'been careful to avoid criticism of the American airlines' and that his 'Ministerial colleagues had followed the same line'.

Later in the year, Heath used Keith to extend an invitation to President Ceausescu of Romania (Keith and Ceausescu had hunted wild boar together), and Robert (later Lord) Armstrong, Permanent Secretary to the Cabinet, wrote to Keith on 17 October 1973, saying:

The Prime Minister suggests that you might tell President Ceausescu how much he is looking forward to seeing the President when he comes to this country. Mr Heath is sure that the President will receive a warm and hospitable welcome from the British people.

Nor did Keith hold back from telling those in power what he thought. He wrote to Edward Heath on 20 November 1973 – just as Heath's world was crashing around him under the strain of the oil price hikes, the collapse of his incomes policy and the threat of a miners' strike – saying:

Thank you very much for including my wife and myself in your luncheon at

Chequer's on Saturday last. Quite apart from the obvious pleasure in being present, I was able to have what I think was a helpful talk with President Pompidou and M. de Beaumarchais.

On an entirely different subject, I and others are very concerned at the poor image of the City. This is partially due to the activities of a limited number of people – not, in most cases, members of the real City – who bring it into disrepute, also to the City's complete inability to communicate – an area in which it steadfastly refuses to help itself. The City's poor image does it no good and is, of course, unhelpful to you and to the Conservative party.

There are many amusing anecdotes of Keith's one-liners, some of them perhaps apocryphal. One of the most famous is his referring to the National Enterprise Board, effectively his bosses, as 'a financial contraceptive'. Another, which was first attributed to Brian Johnston, the well-known cricket commentator, was his reply to the Australian customs official who asked him if he had a criminal record. He said:

I didn't realise it was still a requirement.

Yet another was his query when it was suggested that he should go and see the Shah of Iran to try and cement a possible order:

Has he met me yet?

On one trip back from the Middle East in the company's Gulfstream, the aircraft was diverted from East Midlands to Birmingham airport because of bad weather. The customs official took one look at the presents on board, demanded the keys from the pilot, Cliff Rogers, and impounded the aircraft.

In April 1977, Keith took his fellow director Arnold Weinstock to lobby the Prime Minister, James Callaghan. They were greeted by Callaghan, who said:

Now what do you two scoundrels want?

When the author interviewed T. Wilson, the legendary boss of Boeing during the 1970s, he said of Keith:

He charmed the underdrawers off all of us.

There had never been any doubt about his physical courage which he demonstrated when a railway porter dropped his cased pair of Purdeys onto the railway line. Keith jumped onto the line in front of the approaching train, put the guns back on the platform and climbed back himself. We don't know what he said to the porter.

But, for all the amusing anecdotes, there was no doubting Keith's ability to go right to the top when necessary. He wrote to the Prime Minister, Harold Wilson, in May 1975:

I am writing to ask if, during your current visit to Washington, you would touch upon an issue with President Ford which is of vital importance to the long term future of Rolls-Royce.

He went on to outline the company's plans to collaborate with Pratt & Whitney and state that there could be Justice Department problems under American anti-trust regulations. He continued:

This is an area in which the Executive has the final say. I would, therefore, be exceedingly grateful if you could mention to the President [Gerald Ford] how interested you are in the project, and how very much you hope that the US Administration will feel able to give it a fair wind.

And he was still pressing the case for Rolls-Royce when Margaret Thatcher became Prime Minister in May 1979, writing to her in June of that year:

I know you have many issues to discuss with the Japanese Prime Minister in the course of the forthcoming summit. I hope, however, that you will be able to mention to Mr Ohira or to Mr Esaki that you are aware of our ongoing discussions with the three leading aerospace manufacturers and the Japanese Government [discussions that would eventually bear fruit] – details of which are in your brief.

Although we wish to make haste slowly, it is very important to ensure that it is with us and not with the Americans that the Japanese build their next engine. An expression of your Government's interest in a successful outcome of our discussion would be most helpful.

The 1970s was one of the most difficult decades of the twentieth century in which to try to run a manufacturing business in Britain. Many words have been written on the shortcomings of both management and the workforce in British industry in the period after the war. Let us just say here that the change of Government in 1970 – when the Conservative Party led by Edward Heath replaced Harold Wilson's Labour Party – was supposed to bring an end to the collectivist approach that had been pursued, unsuccessfully in the view of most people, in the second half of the 1960s.

Almost as soon as the new Government was elected in June 1970, the dockers went on strike and, to the surprise of many, the Minister of Labour, Robert (now Lord) Carr, refused to intervene, leaving the dockers and dock employers to fight it out. The logic was that the interventions during both the Conservative administrations of the 1950s and those of Labour in the 1960s, had invariably led to a surrender to trades-union muscle and an inflationary settlement.

This non-interventionist approach had been enshrined in a document produced at a Conservative Conference at the Selsdon Park Hotel in January 1970 and initially, once in power, the Conservative Party had stuck to its guns. Such collectivist structures as the Price and Incomes Board and the Land Commission were abolished, and cuts in public expenditure of £330 million were made for the year 1971/2. There were cuts in subsidies to council house rents as well as in expenditure on school milk. However, by the middle of 1971 the Government's nerve cracked. As we know, the policy of non-intervention in industry was reversed in early 1971 when the Government could not allow Rolls-Royce to go out of business. No sooner had the Government taken Rolls-Royce into State ownership than it faced another industrial crisis, the collapse of Upper Clyde Shipbuilders. Two Communist shop-stewards, adept at publicity, organised a work-in and the Chief Constable of Glasgow warned of possible violence if the yards were closed. Again, the Government swallowed its principles and a grant of £35 million was given to the company.

The industrial chaos continued through 1971 and into 1972. Inflation and unemployment rose at the same time, giving birth to a new phenomenon called 'stagflation'. By the time of the budget in April 1971, manual workers' wages were rising at a rate of 15 per cent, and the more powerful unions in the docks, mines and motor industries seemed to be able and willing to dictate large increases in pay and benefits through industrial action.

An Industrial Relations Bill was enacted in 1971, designed to curb unofficial strikes. It was not dissimilar to the attempt by the previous Labour administration to bring order out of anarchy in Barbara Castle's White Paper 'In Place of Strife', which had foundered on the rock of trades-union power and intransigence. The 1971 Bill set up an Industrial Relations Commission and a National Industrial Relations Court that possessed the power to order a pre-strike ballot and also to enforce a cooling-off period of sixty days in industries critical to the economy or the community. Intended to be a strong weapon in the war against inflationary wage demands, in practice it proved almost wholly ineffective, mainly because of the difficulty of getting trades unions to register under the Act.

'THE MINERS CANNOT STOP THE COUNTRY'

The most severe defeat experienced by the Government came in early 1972 when it caved in to the miners. Furthermore, the flying pickets organised by a new-style miners' leader from Yorkshire, Arthur Scargill, made both the Government and the police look clumsy. On 9 January 1972, a national coal strike began, the first since 1926, after the union had rejected the employers' offer of an 8 per cent wage increase. After successful picketing of many key coal depots and power stations, the Government appointed a Committee of Inquiry under a High Court judge, Lord Wilberforce. Already known for having given the power workers a 20 per cent increase the year before, Wilberforce soon recommended a very large increase for the miners. However, the National Union of Mineworkers (NUM) rejected this offer by twenty-three votes to two. After further negotiations and concessions, the miners won a total earnings increase of between 17 and 24 per cent, and the strike was eventually called off on 21 February. *The Economist*, which had encouraged the Coal Board and Government to take a tough line, saying, 'The miners cannot stop the country in its tracks as they once could have done', was seen as too optimistic. The miners seemed to have come pretty close to halting the country.

And the respite of boom conditions that the dramatic easing of credit produced by 1972 lasted only until the autumn of 1973.

As we have already seen in Chapter Six, the rate of inflation had started to accelerate in the 1960s as successive Governments, Conservative and Labour, pursued a policy of maintaining full employment at all costs. But the results were just a foretaste of what was to come in the 1970s. Anthony (now Lord) Barber, the Chancellor of the Exchequer in Heath's Government (tragically Heath's first choice as Chancellor, Iain Macleod, had died shortly after the election), pumped money into the economy as never before in 1971 and 1972, following the news that unemployment had broken through the one-million barrier. (What would a Government in 1998 do to have unemployment at one million, but in 1972, after thirty years of full employment, it was considered an appalling number.) Unfortunately, Britain's expansion coincided with a world boom largely on the back of an inflationary boom in the United States – caused by the Johnson and Nixon administrations paying for the increasingly expensive involvement in the Vietnam War not by raising taxes but by printing money.

In 1971, the US turned from being an exporter of oil into a net importer; and in October 1973, when yet another Arab–Israeli conflict (the Yom Kippur War) broke out, the Arab oil producers chose the moment to quadruple the price of oil and to cut back production. The effect on world trade was little short of disastrous. Both industry and consumers had become extremely profligate in their use of oil, which had been getting cheaper and cheaper in real terms over the previous decade. Now it was suddenly expensive and in short supply. In truth, it was not in short supply but some panicked, others speculated and the price rocketed. It was as though the whole world had received a massive increase in its taxes and the effect was the same. Tens of billions of pounds, dollars, Deutschmarks, francs and lire were taken out of the world economy and put in the Arabian desert. Until they could be recycled into the system, the world was going to suffer.

And suffer it did, with Britain, because of its continuing structural weaknesses, suffering more than most. The Barber Boom had not worked. British manufactures had not invested as much as had been hoped and the two main results of the expansion in credit had been an explosion in property prices – largely on the back of borrowed money, first from secondary banks and then in their wake from the leading high-street banks – and a rapid increase in inflation. These fuelled resentment from the wage and salary earners, who benefited little from the rising property and stock-market prices but felt the impact of rising prices in the shops.

In an attempt to choke off inflation, Heath's Government tackled the symptoms – rising prices, dividends and earnings – without tackling the cause – too much money. In the autumn of 1973, just before the oil crisis, the Government introduced stage one of an incomes policy that would allow index-linked rises if inflation rose above 7 per cent. Because of the oil price hike, inflation rose quickly above that level and the automatic pay rises gave a ratchet effect pushing it higher and higher. In the meantime, the National Union of Mineworkers had learned the lesson of craven collapses when faced by union pressure by both Governments since the War, and especially of the Heath Government when faced with its own demands in the early part of 1972.

The NUM submitted a large pay claim and imposed an overtime ban on 8 November. Heath panicked and declared a state of emergency on 13 November. A month later he declared a three-day week to preserve fuel, and though negotiations with both Arabs and the miners continued, the only results were the continuation of high oil prices and the actuality of a miners' strike rather than just the threat. In the end, and for some people three weeks too late, Heath called an election with the implied platform of 'Who governs the country, the Government or the unions?' The electorate decided it was not sure who should govern the country – the Tories who at long last had shown some signs of standing up to the unions, or the Labour Party who might at least get the miners back to work. It gave neither party an overall majority but more seats to Labour. Heath tried to negotiate with the Liberal leader, Jeremy Thorpe, to form a coalition but Thorpe would not play and Harold Wilson formed his third administration.

The Labour Government gave in to the NUM and when the rest of the unions followed in the headlong rush, inflation soared to 25 per cent (anyone who was paid monthly in arrears was losing out quite sharply). While inflation was soaring and everyone tried to keep pace, financial institutions and to a lesser extent manufacturing businesses were suffering severely from the financial squeeze that had finally been imposed in the autumn of 1973. Nearly all the secondary banks went into liquidation or, if it was deemed necessary to maintain confidence, were rescued by the main banks prompted by the Bank of England. Even the National Westminster Bank was forced to make a formal denial that it was in trouble. If the new Chancellor, Denis Healey, had not introduced a corporate tax-saving measure – stock relief – many manufacturing companies would have failed. On the Stock Exchange, prices fell throughout 1974 and the *Financial Times* 30-share index plunged to 147 in early January 1975 (it had been over 500 as long ago as 1968 and indeed had reached 146 at the end of the war, thirty inflationary years earlier).

THE COMMERCIAL SURVIVAL OF THE COMPANY

Under these circumstances, trying to run a business whose products had a very long life cycle – from conception through development to sale and manufacture – threatened to become impossible. Inflation and strikes, with which the 1970s became synonymous, became a very unpleasant fact of life as they fed off each other. On 12 December 1974, Keith wrote to the new Secretary of State for Industry, Anthony Wedgewood-Benn:

On 14th August I sent you 'the best shot' we could make, in the present circumstances, at a Five-Year Forecast!

Since that date, we have had continuing inflation, the Budget and our Scottish strike. Their combined effect has been to make both profit and cash-flow significantly worse than was forecast in the memorandum attached to my letter.

One of the attempted methods of dealing with inflation was to introduce a law prohibiting pay rises above a certain level. It was a hangover from the collectivist approach and, perhaps surprisingly, was introduced by the Conservative Government of Edward Heath. The Counter-Inflation (Temporary Provisions) Act of 1972 prohibited any salary increases whatsoever from 6 November 1972 until 1 April 1973 when the Counter-Inflation (Price and Pay Code) Order 1973 came into operation, limiting increases to £250 a year. As a Government-owned company, Rolls-Royce was forced to observe this law not only in spirit but also to the letter. It did not make it easy to retain or recruit the best people – especially when others used devices such as company cars to flout it with impunity.

The consequences for a large manufacturer and exporter of this inflationary merry-go-round were soon made manifest and were explicitly and

cogently laid out in a letter to the Prime Minister, Harold Wilson, from Sir
Kenneth Keith on 6 April 1975:

I feel compelled to write to you about the impending wage claim from our Glasgow
factories and the effect that this claim will inevitably have upon the company, in the
context of our current financial situation.

In 1974, the company earned what could be regarded as a reasonable profit, subject
to final provisions, but one certainly not at as high a level as is desirable in order to
provide for a secure future – let alone give the shareholder an adequate return on his
investment.

In 1975, we have budgeted for a much reduced level of profit, notwithstanding that
this is related to increased sales. This deteriorating situation is due to the fact that a
large part of our commercial business is in direct competition with the two big
American aeroengine companies. The United States is enjoying a rate of inflation very
considerably lower than that of this country, and in order to remain competitive, we
have had to permit our profit margins to be eroded. Our costs have now escalated to a
level where we are ceasing to be either competitive or profitable. It is not possible in
current economic circumstances to recommend to the Government that we should
embark on longer-term new projects with any certainty that they will produce a
profit.

As you are well aware, we entered into a settlement in Scotland in November last
which clearly breached the Social Contract. Our reason for doing so was that we were
about to ground the U.S. airlines through lack of spares, and to disrupt Lockheec's
delivery programmes through lack of engines. We had no option but to settle, in
order to safeguard the long-term future of the company.

Today, we are faced with a new wage demand which, indeed, we were aware of
when we made the November settlement, and which was termed by the Scottish
unions an interim settlement. The circumstances, so far as the airlines and Lockheed
are concerned and indeed all our other customers, remain unchanged. The result of
the settlement, read across the company and the continuing high rate of inflation, has
made the company even less commercially viable than it was six months ago.

We have reason to think that the claim which we are likely to face is for some £10
to £12 per week, with fringe benefits; and if this is read across the company, it would
add some £30m to our costs. So far as the last settlement is concerned, we have not
got the productivity in Glasgow which unions agreed to. The situation in Bristol is
messy – to say the least of it – but in Derby and Coventry the work force, which is
well-led on both the management and union side, has abided by its agreements. When
we receive the new demand, we shall ask for time to consider it.

The policy alternatives, in broad terms, are to stand firm and say that we cannot
entertain another claim before November. This will almost certainly produce another
strike, with an even more immediate effect on our customers, and our reputation,
than the stoppage last autumn.

Secondly, to try to settle for something in the £3 to £4 a week area – which is, of course, a further breach of the Contract.

Thirdly, to give way – which will merely pave the way for another claim in November. This would be disastrous, and would ensure that the company either went bankrupt for the second time, or became a permanent annual charge on the Exchequer.

Your first concern, Prime Minister, must obviously be for the Contract. Mine is for the commercial survival of the company. The situation in Rolls-Royce is really the national situation on a much smaller scale. I am well aware of the fact that, in view of the proportion of our output which we export, we would be temporarily helped by a major devaluation of sterling, but, like you, I fully appreciate that this is only a temporary palliative and if this is the course which the country is obliged to take to get out of its troubles, the last stage will inevitably be worse than the first.

I am writing this letter for your information and I am sending a copy to those of your ministers whom I think are most concerned.

The company will act responsibly and will keep in the closest touch with officials and ministers.

May I conclude by saying that the Cabinet and the board have a close identity of interest. If we are going to face serious industrial trouble, I very much hope that we shall all stand together and present a united front – you and your Ministers doing all they can to help in order to bring about a sensible solution.

However, the Government decided not to intervene and on 24 March, Tony Benn replied on behalf of Harold Wilson, telling Keith that the Government looked to the company

. . . as to all other employers, to negotiate within the TUC guidelines and not to make any settlement in excess of them.

Keith turned to Gordon Richardson, the Governor of the Bank of England, writing on 25 March:

I am enclosing a copy of a letter which I sent to the Prime minister on 6th March, which is self-explanatory, together with a copy of his reply through Benn, dated 24th March.

My letter sets out the situation which faces Rolls-Royce – and indeed is facing the country. We have now received a wage claim for an additional £12 per week with a demand for a 35-hour week, together with indexation.

We are bound to be involved in strike action, which, in view of our shortage of complete engines and spare parts due to the three-day week and the November strike, must have a very serious effect on Lockheed and the American airlines, some of whose finances are in bad shape.

I have just returned from the United States, where I took the opportunity to warn certain of our customers and Lockheed and its bankers of the problems we face. They, of course, are completely unable to understand how we could be faced with a further strike in view of the fact that we only settled the last one in November and on terms which were much criticised by the Government!

In view of the very serious consequences of another major Rolls-Royce strike to the company and country, I had hoped that ministers would have seen fit to intervene. Regrettably, and after much discussion, they decided not to do so.

My reason for writing to you is that I think you ought to know what is going on. I would welcome a general chat with you some time in the near future.

On 2 April, Keith wrote a blunt letter to all employees, setting out the position.

Costs have risen much more rapidly than those of our competitors, Pratt & Whitney and General Electric . . . no government, whatever its political complexion, is going to make aero engines which we cannot sell . . . our customers have had their operations dislocated by the bankruptcy, the three-day week, the recent major strike in Scotland, and a number of minor strikes – not to mention strikes in the plants of some of our sub-contractors and suppliers . . . almost all airlines have serious financial problems . . . This makes them all the more touchy, if modern and expensive aircraft are grounded by another industrial dispute in Rolls-Royce . . . certain members of our work force think they have the company over a barrel. The facts of life are that they have themselves, as well as the rest of us, over a barrel.

During one of the many strikes of the late 1970s when those in union leadership pulled their men out for three days in the week, Managing Director Dennis Head said, 'We can't run a business like this' and locked them out for the other two. If he hoped that other engineering companies would support him he was disappointed and was forced to say:

We were left with our asses hanging out the window.

'SO LONG AS I AM NOT BUGGERED ABOUT'

Sir Kenneth Keith had been persuaded to take the chair at Rolls-Royce under a Conservative Government by John Davies, a former businessman, and when Davies became ill and retired, he was replaced by another former businessman, Peter Walker. When the Conservative Government called a swap election in February 1974, it lost power and was replaced by a Labour Government, many of whose members had expressed anti-capitalist and anti-big-business views over the years. The new Prime Minister, Harold Wilson, was a pragmatic individual – indeed, few were more so – and he knew that to hold his Party together some left-wingers had to be invited into the Cabinet. He chose as his Secretary of State for Trade and Industry Anthony Wedgewood Benn. Formerly Lord Stansgate, he renounced his title to become the Rt. Hon. Anthony Wedgewood-Benn. However, in his estimation this was still too élitist and he moved quickly to Tony Benn. He was a very different proposition from Davies and Walker, and Keith said later that he found him very difficult. When Keith complained to Harold Wilson, the Prime Minister asked, 'How often do you have to see him?'

'About once every six weeks,' replied Keith.

'What about me?' said Wilson 'I have to listen to him three or four times a week.'

When, presumably because the chorus of criticism from leading businessmen reached deafening levels, Harold Wilson removed Benn from Trade and Industry and replaced him with Eric Varley, Keith wrote:

You will be glad to know that relations with the Department of Industry have greatly improved since – and I think as a result of – your last ministerial changes.

A meeting between Keith and Benn was arranged very soon after Labour came to power and if Keith found Benn difficult, the antipathy was reciprocated. This is what Benn said of their first meeting after Labour was returned to power in February 1974:

Friday 22 March
At 10.45, Sir Kenneth Keith, Chairman of Rolls-Royce, accompanied by Sir William Nield, came in with the Secretary, Frank Beswick and David Jones present, to do a tour d'horison of the whole Rolls-Royce picture. I have known Keith for some time, a tall, arrogant man who thinks that because of the critical press comments of me, he can simply lay down the law.

He began from notes. 'I want to tell you, Minister, the circumstances in which I was appointed. I was approached by Sir William Armstrong in 1972 who asked me if I would take on a job in the national interest. I said, "Yes, but why do you ask without telling me what it is?" He said, "Because the Prime Minister [Heath] cannot have his request refused and won't put it if it is likely to be refused." Then Sir William told me it was Rolls-Royce. I went to see the Prime Minister and he said, "This is a most important job. Will you take it on?" So I answered, "Yes. I will take it on so long as I am not buggered about by junior Ministers and civil servants and officials." And that is the basis on which I accepted it.'

He went on to describe his attitude to the future of the company, to the need for the RB-211-524, to the fact that Bristol was a great headache with so many strikes. He said he was looking for new hardware, travelling the world to try to help Rolls-Royce and doing it without a penny's recompense. [This was not strictly true. Keith was not paid a salary but his employer, the merchant bank Hill Samuel, received a substantial annual fee.] He made some reference to 'that ass Scanlon' and so on. He spoke for about half an hour and I listened carefully. It was a most offensive presentation of the case.

When he finished I said, 'Well, since you have put your cards on the table and told me how you became Chairman of Rolls-Royce, let me tell you how I became Secretary of State for Industry. In February, I had a marginal constituency, I got the support of the trades unions in Bristol and thousands, many of them aircraft workers, turned out on polling day and returned me to Parliament, and the Prime Minister asked me to become Secretary of State for Industry.

'I don't know how long this minority Government will last but while I am in charge I will not accept chairmen of nationalised industries indicating to me that they won't be mucked about by junior Ministers and civil servants: Rolls-Royce is a nationalised company and must be accountable for what it does. On the trade union side, I know the Bristol Rolls-Royce workers very well and I can only tell you that plant there has been very badly managed indeed. Anyone with any sense would know that the real problem in Bristol is the anxiety the workers have experienced arising from the uncertainty of the Concorde project. They have not been involved in decisions until the decisions have already been made.'

I think all this came as something of a shock to Keith who spoke at best as if he were an army officer talking to the troops and at worst as the most arrogant right-wing Tory boss. At one stage when Keith said, 'I can't bother with politics, I am in business . . .' Frank Beswick intervened and I said, 'Well, politics is politics and while we retain the power to change our Government by democratic means, we expect industry to take account of decisions that are reached.'

It was a bitter exchange and at the end he said, 'I hope we can get on well together with you as my Minister.' I added, 'And the owner of your company.' So he said, 'But not my employer.' I said, 'I am inclined to say to you what my sergeant said to me in the war: "You play ball with me and I will play ball with you."'

I was extremely angry, I must confess. My Private Secretary said he never realised how easy it was to become a socialist until he heard Keith.

Others might add that they did not realise how politicians could blow this way and that until they met Benn. His Labour Party colleague, Harold (later Lord) Lever, said of him:

I have heard Tony address two contrary arguments to the Commons and be cheered for both . . . I told him if ever I was accused of a crime I'd like him to plead for me.

We should also note that Keith's view of the National Enterprise Board (NEB), which he famously referred to as a 'bureaucratic contraceptive', was shared by many including the millionaire socialist in the Wilson Cabinet, Harold (later Lord) Lever, who told the author Peter Hennessy:

I did not believe in the so-called industrial strategy . . . all this crap about Benn and his industrial planning . . . the NEB. I knew it was going to be a dud.

Apart from this general dislike of each other, there was some disagreement about the continuing labour problems at Rolls-Royce's Bristol factories. It did not help that Benn had been the Member of Parliament for Bristol South East since 1963. He felt that Rolls-Royce was very Derby-orientated and inclined to ignore the interests of the workers in Bristol. In the words of the Secretary of State's under-secretary:

Sir Kenneth Keith agreed that uncertainties about future employment were at the heart of the Bristol problem but this could not be solved unless Rolls-Royce obtained orders. He believed Mr Whitfield was doing a good job at Bristol, and this view was confirmed by what he had heard from the shop floor.

By the time Keith met Benn on 10 July 1974 to discuss Rolls-Royce's 1975–79 five-year plan, the general situation had darkened and the Plan showed that Rolls-Royce would be bankrupt in 1975 unless substantial new capital was injected or expenditure cut, or both. The economic clouds were darkening both generally – the oil crisis and inflationary trends – and especially, with Government expenditure cuts on Concorde, delay in support for the RB 211-524 and reduced military expenditure.

The delay in capitalising the company properly since the bankruptcy was pushing the Rolls-Royce board to the point of calling in the receiver again and Benn was told that:

There has been a three and a half year delay since the bankruptcy in capitalising the new Company: indeed we have had to contribute £37 million cash to pay the Receiver for our assets.

The effects since last October of inflation, oil and economic crises, the 3-day week and government expenditure cuts on our markets, costs and workload inevitably mean that our financial prospect is much worse and we are faced with a negative cash flow of £100m in 1975 with further deficits in following years.

We have no assurance that we shall be properly capitalised before 1975 and consequently this year under the Company's Act we shall be obliged to give notice that we are not entitled to continue trading unless we are given firm and formal assurances or proper capitalisation and funding from 1975 on. This financial situation also affects our credibility as a supplier.

By the middle of 1975, with the world economy in recession thanks to the oil crisis, Rolls-Royce produced a long-term plan in an attempt to galvanise the Government into committing itself fully to Rolls-Royce's plans for the future. Keith was becoming very frustrated, as is clear from this comment in a letter to Eric Varley on 30 October 1975.

During my chairmanship, the company has prepared two five-year forecasts and, under the arrangements entered into in 1971, has sent copies of these forecasts to HMG. Little useful discussion has so far taken place and no clear understanding of HMG's attitude towards the company and its future has yet to be established.

I have therefore had little evidence that HMG appreciates the company's potential as a national asset and the problems involved in exploiting its potential, and much to the contrary. I have also felt inhibited by the lack of freedom I have been accustomed to in the private sector, but at the same time my ultimate 'masters' have been remote and inscrutable – the worst of both worlds. This letter and appendices will, I hope, help to remedy that situation.

CHAPTER NINE

A LAUNCH ENGINE AGAIN

THE 524
THE BOEING 757
MOSTLY IN DOLLARS
VERTICAL TAKE-OFF AND LANDING
THE HARRIER AND PEGASUS IN ACTION
ALLIANCES
THE TORNADO
HELICOPTER ENGINES

THE 524

ROLLS-ROYCE, IN ITS IMMEDIATE RECOVERY in the early 1970s, concentrated on getting the RB 211 right and delivering the 555 engines ordered by Lockheed. However, at the same time, it knew that the uprated RB 211, the RB 211-524, which would be certified at 50,000 lb thrust at the end of 1975, was a great improvement on Pratt & Whitney's JT-9D, the engine currently being used by Boeing on the 747. The 747 application was particularly important to Rolls-Royce because it was complementary to the Lockheed L.1011 (TriStar), although the B747 had 50 per cent greater passenger capacity and could fly 1,000 nautical miles further than the L.1011-250. And the potential market was very large. Worldwide traffic forecasts by Rolls-Royce, Boeing, Lockheed and McDonnell Douglas showed a good measure of agreement and Rolls-Royce's forecast for the 747 was that 1,000 would be required by 1989 and that 700 had still not been ordered.

Rolls-Royce estimated that an RB 211-524-powered 747 would have at least five per cent specific air range advantage over the next best competing installation. This improved efficiency would translate into a 2·5 per cent lower direct operating cost. This was a very significant operational advantage calculated to be the equivalent of a 50 per cent reduction in engine price (KLM, the Dutch national airline, had changed to the GE CF6-50-powered version of the 747 for their follow-on orders for a predicted advantage in fuel consumption of only 2 per cent over the Pratt & Whitney version, i.e. for only a 1 per cent lower direct operating cost).

Furthermore, 5 per cent better fuel consumption also meant a lower take-off weight for a given mission. This would mean that the RB 211-524 with a 50,000 lb thrust would give the equivalent of 52,500 lb thrust for an engine without such fuel advantage.

The Rolls-Royce sales team made a great effort to convince Boeing that it should consider the RB 211-524, and Managing Director Ken Wilkinson, formerly Managing Director of BEA, would have been delighted to receive a letter from E.H. ('Tex') Boullioun of Boeing on 18 October 1973 saying:

The purpose of this letter is to reaffirm our proposal to offer Rolls-Royce engines on our 747 airplanes ... [W]e believe that the RB.211-524 is well suited to our 747-200 airplane and also to the 747 SP and 747 SR models . . . It appears that a common engine and nacelle package can be adapted to each of these 747 models.

At a meeting, key executives from British Airways and four representatives of Rolls-Royce (Sir Kenneth Keith, Sir William Nield, Dennis Head and Ernest Eltis) discussed British Airway's plans for future purchases of Lockheed TriStars and Boeing 747s.

The meeting had been called at the suggestion of David (later Sir David) Nicolson, the Chairman of the British Airways Board, and Sir Kenneth Keith to discuss British Airways' aircraft requirements up to the 1980s and the part the RB 211 engine might play in those requirements. Nicolson made it clear that BA would make judgements about aircraft and engines on purely commercial grounds. If it felt that General Electric or Pratt & Whitney engines were more suitable, it would specify them. He anticipated that British Airways would require forty-four wide-bodied aircraft by 1982, of which half would be TriStars and half 747s. This was a minimum figure and the maximum was expected to be seventy aircraft, of which thirty-eight would be TriStars and thirty-two 747s. British Airways planned to give a firm order to Lockheed for six of its option of nine TriStars and, at the same time, tell Boeing that all its 747s from 1977 onwards should have Rolls-Royce engines. (BA's early 747s had been powered by Pratt & Whitney JT-9Ds. One of the pilots was the present General Secretary of the Rolls-Royce Enthusiasts Club, Peter Baines.) British Airways would require an engine with 50,000 lb thrust initially, with further development to provide 53,000 lb by 1980 and whatever future increase, perhaps 55,000–56,000 lb, required by Boeing for the stretched version of the 747 by the mid-1980s.

Ernest Eltis confirmed that whereas the basic RB 211 was providing 42,000 lb thrust for the TriStar, the RB 211-524 would provide 48,000 lb (in fact, it was proving more successful than expected and could be rated at 50,000 lb) and was aimed at the uprated TriStar and the A300 B10. Rolls-Royce was also developing an RB 211 for Boeing giving 50,000 lb, which would be uprated to 53,000 lb by 1980. This would be for the 747-200 and would also be appropriate for the A300 B2/4.

Rolls-Royce needed Government permission to proceed with its development of the RB 211-524 and this was extremely slow in coming.

At last, on 19 June 1975, Keith received a letter from Eric Varley, who had succeeded Tony Benn as Secretary of State for Industry, which gave Rolls-Royce the necessary permission to develop the RB 211-524. And over the following two years, the company strove mightily to secure an order from an airline to join British Airways in specifying the engine for its purchase of wide-bodied aircraft.

Pan American *World* Airways (Pan Am had changed its name in 1950), at the time perhaps the United States of America's leading airline, was considering buying twelve wide-bodied aircraft, an order worth $540 million, and all three US aircraft manufacturers were eagerly competing for the business. Rolls-Royce hoped that Lockheed would secure the contract because it was almost certain that the engine specified would be the RB 211-524, an order worth about $120 million. Pan Am's prestige was such that other airlines such as United, American and Qantas would probably follow suit. Rolls-Royce calculated that a favourable Pan Am decision would lead to further sales of 200 engines in addition to the direct sale of 100 engines.

And the loss of the order to General Electric would mean that the Lockheed TriStar would fly for the first time without a Rolls-Royce engine. It was very important for Rolls-Royce to fight off the General Electric and Pratt & Whitney competition. Richard Turner, currently Group Marketing Director of Rolls-Royce, spent a whole year in New York to make sure that the order with Pan Am was secured. However, it was a close-run affair and provides a classic example of how competitive the market was. John Newhouse wrote extensively about the battle to secure the Pan Am order, and much of what follows comes from his book *The Sporty Game*.

Pan Am had narrowed its choice down to the DC-10 and the TriStar, and by the end of 1977 favoured the TriStar. The choice of engine was apparently obvious. The TriStar had only ever flown with Rolls-Royce's RB 211. To change to Pratt & Whitney or General Electric would mean redesigning certain features and recertifying the aircraft. There would inevitably be substantial extra cost. John Borger, the powerful Vice-President of Engineering at Pan Am, recommended the RB 211.

However, Lockheed was not sure that Pan Am, the oldest American flag carrier and one that previously had only bought Pratt & Whitney engines, would buy a British engine and it therefore offered Pan Am the TriStar with a

Pratt & Whitney engine. To complicate the situation further, Pan Am was suffering financially and was looking for supplier financing. Lockheed was willing to help as much as it could but was not in robust financial health itself. Pan Am turned to Rolls-Royce.

On 15 March, not long before the board meeting at Pan Am when the vital decision would be made, the Vice-President of Finance at Pan Am, J. Kenneth Kilcarr, called a meeting with Rolls-Royce and told those present that Pan Am had an 'operational preference . . . for the General Electric engine on the Lockheed airplane' but continued that its 'operational preference is substantially small so that it could be offset by a superior financing package'. Kilcarr then told the shocked Rolls-Royce team that General Electric had offered a financial package over fifteen years, which covered not only the engines but also the aircraft, and that Rolls-Royce would need to offer something similar but with a lower rate of interest. In stark terms, the conditions were:

100 per cent financing, a term of 15 years, no involvement of Pan Am's own credit and a commitment by Rolls-Royce by Wednesday 22 March [seven days away] to cover the initial 12 aircraft order.

Rolls-Royce was also told that its chances of selling engines for use on Pan Am's 747s would depend on its success with the TriStar order.

Here was a classic case of cut-throat competition, of the type we saw earlier when the initial battle between Lockheed and Douglas was raging at the time of the introduction of the wide-bodied aircraft in the 1960s.

Pan Am, perhaps already showing some of the financial strains which led to its restructuring at the end of the 1980s, was exploiting the keenness of General Electric, Pratt & Whitney and Rolls-Royce to secure the order for the engines on its TriStars.

Rolls-Royce would have to decide whether it should grit its teeth and match the required terms or let go this possibility of what would be a very significant order. Keith had no doubt that he must secure the order and flew to New York to negotiate with Pan Am in person. He was prepared to match, but not exceed, the terms offered by General Electric. The deal was closed.

Pan Am itself said at a press conference after announcing its decision:

In addition to the Rolls-Royce engine, Pan-Am also evaluated the TriStar with Pratt & Whitney's JT9D-20C and GE's CF6-50C2R. The Pratt & Whitney candidate was particularly attractive because of some commonality with existing powerplants. However, in the final analysis the Rolls-Royce engine was considered superior because the propulsion system had been specifically tailored to the TriStar requirement . . . Rolls-Royce's previous work on the TriStar program also provided a greater confidence level that the tight 1980 delivery schedule can be met.

There were two happy outcomes as far as Rolls-Royce was concerned – the vital order was secured, and Richard Turner was awarded an OBE for his efforts in securing it.

THE BOEING 757

In spite of its huge success in building a series of jet airliners since the 1950s, Boeing had not always been as successful in making money out of them. The author interviewed Thornton Wilson (known to everyone as 'T'), President of Boeing in the 1970s, who told him shortly before he died in 1999:

At that time [the early 1970s] Boeing had sold about twenty billion dollar's worth of commercial airplanes and hadn't made any money . . . I had never seen us come close to budget on any plane . . . We didn't on the 707, we didn't on the 727 although we came closer, we were an absolute basket case on the 737, and we had great difficulties at the start on the 747 program.

Nevertheless, by the mid-1970s, the company had made money on the 727 and was still doing so on the 707; additionally, both the 737 and 747 were beginning to be profitable. As was its wont, Boeing took another calculated gamble and decided to build two new aircraft, the 757 and the 767, simultaneously. The 757 was designed to exploit the six-abreast body to its optimum extent. It was 20 per cent larger than the hugely successful 727 (by this time 1,500 727s had been sold), and would be quieter and more fuel-efficient. Boeing expected to sell up to 1,000 aircraft in the 160–180 seat capacity and another 1,000 in the 200–230 capacity.

The initial studies of the 757, code-named 7N7, produced a twin-engine design aiming at a passenger load of about 150. Rolls-Royce was hoping to offer the JT-10D, the engine that it was developing jointly with Pratt & Whitney. As the project developed, the aircraft size and weight increased to enable it to carry 200 passengers. At this point, GE entered the competition with a reduced fan-diameter CF6 engine called the CF6-32. Rolls-Royce realised it would make more sense to re-fan the RB 211 rather than keep stretching the proposed JT-10D. Rolls-Royce therefore withdrew from the JT-10D collaboration and developed the RB 211-535. This two-engined aircraft placed greater emphasis on the reliability of the engines, and a derivative engine offered lower engineering risk, with shorter timescales and reduced launch costs. (We saw what Stewart Miller had been advocating in 1972.)

The RB 211-535C, at 37,400 lb thrust, allowed the use of conservative turbine temperatures, although the existing -22B HP spool was slightly larger than the optimum and meant that weight and performance penalties would be incurred. As we shall see, it also allowed Pratt & Whitney to come back strongly with their PW2037, effectively an uprated JT-10D, which featured an ambitious thermodynamic cycle and advanced component technology which promised an 8 per cent specific-fuel-consumption advantage over the RB 211–535C.

Ernest Eltis with Geoffrey Wilde went down to Keith's Hill Samuel office in St James's Square and persuaded him that the company should develop the RB 211 derivative for the 757. The logic was such that Keith was persuaded.

On 28 April, he wrote a long letter to Eric Varley in which he outlined the importance of the project, as well as its profit potential. He began:

The following brief observations (from paragraph 7 onwards) set out the considerations which in my board's opinion leave us with no viable or sensible alternative than to finalise as rapidly as possible the establishment of the RB.211 as the launch engine in the Boeing 757 aircraft.

A decision to this effect is critical for Rolls-Royce and we need to move very quickly if we are to get the important advantage of being the market leader.

I cannot state too seriously how strongly my board hold this view, as being the only viable route for Rolls-Royce. Without the commercially right decision, I should fear for the long-term security of the company and the jobs of many thousands of its workers.

At a moment when the profitability of the company is on the way to being restored, and morale can reasonably be claimed to be high, it would be specially injurious to set off on some speculative track in which none of us has any confidence. The mood so created would reflect throughout the company.

Keith went on to say that the 757 gave Rolls-Royce the opportunity of providing a launch engine for a Boeing aircraft for the first time in its history and wrote:

This would lead us to a very high penetration of the market, possibly making the 535 the JT 8 of the 1980s. On the assumption of achieving slightly over 40 per cent market penetration, the total 535 programme would produce almost £5 billion of business.

The 757 would be the first Boeing aircraft to be launched with anything other than a Pratt & Whitney engine.

Keith also pointed out that the British Airways's stated fleet plan required a 757-type aircraft. (Rolls-Royce worked hard to persuade British Airways to buy the 757 specifying the Rolls-Royce RB 211-535.) And he also played the 'national importance' card, stating:

The project would be profitable by any conventional commercial criteria, but equally important, at a time of depressing national and international outlook for trade and employment, it would provide export potential and employment for the U.K. at a level rivalled by few new investment projects which had become available to the U.K. in recent years.

If the RB 211-535 became the launch engine for the 757 the Rolls-Royce programme would provide employment for 12,000 to 14,000 in the U.K. through to the year 2000.

The engineering and production of the share offered to British Aerospace would provide 9,000 jobs in that company alone up to the turn of the century. Other suppliers in the areas of undercarriage and systems would also be substantial.

On 31 August 1978, Eastern Airlines and British Airways announced their orders for the Boeing 757 powered by Rolls-Royce RB 211-535 engines. GE, who had won orders from a couple of airlines, abandoned their CF6-32 and paid compensation to the customers who had selected it.

MOSTLY IN DOLLARS

Rolls-Royce's attempts to rebuild itself as one of the three major aero engine makers in the world, were not helped by the oil price hike induced recession of the mid-1970s. In preparing the Budget for 1977, the Aero Engine Division accepted that the depression in the market for both wide-bodied civil aircraft and military aircraft that had begun in 1975 was deeper and longer-lasting than had been anticipated.

Compounding the problem was the continuing high level of development expenditure – maintaining the RB 211-22 in airline service and developing the -524B and -524D versions. We have already seen how Rolls-Royce was developing uprated versions of the RB 211-524 to give thrusts higher than 50,000 lb.

As Dennis Head, the Managing Director of the Division, put it:

These RB211 costs are extremely heavy but the whole future of the Division is dependent on these engines and curtailment of this work cannot be contemplated. Other new engine programmes have been cut back from previously forecast levels.

And the effect on cash-flow was severe:

There is no immediate prospect of the Division becoming self-supporting in terms of cash.

The majority of profits were still coming from the established Derby engines – the Spey including income from China of £7 million (we saw how Sir Stanley Hooker led a delegation to secure the licence), the Dart, the Avon and other older models – and much of the profitability was due to the supply of spare parts. By contrast, the RB 211 was still showing losses, though the Division was hopeful:

There is every reason to believe that we shall in due course find the RB211 contributing proportionately more substantial profits [than the Spey, which had also been a substantial financial burden in its development days].

Compounding the short-term financial burden was the decline in development work carried out on behalf of the Military – from 18 per cent of turnover in 1975 to 11 per cent in 1977. The four major Government projects – the Adour, the Olympus 593, the RB 199 and the Gem – were all passing their peaks and not being replaced by new projects. While this development was healthy in the long term as such business was low margin, in the short term it was a burden as the overhead costs took time to cut back.

Repayments to the Government on sales of the RB 211 were a continuing burden. By the end of 1977, 545 of the original programme of 555 engines had been sold and in three years, from 1975 to 1977, £69 million had been paid to the Government. A further £10 million would have to be paid to conclude the agreement made at the time of the receivership.

And while this pressure from recession was squeezing the company, the Labour Government was exerting its own attempt to squeeze as well. Keith reported to his board that he had been unable to respond constructively to the Secretary of State's request that the company's cash requirements be substantially reduced.

Rolls-Royce's long-term strategy at this point was to continue developing and exploiting the RB 211 family of engines; launch an engine for the significant new intermediate commercial aircraft; develop the military engines (especially the RB 199, Pegasus and Adour) to achieve higher ratings suitable for new applications; develop and demonstrate technology for new low-cost engines to meet emerging market requirements in the 5,000 and 18,000 lb thrust classes; and to continue to exploit industrial applications for Rolls-Royce engines.

Specifically, Rolls-Royce expected continued demand for wide-bodied jets and while the RB 211 was still the only engine used in the Lockheed TriStar, the company hoped to capture new customers for the Boeing 747. It saw a great opportunity for the RB 211-535 in the 180–200 seat aircraft range. It expected to continue selling the RB 199 for military aircraft. It hoped that the Gem could be sold to Sikorsky as well as Westland and that engines being developed could be sold in the 'Bizjet' market.

By the end of 1977, the gloom seemed to be lifting a little. Ralph Robins, Commercial Director, reported to the board on 30 November that there were hopes of 'increasing opportunities for further RB 211-22B sales in both standard and derivative versions of the TriStar'. On the military front, significant opportunities were opening up in the Middle East and Africa so that:

The whole area [was] a high priority target in our military engine strategy with opportunities in Egypt, Sudan, Algeria, Saudi Arabia and the Gulf.

Rolls-Royce was straining every muscle to win the expected Pan Am order, and Richard Turner had been positioned in New York until a decision was made. British Airways had recently ordered its eighth RB 211-powered 747 and was reporting satisfaction with those in service. Saudia (Saudi Arabian Airlines) was probably going to buy more TriStars (they did ultimately launch the RB 211-524-powered version of the TriStar) and was considering five 747-200s. As Robins said: 'There is no doubt that Saudi represents a continuing major business opportunity for us.'

In January 1980, Robins outlined the company's marketing strategy as being the continued penetration of the large fan market from 1982 onwards in an attempt to maintain a steady production load of 250-plus engines throughout the decade. A major factor would be Rolls-Royce's ability to sell the RB 211-535 engine, although continued success in the TriStar and 747 would also be very important.

Sterling, which had started to rise sharply as it came to be viewed as a petro-currency following the latest oil price hike in 1979, was making Rolls-Royce prices increasingly uncompetitive.

Rolls-Royce had always been a very large exporter. Indeed, as Sir Kenneth Keith pointed out to Harold Wilson in a letter in October 1975:

In the first seven months of this year, Rolls-Royce exported some £218 million of engines and spares which compares with a total of £207 million of aircraft and parts, instruments, avionics etc., for the whole of the remainder of the British aerospace industry.

Following the breakdown in 1972 of the exchange rate system set up at the Bretton Woods Conference (in Bretton Woods, New Hampshire) in 1944, the major currencies of the Western world fluctuated much more sharply than at any other time since World War II. Both the English pound and the American dollar depreciated steadily against the Deutschmark, Swiss franc and Japanese yen. This helped British exporters, though the beneficial effect was largely eroded by the much higher rates of inflation in Britain.

The key exchange rate for Rolls-Royce was that of the American dollar and in 1979, much to the company's consternation, the dollar began to depreciate not only against the stronger European currencies but also against sterling.

Because of the dominance of the USA in the world aircraft industry, nearly all transactions were – and still are – in US dollars. A depreciating dollar was therefore bad news for Rolls-Royce. The second oil crisis of the decade meant that countries with oil – and Britain was one of those – were counted as beneficiaries and their currencies began to strengthen. This did not suit Rolls-Royce at all and Neil Collins, City Editor of the *Evening Standard*, wrote under the headline 'The $ makes whispering Rolls wince':

No company is watching the dollar's dismal performance in foreign exchanges more closely than Rolls-Royce, the aero-engine giant whose business is largely with the U.S. plane-makers.

The company frankly admits a 1% drop in the dollar wipes £1 million off profits – a matter for major concern when the dollar has fallen by over 14% since January 1.

'The weak dollar is having an unfavourable effect on profits,' finance director Peter Molony told me today . . .

City conclusions are that the State-owned outfit – nothing to do now with the famous cars – must end the year with a hefty loss.

RR – one of our top 10 export earners – made profits of only a slim £12 million – cash earned mostly in dollars.

And there was to be no let-up in the roller-coaster ride of the pound–dollar relationship. The pound, devalued from $2.80 to $2.40 in 1967, had slid gradually down to $1.70 by 1977, but in 1980 it rose all the way back to $2.40, before plunging down to near-parity by January 1985. For a large exporter, it made forecasting and planning very difficult.

The late 1970s was a very successful period for Rolls-Royce in winning orders but, as we saw in the late 1960s, winning orders was only part of the battle. Delivering them on time at the right price and with costs under control was the key to success.

On 30 July 1979, Keith wrote to the new Secretary of State for Trade and Industry in the new Conservative Government, Sir Keith Joseph (the Conservatives under Margaret Thatcher had won the General Election in early May). He pointed out the problems Rolls-Royce faced.

As I mentioned when we met previously, the success there has been in sales of RB.211 engines, with consequential increases in workload, and the strengthening of the pound, have meant that the financial needs of the Company are considerably larger than estimated a year ago for the then five year forecast or for the later 1979 budget, both of which being reported to your Department through the National Enterprise Board . . . The exchange rate and rise in inventory dominate the Company's finances. The RB.211 competes with General Electric and Pratt & Whitney, both of whom sell aggressively in dollars and incur most of their costs in that currency.

The airlines are in the middle of a major re-equipment phase. The airframe and engine choice that they make now will determine follow-on orders and the all-important sales of spares for the next 20–30 years.

The Company has taken this unique opportunity to expand its position in the Lockheed 1011, to enlarge its bridgehead on the Boeing 747 and to consolidate its position as the lead engine powering the new Boeing 757.

Robins reported:

There is little evidence that the fierce competition between GE and Pratt & Whitney is abating; indeed in the recent competition for the Air France order it became even more extreme, and this trend is expected to be continued over the engine competitions for ANA and TWA's 767s. We therefore face a considerable problem if called on to match this competition at today's sterling/dollar rate.

An indication of the fierceness of the competition was given by Don Pepper's report to the board where he noted that Pratt & Whitney was offering a fuel efficiency guarantee to all current potential customers that its PW2037 would be 8 per cent better than the RB 211-535C or its derivatives. Frank Borman of Eastern had told Pepper this could mean that Delta, Eastern's main competitor, would receive a subsidy of no less than $36 million a year and that, whereas Eastern wanted the RB 211-535E4, such a subsidy was very attractive. There is no doubt that after Boeing had launched the 757 with Rolls-Royce's RB 211-535C, Pratt & Whitney attacked very strongly with its fuel-efficient PW2037. Rolls-Royce countered with the RB 211-535E4. This engine, at 40,100 lb thrust, contained many of Rolls-Royce's most advanced technologies including the wide-chord fan and integrated nozzle/nacelle.

One of the most significant advances made by Rolls-Royce on this engine was the introduction of the hollow, wide-chord fan blade. Previous fan blades of solid titanium construction were restricted in chord by a limit in blade weight – necessary to ensure successful containment in the event of failure. These relatively narrow blades were aero-elastically unstable, requiring a snubber or damper to prevent flutter, thereby reducing efficiency and capacity.

The HP nozzle guide vane on the RB 211-535E was the first to benefit from 3D aerodynamics in order to reduce the secondary end-wall losses. Furthermore, developments in materials, particularly the widespread use of titanium in the HP compressor and carbon composites in the nacelle, enabled the thrust-to-weight ratio to be superior to that of the PW2037. Titanium accounted for over 30 per cent of the engine by weight. The 535 was also the quietest engine in its class and was the only engine allowed to power aircraft with over 100 seats into the Washington National Airport during the night-time curfew. It may not quite have matched the PW2037 for fuel efficiency but it was more reliable, staying on the wing between services for some 15,000 hours. Fortunately, the oil price hike of 1979 faded, and reliability became the name of the game.

Nevertheless, Robins was optimistic about the overall RB 211/747 programme. Problems between Boeing and Saudia had been resolved, and the contract for six RB 211-powered 747s, for delivery in 1981, was firm. If Rolls-Royce could be successful with Air New Zealand, Swissair or Pan Am, the RB 211 could achieve a respectable 747 share.

VERTICAL TAKE-OFF AND LANDING

Whereas the military aviation market had been everything to Rolls-Royce up to the end of World War II and indeed almost so in the years immediately after it, as we have seen, the civil aviation market, especially in the USA, became increasingly important in the decades following the war.

Nevertheless, military business remained extremely valuable to Rolls-Royce and there were plenty of outbreaks of violence around the world to remind governments of the need to sustain their military strength. The immediate post-war challenge was in Asia.

When World War II was finally won and the Fascist dictatorships had been defeated, the democratic nations soon faced a new enemy, Communism. The British Prime Minister, Winston Churchill, and the US President, Harry Truman, were equally determined that the social system under which individuals enjoyed liberty must be defended. In 1949, Truman said:

Democracy maintains that the Government is established for the benefit of the individual and is charged with protecting the rights of the individual and his freedom . . . The actions resulting from the Communist philosophy are a threat to the efforts of free nations to bring about world recovery and lasting peace.

The first rattling of sabres came in 1948 when the Soviet Union blockaded Berlin but the first real threat to peace came from Asia when the Communist North Korea invaded South Korea on 25 June 1950. Truman in particular felt that the democracies could not hold back, saying:

The Kaiser and Hitler, when they started their great wars of aggression, believed that the United States would not come in. They counted on being able to divide the free nations and pick them off one at a time. There could be no excuse for making that mistake today.

Truman, according to his daughter Margaret, feared 'this was the opening of

271

THE MAGIC OF A NAME

World War III'. If the Communists were not resisted, 'no small nation would have the courage to resist threats and aggression by stronger Communist neighbours'. The Prime Minister of the People's Democratic Republic of Korea and Supreme Commander of the North Korean armed forces, Kim Il Sung, had seized power after a spell in Moscow learning Communist ideology and techniques. There was no doubt who was behind him. (His rule in North Korea was to last over forty years, outliving even the Soviet Union, his initial supporter.)

The threat of resistance from the United States, Britain and other United Nations countries did not deter the North Koreans who were faced by a South Korean army of barely 25,000 men. After a visit to South Korea, General MacArthur recommended full American commitment. Truman acceded, and, on 1 July, the first United States troops, under the flag of the United Nations, landed at Pusan in the south of the peninsula.

Other nations rallied to the cause of freedom. On 5 July, the British Prime Minister, Clement Attlee, told the House of Commons:

I think that no one can have any doubt whatever that here is a case of naked aggression. Surely, with the history of the last twenty years fresh in our minds, no one can doubt that it is vitally important that aggression should be halted at the outset.

Mobilisation orders went out to several thousand reservists throughout Britain and those doing their National Service were warned it might extend beyond the usual two years. Many wondered whether the atomic bomb might be used again, but British military chiefs favoured conventional air attacks. Nevertheless, when President Truman answered questions put to him on the subject rather ambiguously, a *United Press* bulletin announced on 30 November 1950:

President Truman said today that the United States has under consideration use of the atomic bomb in connection with the war in Korea.

The United Nations forces gradually pushed back the North Koreans but the war escalated as the Chinese became involved and, in November 1950, Soviet-built MiG-15 jet fighters appeared, flown by both Chinese and Soviet pilots. By the end of November, United Nations forces were in retreat. MacArthur cabled Washington, warning that the United Nations Command was 'facing the entire Chinese nation in an undeclared war'.

By December, the number of Chinese troops inside North Korea was estimated at between 400,000 and 450,000. The United Nations forces retreated and retreated, 300 miles in thirty days.

Truman knew that another world war was close. Indeed, if he had followed the advice of General MacArthur, there is no question that the Soviet Union as well as China would have become fully involved. MacArthur wanted the US Air Force to drop between thirty and fifty atom bombs on Manchuria and the cities of mainland China.

The nuclear option was not used and the United Nations forces gradually fought back until an armistice agreement was eventually signed on 27 July 1953 at Panmunjom. Within two weeks the United States signed a peace treaty with South Korea, and for over forty years North and South Korea pursued different ideologies. By the 1990s the South was enjoying considerable prosperity while the North languished in poverty. There is scarcely a better example of the results of capitalism as opposed to communism.

The effect on Rolls-Royce of the Korean War was seen mainly in a rapid expansion of the production of the Avon engine, which powered the Canberra bomber. Demand was such that Bristol, Napier and the Standard Motor Company were licensed to build the engine. It also led to the opening of Rolls-Royce's East Kilbride factory. Out in Korea, Supermarine Seafires with Griffon 88 engines and Hawker Sea Furies with Bristol Centaurus engines were used by British forces, as were Fairey Firefly aircraft with Griffon 74s.

In his book, *The Korean War,* Max Hastings wrote:

From the first day of the Korean war the importance of fighter-bombers in a close support role was beyond doubt. The Yak piston-engined fighters of the North Korean air force were cleared from the skies within a matter of weeks, and the USAF's Mustangs [powered by Packard-built Merlins] together with carrier-based American Corsairs and British Seafires and Sea Furies, played a critical tactical bombing role.

Later in the war when the Russian-built MiGs appeared, the piston-engined aircraft were, of course, completely outclassed. The Fireflies were instructed to fly at only 125 knots if they encountered MiGs. This was reckoned to be too slow for the MiGs to make contact.

Following this 'hot war', what became known as the Cold War settled into its forty-year lifespan and the concern of the North Atlantic Treaty Organization (NATO) planners was the firepower of the Soviet Union. They saw the Soviet Union building, and possibly destroying, thousands of tactical nuclear weapons that could be delivered either by manned aircraft or by ballistic missiles. Conventional airfields would be sitting targets and dispersed sites seemed to be the answer. From those sites, aircraft that could take off vertically would be able to escape the attacks and would still be a threat to the enemy.

The advance in technology, which saw the speed of jet fighter aircraft increase from 500 miles per hour in 1945, brought problems as well as solutions. By the mid-1950s, virtually all NATO fighters needed runways of up to two miles from which to operate. Not only could such runways not be built close to the battlefield, but also wherever they *were* built they provided a large and easily identifiable target. As Stanley Hooker recalled in his book *Not much of an Engineer*, this problem was identified at the highest levels, notably by

... Col Johnnie Driscoll, USAF, head of the MWDP [Mutual Weapons Development Programme] office in Paris. MWDP was a US organization set to organize projects within NATO, and bring in additional nations who had weak industrial strength. The NATO Supreme Commander was Gen. Lauris Norstad, also of the USAF, and he was very interested in light fighters, partly because they could be dispersed widely away from the vulnerable airfields and partly because they were within the industrial capability of many nations, including the rebuilt Italy and West Germany. Driscoll therefore convened a meeting of NATO aircraft companies where they were given details of a requirement for a light tactical strike fighter weighing around 8,000 lbs and able to take off and land in less than 2,000 ft, in what Driscoll called 'cow pastures'.

The challenge to produce an aircraft that could take off vertically and trans-fer to horizontal flight, and then achieve speeds and manoeuvrability that would challenge conventional fighter aircraft, was an extremely formidable one. Perhaps the easiest way was to install a powerful engine and tilt the aircraft on to its tail. This was tried by the Germans before the end of the war in the rocket-powered Bachem Ba349 Natter. The pilot bailed out when it ran out of fuel. It was also tested by the Ryan X-13 Vertijet with a Rolls-Royce Avon turbojet of 10,000 lb thrust in 1956. The Ryan X-13 launched from, and landed back on, a tilting platform. By the early 1960s this approach – vertical attitude take-off and landing or VATOL – had been abandoned because of piloting problems in transition and through lack of an overload short take-off capability in which wing lift could augment jet lift. (David Huddie told the story many years later that when he went to a reception at No. 10 Downing Street for the astronaut Neil Armstrong, Armstrong hugged him and shouted 'Dave, how are yer?' Huddie was taken aback until Armstrong said, 'Don't you remember me? I was a project test pilot on the Ryan X-13.')

The trick was to find a power plant that would provide extra jet lift with little weight increase, and there were two ways of achieving this. One would be to link the propulsion engine to a device that would increase its thrust by allowing it to deal with a much larger mass flow of air. General Electric went down this route and installed a thrust magnification device (in the form of buried fans) in the Ryan XV-5A Vertifan, but the volumetric and weight penalties were such that the project was abandoned.

The other method was to develop 'dedicated' lift engines and this was the approach adopted by Rolls-Royce. It was attractive because the lift engines, which only had to operate for a few minutes at a time, could be made extremely light and could disregard the level of fuel consumption. This method was chosen by the Air Ministry as the way ahead, and Rolls-Royce and Short Brothers, whose aircraft the Short SC.1 was chosen as the test aircraft, received the whole of the Government support available for VTOL aircraft and engines.

We have already seen the contribution made by Dr. A.A. Griffith to the development of the jet engine, particularly in connection with axial compressors. He also made a considerable contribution towards VTOL concepts. He is best known for his idea of employing numbers of small specialised lift engines, but prior to that he had also considered an approach using the deflection of propulsive jets. As early as 1941, he discussed, in an internal Rolls-Royce paper, the potential offered by the higher power-to-weight ratio of the jet engine relative to the piston engine, and floated the idea of a 'jump take-off' and 'thistledown landing'. The 'multiple small engines' idea came to him later, primarily for use as disposable engines in relation to guided weapons.

In 1950 and 1951 Griffith conducted several studies on military aircraft with a number of small engines which deflected the jets for take-off and landing. He came to the conclusion that, apart from the operational aspects, jet lift could offer important advantages in allowing the wing and main propulsion system to be designed essentially for flight without the compromise necessary for take-off and landing requirements.

In 1952, he produced his first scheme with separate engines to be used solely for take-off and landing. He proposed a scheme using two Nene engines with their jets directed downwards and with appropriate compressor bleed for control nozzle arrangements. The rig constructed was called a Thrust-Measuring Rig but quickly became known as 'the flying bedstead' and began 'flying' in July 1953. It was a steel-tube frame standing on four small swivelling castors and carrying two Nenes, two fuel tanks and a pilot seat. The engines directed their jets towards each other, one with its jetpipe turned down through 90 degrees directly in the centre of the rig and the other with its pipe bifurcated and turned 90 degrees down on either side of the other engine's jet pipes so that its thrust, or loss of it, caused no pitch or roll. It was extremely unstable, with its attitude controlled by four reaction jets on long arms, known as 'puff pipes'.

The following year, Rolls-Royce began work on the first small engine designed specifically for lift. Named the RB 108, five were installed in the Short SC.1, which first flew on 2 April 1957. Four were mounted vertically in crosswise pairs inside a central engine compartment to provide lift. The thrust from each pair could be swivelled through a total angle of 37 degrees

(25 degrees rearward, 12 degrees forward vectoring of the thrust) about the vertical to assist take-off and landing transitions. The fifth engine was installed horizontally at the tail for propulsion.

The *Daily Telegraph* wrote:

The Short S.C.I, Britain's first vertical take-off jet plane made its maiden flight yesterday from the Ministry of Supply experimental airfield at Boscombe Down, Wilts.

'Normal take-off technique' was used and the aircraft was tested conventionally in forward flight only, said the company last night. Mr Tom Brooke-Smith, 38, the firm's chief test pilot was at the controls and had the machine airborne for about 15 minutes. Later he said, 'I was very pleased with the flight. Everything went well. In fact, it exceeded my expectations.'

Short's stated: 'This flight marked the start of a phase of development expected to continue for some time before the aircraft proceeds logically to unrestricted vertical take-off and transition from hovering to forward flight.'

Over the following decade, this aircraft contributed a great deal to the under-standing of multi-engined V/STOL aircraft. It laid the foundations for the Dassault Balzac and Mirage IIIV.

Griffith was quoted as saying that he expected vertical take-off airliners would be flying on passenger routes by 1965.

Development of a second-generation lift jet, the RB 162, began in 1959 and in November 1961 Rolls-Royce was authorised by the UK, French and West German Governments to continue its work for the Mirage IIIV and for the Dornier DO.31. In 1964, a second phase of this development was authorised to produce the scaled-up RB 162-31 and RB 162-81 versions. Eighteen flight engines at 6,000 lb thrust were delivered for the three VAK.191B proto-types. These engines pioneered lightweight construction and systems using glass fibre in the compressor and a titanium turbine disc. One version of the engine, the RB 162-86, received a full transport category type certificate from the British ARB in 1971 for booster use in the Hawker Siddeley Trident airliner.

However, there were many problems with the multi-engined configurations, not least the complex control systems and the ground–aircraft interactions, in spite of the high reliability achieved in the specialised lift engines. Cost was also a big factor. Since engine costs tended to increase as the square root of the thrust, the thrust achieved by a small number of engines would be more expensive than that achieved by a single engine. Maintenance was also going to be more expensive. Other complications arose, such as the need to operate from grids due to ground erosion and reingestion – caused by perma-nently deflected lift jets and suck-down complex inlet momentum drag problems affecting transition.

David Davies joined Rolls-Royce in 1944. He was directed to Barnoldswick where he was supervised by Fred Morley who, at intervals, asked him his name.

'Omre', replied Davies.

'Haven't you got another name?' said Morley.

'David.'

'Well, from now on you're f---ing David!' replied four-letter-word Fred.

Davies worked on most of the company's engines after World War II (with particular reference to design, stressing and systems) before becoming chief designer on the Spey family of engines, and was deeply involved in Rolls-Royce's attempts to perfect a VTOL system. He maintained that Rolls-Royce Derby ignored the alternative vectored approach largely because, for supersonic-type fighter aircraft, the thrust boost from the application of reheat was essential for overall effectiveness.

The Derby engine design and the Hucknall installation teams worked very closely with Dornier, supported by German funding, to install sixteen lift engines on the German DO.31 transport aircraft. The subsequent flight test programmes confirmed the project's feasibility but also exposed many practical problems. The German Government was more interested in initiating programmes that could lead to the definition of a fast-response interceptor aircraft because of the closeness of Soviet forces to the German borders.

There were three other significant programmes fully funded by the German Government. The first was the RB 145 engine with reheat for the EWR VJ101C. This was a development from the RB 108, requiring the addition of a high-thrust boost reheat system to provide adequate thrust for the supersonic VJ101C. The four engines were installed in two rotatable pods located at the wing tips.

The engine and pod development programmes proceeded satisfactorily to a flight clearance standard, much to the credit of the Derby design development and Hucknall teams involved. Supersonic speed was achieved in the flight tests that followed. However, at the time of these developments there was a spate of fatal crashes of Lockheed Starfighters in Germany, reinforcing the views of many that single-engined, high-performance aircraft would not be acceptable in the future. This view, somewhat reinforced by the crash of one of the VJ101C aircraft, led the German authorities to terminate their investment in the VJ101C.

The next programme was the RB 153-61 engine for the VJ101D interceptor aircraft. This aircraft would have two bypass propulsion engines fitted with maximum-boost reheat systems, thrust reversers and deflectors, and a small number of lift engines. The high-pressure core of the definitive propulsion engine, the RB 153-61, was a scale-down from the Spey, but the design of the HP turbine disc, shafting and bearings in the HP turbine area proved very

THE MAGIC OF A NAME

difficult. A satisfactory solution was finally achieved but the problems of scaling down from highly stressed components had been highlighted.

Next came the RB 193 which was in part a return to the idea of four lift engines and conventional thrust. This time the engine had a thrust reverser for short take-off and landing (STOL). Rolls-Royce decided that the civil version of the Spey was a good starting point but it was too big and had to be scaled down. At this point, the German Government came back with several new suggestions and sent great numbers of designers to Derby.

However, yet again, they seemed to lose heart. Nevertheless, the engineering work on the RB 193 for the VAK.191 gave Rolls-Royce invaluable technology-base information which greatly helped in developing the Spey TF41 which was sold in large numbers to the US armed forces. The six-stage HP compressor development also provided the base-line HP compressor for the RB 211 family of engines. Derby's work on VTOL now lost priority as the problems of the RB 211 destined for the Lockheed TriStar loomed larger and larger. Bristol, with its successful Pegasus, took over the VTOL work for Rolls-Royce after the merger of the two companies.

Reluctantly we would have to conclude that Rolls-Royce spent a lot of time, effort, resources and money on VTOL and with very little to show for it. Fortunately, most of the money came from the British, French and German Governments, especially the German Government. Ernest Eltis was to say later that Pearson, the company's Chief Executive, was probably unhappy about this use of resources.

In the meantime, the solution to the thrust/weight problem, or at least the beginnings of it, was being found by a French engineer, Michel Wibault. Wibault had been encouraged and financed by Winthrop Rockefeller. He conceived the idea of using a high-powered turboshaft to drive centrifugal blowers mounted in the sides of the fuselage. Through turning their casings these compressors could provide either jet-lift or propulsive thrust. Searching around for a suitable engine, Wibault selected the 8,000 shp Orion designed by Bristol.

Wibault's initial design was complicated and did little to solve the weight problem but, although he was rebuffed by the Pentagon and the French Ministry of Defence, the head of the Mutual Weapons Development Programme, Colonel John Driscoll, could see the potential, and passed Wibault on to Stanley Hooker at Bristol. In turn, Hooker put Gordon Lewis in charge of developing Wibault's ideas into something more practical. In a memo of 2 August 1956, Lewis suggested to Hooker that Wibault's four centrifugal blowers should be replaced by a single axial-flow fan, mounted co-axially with the Orion engine, with the exhaust through a rotatable nozzle on either side of the fan casing.

At this stage, the design still included a reduction gearbox, but this was soon

In February 1971, the receiver, Rupert Nicholson, had
breakfast at Duffield Bank House (which had been
used since the mid-1930s by Rolls-Royce to entertain guests)
with Secor Browne, the chief of the American Aeronautics
Registration Board, who had been sent over by
President Nixon. Nicholson said later that: 'The talk
was very reassuring to me.'

An early Eastern Airlines TriStar at Logan Airport, Boston.

OPPOSITE TOP: Sir Kenneth Keith or 'K squared' as he was popularly known. He was appointed Chairman of Rolls-Royce in October 1972 and was prepared to serve, provided he was not 'buggered about'. He raised the morale and used his contacts ruthlessly to help put the company back on the map. Described by Prime Minister Callaghan as a 'scoundrel', he could nevertheless 'charm the underdrawers off all of us', according to the boss of Boeing, T. Wilson. When he took over, the company's future was still in doubt; when he handed over, nine years later, Rolls-Royce was definitely back as a world player. Here, he is greeting Sir Denis Spotswood, Marshal of the RAF, in Derby in 1972. On the right is Geoff Fawn, Managing Director of the Aero Engine Division.

OPPOSITE BOTTOM: Keith with Dennis Head, who succeeded Geoff Fawn.

TOP: An RB 211-524. Rolls-Royce was successful in convincing
Pan Am it was the engine for their TriStars.

BOTTOM: A British Airways 747, powered by RB 211-524s, over
Kai-Tek Airport, Hong Kong.

The Boeing 757

TOP: Rolls-Royce was successful in convincing British Airways and Eastern Airlines to buy their 757s specifying RB 211-535 engines. It was the first British Boeing aircraft to be launched with anything other than a Pratt & Whitney engine. Pratt & Whitney offered their new PW2037 as an alternative, claiming performance advantage. Rolls-Royce responded, adding their most advanced technology to produce the 535E4 shown here.

BOTTOM: The prototype 757 calls at East Midlands Airport for the benefit of Rolls-Royce employees, on its way back from Farnborough to Seattle.

The Ryan X-13 launched from, and landed back on,
a tilting platform.

A Thrust-Measuring Rig that quickly became known as 'the flying bedstead'. This was Rolls-Royce's first scheme to test A.A. Griffith's V/STOL ideas. It first flew, tethered, in 1953 and is seen here flying untethered at Hucknall a year later.

The Short SC.1. Five Rolls-Royce RB 108 engines were installed. Four were mounted vertically in crosswise pairs inside a central engine compartment to provide lift. The fifth was installed horizontally at the tail to provide propulsion. It first flew on 2 April 1957 and would contribute a great deal to the understanding of multi-engined V/STOL aircraft.

OPPOSITE TOP: The Mirage IIIV. Development of Rolls-Royce's second-generation lift-jet, the RB 162, began in 1959 and in November 1961 the company was authorised by the UK, French and West German Governments to continue its work for the Mirage IIIV.

OPPOSITE BOTTOM: The VFW VAK.191B V/STOL tactical reconnaissance fighter. One RB 193 engine was used for propulsion and to augment the lift provided by two RB 162 lift engines.

TOP: The P.1127. Two of this revolutionary vectored thrust design aircraft were ordered in 1959. Powered by the Bristol Pegasus, its first tethered hover took place on 21 October 1960 and the first untethered hover on 19 November. The first conventional take-off and landing was on 13 March 1961.

BOTTOM: Stanley Hooker, assisted by Gordon Lewis, evolved the concept of the Pegasus, in which a single engine performs the roles of both lift and propulsion. Its simplicity led to the adoption of the concept as the way forward for V/STOL.

The prototype XP.831 of the P.1127 on HMS *Ark Royal*.
It was the first V/STOL aircraft at sea.

The McDonnell F4 Phantom, powered by the
Rolls-Royce Spey.

OPPOSITE: HMS *Invincible*, powered by Rolls-Royce Olympus
engines, provided the air power with its Harriers, powered
by Rolls-Royce Pegasus engines, in the Falklands conflict in
1982. Commander ('Sharkey') Ward, DSC, wrote later:
'The Mirage IIIs and other aircraft had been outclassed by the
Sea Harrier on day one of air combat and, as a result, all
Argentine pilots were instructed to keep out of the way when
Sea Harrier was around.' Also in the photograph is a
Sea King helicopter powered by Rolls-Royce Gnome engines.

TOP: The Rolls-Royce Adour – a Rolls-Royce-led design with Turbomeca of France as the partner. Rolls-Royce built the combustion chamber and 'hot end' of the engine, while Turbomeca produced the compressor and gearbox.

BOTTOM: The Jaguar – the Ecole de Combat d'Appui Tactique (Combat Training and Tactical Strike) aircraft built by Breguet supported by BAC, and powered by two Adours. A great success, over 600 aircraft were sold.

OPPOSITE: The Adour also powered the British Aerospace Hawk (TOP) and the US Navy deck-landing version developed by McDonnell Douglas, the T-45 Goshawk (BOTTOM).

The Panavia Tornado (TOP) – Rolls-Royce beat off fierce
competition from Pratt & Whitney and provided the three-spool
RB 199 (BOTTOM). The Tornado, as a Multi Role Combat
Aircraft, was perhaps a compromise but the RB 199, after some
early problems, achieved the targets of fuel economy and
the ability to accelerate from low to high speed in a few seconds,
while enabling the Tornado to remain in the air for long periods.
Shown in the photograph are Tornados from British, German
and Italian defence forces.

seen to be undesirable because of its weight and cost penalties. One solution was to develop a low-speed power turbine for the Orion, but this was not practical as the Orion already had two spools and could not take a third without extensive and expensive redesign. Lewis and his team therefore replaced the Orion with the single-shaft Orpheus, which used the low-pressure spool from the Orion as its compressor. A new two-stage power turbine was added to the Orpheus, designed to run at the same speed as the Olympus fan, which it would drive by means of a shaft turning inside that of the Orpheus. Stanley Hooker became very excited at the prospects, as he wrote later:

At this juncture Sir Sidney Camm took a hand. [By this time, Hawker was becoming concerned about what would replace their very successful Hunter fighter.] He had been watching the Derby proposals with growing disbelief, and suddenly sent me a one-line whip: 'Dear Hooker, What are you doing about vertical take-off engines? Yours, Sydney.' I sent him our first BE.53 (Pegasus) brochure, and at once busied myself in Olympus work. It was thus a shock when Camm telephoned me a few weeks later, asking 'When the Devil are you coming to see me?' I replied 'As soon as you like, of course; but what is the subject?' And he said 'It's vertical take-off; I've got an aeroplane for your BE.53.'

We set off hot-foot for Kingston, where Camm showed us a drawing of the P.1127. It looked very like the Harrier of today and we were thrilled when Camm told us that, despite the inability of the Ministry to show much interest (because manned military aircraft were out), he intended to go ahead and fly a prototype. The Hawker Siddeley Board supported him, and put up the money, just as it had done in 1936 in tooling up to make 1,000 Hurricanes. In 1936, however, at least the company knew the Hurricane would be ordered eventually, whereas in 1958 there was absolutely no reason at all to think that the P.1127 would be ordered, even as a prototype.

On this historic visit to Kingston we met Roy Chaplin, whom I knew from Hurricane days pre-war, and Ralph Hooper and John Fozard. The latter were brilliant young designers who, after the death of Sir Sydney and the retirement of Chaplin, were to bring the revolutionary P.1127 through the intermediate stage called the Kestrel, to service as an operational weapon system for the Royal Air Force in early 1969 as the Harrier. To them and to Chief Test Pilot Bill Bedford must go the credit of producing the first practical VTOL jet.

Hooper's key contribution was the configuration of the rear nozzles close to the turbine exhaust, which moved the thrust centre forwards and enabled satisfactory balance to be achieved. And as Bristol now had a definite aircraft to work on, it developed the engine to a stage where it replaced the Olympus-based fan with a new two-stage trans-sonic fan of increased mass flow, designed from the outset to rotate in the opposite direction to that of the high-pressure spool. This fan gave the engine, by this time called the

Pegasus (there was a tradition at Bristol that an engine was not named until it became a committed project), better fuel consumption, allowing a reduction in the fuel carrying capacity of the Hawker P.1127. At the same time, Hooper moved the engine-driven accessories to the top of the fan casing, which meant the frontal area of the fuselage could be reduced.

And Hooker gave credit to Camm for his contribution:

Camm, remembering his 'Bifurcated' jet pipe in the Sea Hawk, suggested bifurcating the jet pipe on the Pegasus and using a second pair of left/right nozzles rotating in unison with the first pair. A further desirable feature was to make the HP and LP spools rotate in opposite directions, thus almost eliminating the engine's gyroscopic couple. This is a very important objective for a V/STOL aircraft, enabling it to hover under perfect control.

Hooker pointed out that there were many new features on the Pegasus.

[It was] one of the first engines to have no inlet guide vanes, the first part of the engine encountered by the airflow being the rotating fan blades. When we added the third stage, we found that we could overhang this fan ahead of the front bearing like a propeller as is today the practice on all turbofan engines. This enabled us to do away with the front bearing and all fixed radial struts or vanes upstream of the fan, and thus also to eliminate the hot-air supply previously needed to prevent ice forming on them. Inlet guide vanes were added to all early axial jet engines to swirl the incoming air in the direction of rotation of the rotating blades, in order to keep the relative Mach number between the air and blade below unity. With engines such as the Pegasus we had fans whose blades were supersonic over their outer portions, the tips running at Mach 1.3 to 1.5. We suddenly recognised that inlet guide vanes had become part of gas-turbine folklore and were no longer needed.

Another innovation was contra-rotating spools. On conventional aeroplanes the large gyroscopic forces imparted by the engine – which in effect is a spinning top – are not noticed because they can be continuously counteracted by the aircraft control surfaces. On the Harrier the airflow over these surfaces can be zero while hovering, so RCVs [reaction control valves] fed with HP air bled from the engine must be used instead. Any attempt to yaw the hovering Harrier, swinging the nose left or right, would have resulted in a powerful nose-down or nose-up tilt, because gyroscopic forces act at 90° to the axis of disturbance. Contra-rotating the two spools overcomes this, and the resultant gyroscopic force imparted by the Pegasus is exceedingly small.

The Mutual Weapons Development Programme was sufficiently encouraged by progress to provide 75 per cent of the funding necessary for Bristol to produce a small batch of Pegasus engines during 1958. The Pegasus 1, rated at 9,000 lb, first ran on the bench in September 1959. Pegasus 2, at 11,000 lb, ran in February 1960 and was cleared for flight in the P.1127 later that year. The first tethered hover took place on 21 October 1960 and the first untethered hover on 19 November. When asked what the transition from VTO to high-speed horizontal flight felt like, Hawker's chief test pilot, Bill Bedford, replied:

The aircraft felt like a brick on ice!

The first conventional take-off and landing was on 13 March 1961.

And the pressure to succeed was increased in 1961 when NATO initiated design competitions for aircraft which it intended should become standard throughout NATO air forces. The first was for a supersonic lightweight vertical take-off strike and reconnaissance fighter aircraft, and the second for a short take-off and landing tactical transport aircraft, intended both to support the fighter at dispersed sites in the forward battle zone and for general logistical purposed.

These were the first attempts by a large military organisation in the West to pull together all the various strands of research into VTOL and STOL (short take-off and landing) and turn them into a practical aircraft.

In May 1962, the Governments of the UK, USA and West Germany agreed to fund an improved version of the P.1127, and money was made available for the production of nine Kestrel aircraft and eighteen Pegasus 5 engines. The Kestrel trials lasted from March 1964 until March 1965, and enabled an assessment to be made of the practicality of V/STOL combat aircraft. In February 1965, Hawker Siddeley was given a new Government contract for development of a new P.1127 variant. The company initially described this as the Kestrel Development Aircraft, but was discouraged from doing so to prevent the US or German Governments claiming any proprietary rights. It was officially known as the P.1127 (RAF), and in 1967 became known as the Harrier.

The original RAF plan was to have two Harrier squadrons – one in Britain and one in Germany. This would mean sixty aircraft. Subsequently the RAF decided to have four squadrons and the original order for sixty-one aircraft was increased to 114. The Harrier entered service with the Royal Air Force in April 1969 and, as the Sea Harrier, with the Royal Navy in 1979. In 1971 the Harrier was equipped with the Pegasus 10 engine with 20,500 lb thrust and in 1976 with the 21,500 lb Pegasus 11.

THE HARRIER AND PEGASUS IN ACTION

On Good Friday 1982, it was announced that Argentinian forces had invaded the Falkland Islands, or to the Argentinians, Las Malvinas. The British Government, under Prime Minister Margaret Thatcher, had been reluctant to spend the money necessary to offer any sort of defence of the Islands and had even withdrawn the ice-breaker, *Endurance*, which was equipped with missiles and helicopters. This was obviously taken by the Argentinian Government, led by General Galtieri, as a sign of retreat. The general was therefore probably surprised by the furious reaction. The House of Commons met on a Saturday for the first time since World War II and was treated to the bizarre sight of the pacifist leader of the Labour Party, Michael Foot, condemning the usually belligerent Margaret Thatcher for her weakness.

If ever there was a perfect case for referring the issue to the United Nations this was it, but Thatcher adopted the role of Warrior Queen and on 10 April it was announced that a taskforce of 10,000 men would be assembled and sent to the South Atlantic to recapture the Islands. It was a risky venture and would not have been possible a few months later as the last two aircraft carriers in the Royal Navy that were despatched, HMS *Invincible* and HMS *Hermes*, had been destined for the scrapyard. There were amphibious assault ships such as the *Fearless*, but many of the troops were carried in tourist passenger vessels led by the *QEII*.

The logistic capacity of both the Royal Navy and the Royal Air Force was stretched to the limit – especially the RAF, which was expected to provide air cover and defence against missile attack. As Commander ('Sharkey') Ward, DSC put it in his book, *Sea Harrier over the Falklands*:

The surface and sub-surface threat was pretty well catered for, but what could a handful of Sea Harriers do against up to 200 Argentinian military aircraft? There were Mirage IIIs and Vs (faster than the jump jet and they could pull more 'G' as well), Navy Sky Hawk A-4s, Etendards armed with Exocet, Pucaras, and many more. It was a valid question.

Nevertheless, as Ward goes on to show in his book, the Sea Harrier performed outstandingly well.

The Mirage IIIs and other aircraft had been outclassed by the Sea Harrier on day one of air combat and, as a result, all Argentine pilots were instructed to keep out of the way when Sea Harrier was around (this was confirmed in Argentina following the war). Jets sent out from the mainland to bomb or strafe Task Force targets had orders to the effect that if a Sea Harrier CAP [Combat Air Patrol] got in their way en route to the target, they were to jettison their ordnance and return home. One Argentine attack pilot confirmed after the war that he was turned away by Sea Harrier CAPs on four missions before eventually getting through to his target.

Ward did not feel that the Sea Harrier was used as effectively as it could have been, writing:

Had the Command in the Falklands understood the Sea Harrier and its capabilities better, the aircraft could undoubtedly have been used to greater effect and the war might well have been a less costly affair.

In his 'Acknowledgements', Ward wrote:

To Rolls-Royce, who provided the power plant for Sea Harrier in the form of the Pegasus jet engine with its astonishing reliability and performance.

Chief of the Naval Staff, Admiral Sir Henry Leach, said after the war:

Without the Sea Harrier there could have been no taskforce.

Over 2,000 sorties were flown by forty-two Harriers of both the Royal Air Force and the Royal Navy, with individual aircraft flying up to six sorties a day. The reliability was such that aircraft availability exceeded 90 per cent throughout the campaign. Rolls-Royce could be proud of the fact that there were no unplanned engine removals.

The Falklands War was won and, after nearly forty years of defeat and retreat, this victory did wonders for the morale of the nation as well as for Margaret Thatcher's standing in the polls. Her popularity had been plummeting in the severe recession of 1980/1 but it now turned round and soared. She instructed everyone to 'Rejoice! Rejoice!' The *Sun* newspaper indulged itself in distasteful jingoism. It was left to Archbishop Runcie to remind everyone that people, including the enemy, die in wars and that there was nothing to be joyful about. Nevertheless, it gave Thatcher new stature as Prime Minister, allowed her to gain re-election in 1983 and continue the work she had been doing in forcing the country to wake up to the realities of international competition.

There was no doubt that the Harrier proved itself in combat. As it was a subsonic aircraft, it could not match at high altitude the Mirages flown by the Argentinian Air Force. However, at low altitudes the Mirages needed to use reheat and this meant that they would not operate at low level for fear of using too much fuel, rendering them unable to return to their bases in Argentina. At this low altitude the Harrier also showed itself superior in its manoeuvrability.

During the Falklands War, Stanley Hooker received a letter from the Deputy Chief of Staff for Aviation, Lieutenant General W.J. White, US Marine Corps, which said:

On the occasion of the highly successful introduction of V/STOL aircraft into combat, we congratulate you on your keen foresight and your steadfast pursuit of the development of the Pegasus engine for V/STOL aviation.

As you know, the US Marine Corps will soon make the transition to an all V/STOL light attack force with the Harrier II. This serves as the highest testimony to your contribution to the goals, of not only the US Marine Corps, but to those of free world aviation.

However, selling the Harrier in the USA had proved difficult and it was only after a marketing film was shown to the US Marine Corps Deputy Chief of Staff (Air), Major-General Keith McCutcheon, that he sent two pilots to Britain to test it. They made some criticism but, on the whole, their report was positive and the US Marine Corps ordered twelve Harriers in 1970, eighteen in 1971, thirty in 1972, thirty in 1973 and twenty in 1974. It was the first time since the First World War that operational military aircraft built in Britain were sold to the USA. In those days it was the Sopwith Camel and Dolphin. Now it was the Hawker Siddeley Harrier. (The USAAF did operate a photo-reconnaissance squadron of Spitfires in the Second World War and the English Electric Canberra bomber was built in the USA and powered by licence-built Armstrong Siddeley Sapphire engines.)

Early in 1976, McDonnell Douglas began work on a series of aerodynamic improvements to the original Harrier, the AV-8A. In mid-1978, the US Navy contracted for full-scale development of this new version, the AV-8B. British Aerospace, which had worked with McDonnell Douglas from the start of the programme, became the major subcontractor for the AV-8B for the US Navy and the prime contractor for the GR Mk 5 for the RAF. Developed and built during the 1980s, these new Harriers proved themselves in the Gulf War in 1991 where eighty-six Harrier AV-8B aircraft were deployed by the United States Marine Corps. Over 3,380 sorties were flown in eight days with 4,112 combat hours completed. Some 5.95 million lb of ordnance were delivered and, as in the Falklands War, aircraft availability exceeded 90 per cent and there were no unplanned engine removals.

The Commanding Officer of UMA-231 said:

The employment of the AV-8B was just as had been planned and explained for twenty years.

ALLIANCES

Just as collaboration became an important part of the aero engine manufacturers' policy on engines for the civil market, so it did in the military area as well.

Rolls-Royce had dominated the military aero engine market in Europe immediately after World War II when both the German and French industries were shattered. However, by the 1950s, France, with a long and proud aeronautical tradition, was beginning to rival the UK – both in the production of civil and military aircraft (the Caravelle airliner and military aircraft built by Dassault) and of aero engines built by the State-owned SNECMA. Germany, starting again in the 1950s virtually from scratch, was looking for help and, in the early 1960s, joined with France to produce two aircraft. The French Breguet combined with the German Dornier and other subcontractors to build an Atlantic patrol aircraft, the Atlantique, while the French Nord Aviation combined with Weser of Germany to build the Transall troop transporter. Both were powered by the Rolls-Royce Tyne engine built under licence by a European consortium. Neither aircraft was very successful, but the groundwork of co-operation had been laid.

Meantime, the Conservative Government of Harold Macmillan and the Labour Government of Harold Wilson first vacillated and then wielded the knife on support for military aircraft projects in Britain. When the Labour Party came to power in 1964 and found public finances in a weak state, defence budgets were an obvious target if public spending needed to be cut back. First the AW681, the V/STOL troop transporter, and P1154, the supersonic Harrier, were cancelled; followed, as we have already seen, by the TSR2. In total, these were huge blows both to the airframe and the components manufacturers.

The Government's plan was to replace these three aircraft with the F-111 from General Dynamics and the Phantom from McDonnell. The only consolation for Britain was the ordering of more subsonic Blackburn Buccaneers, which were already operating.

Development of the F-111 took so long that the British order was subsequently cancelled, but the Phantoms, powered by Rolls-Royce Spey engines, were delivered. As for future aircraft, the Phantom Committee, which reported in December 1965, called not only for support of the British Aircraft industry but also for collaboration. This begged the question, collaboration with whom? The answer was probably not with the USA, which did not need it, but with Europe – notably France and Germany.

In May 1965, agreement was reached with the French Government on the development of a supersonic attack/trainer, *Ecole de Combat et d'Appui Tactique* (Combat Training and Tactical Strike/ECAT, or later the Jaguar). Breguet, supported by BAC, as SEPECAT would build the aircraft while Rolls-Royce, supported by Turbomeca, would provide the engine, which would be the Rolls-Royce Turbomeca Adour. The two Governments also agreed to develop a sophisticated multi-role combat aircraft, the Anglo-French Variable Geometry or AFVG. BAC, supported by Dassault, would build the aircraft, while Bristol would combine with its old friend, SNECMA, to build the engine that would be the M145G.

The Jaguar was a success with eventual sales of over 600 aircraft, but the

AFVG arrangement was not, as Dassault broke away to develop its own variable geometry fighter, the Mirage IIIG. As a result, Britain, Germany and Italy came together to develop the Multi-Role Combat Aircraft (MRCA), later known as the Tornado. The initial order was for just over 800 aircraft – 385 for Britain, 324 for Germany and 100 for Italy.

In contrast to the deceptions and bad feelings over the AFVG, collaboration on the Jaguar worked harmoniously. The engines were two RB 172/T260 Adours, a Rolls-Royce-led design with Turbomeca as the partner. Rolls-Royce designed and built the combustion chamber and 'hot end' of the engine, and Turbomeca produced the compressor and gear box. Initially, the engine was a two-shaft turbofan of 4,600 lb thrust, boosted to 6,900 lb with reheat. Learning from Concorde that a separate management company was necessary, the two airframe manufacturers – BAC and Breguet – set up a new joint company, Sepecat (*Société Européenne de Production d'Ecole de Combat et d'Appui Tactique*) to co-ordinate engineering design, manufacture, finance and sales. This arrangement worked well and was to be used later in the formation of the three-nation company Panavia, set up by Britain, Germany and Italy to handle the MRCA Tornado.

Typical of the rapport within Sepecat was the selection of the name, Jaguar. Henri Ziegler, head of Breguet, rang BAC and said:

We need a name and it has to be the same in both languages. I suggest Jaguar.

Sir George Edwards agreed immediately, the Motor Car Company granted permission and, within twenty-four hours, sales literature was being prepared for the Paris Air Show of 1965. [William, later Sir William, Lyons had acquired the name for his car company, originally called Swallow Sidecar, from Armstrong Siddeley when they stopped making the Jaguar aero engine.]

According to Charles Gardner in his book *British Aircraft Corporation* (he was publicity manager for BAC from 1960 until 1977), the manufacture of the airframe went very smoothly but:

The main problem which Jaguar had to face throughout its early life was that the Adour engine was not coming on as well as it should. It was down on thrust, up on consumption and there were reheat problems.

The great Lord Hives is credited with the remark that he suspected there was much more to reheat than just piddling fuel down the jet pipe, and so, in the Adour, it was to be.

Eventually a part-throttle reheat system, by which the pilot increased reheat percentage as he opened up, was introduced successfully. Extra power was provided in the export version, the Jaguar International, with its two Adour 804-26 engines.

Whatever the setbacks, the Jaguar first flew on 8 September 1968 and by October had flown supersonically. Three more aeroplanes were in the air early in 1969. As Gardner admits:

This was a rapid programme by any standards for a sophisticated modern supersonic aeroplane; for one which had been the product of a two-nation collaborative enterprise, it was outstanding.

Rolls-Royce were so impressed by Turbomeca and its President, Joseph Szydlowski that they suggested to the British Government that he be recommended for an Honorary KBE. In making this proposal, Rolls-Royce told the Government:

He is a staunch supporter of any efforts to find further fields for this collaboration and has always been willing to find funds to progress them. For example, the development of the Adour which enabled the successful recent contract with India was a joint privately financed venture by Rolls-Royce Turbomeca Ltd; currently a further project for a joint company in Brazil is in progress.

In 1966, Szydlowski and Rolls-Royce formed an equally owned joint company, Rolls-Royce Turbomeca Limited, primarily for the design and manufacture of the Adour engine. This engine powers the Jaguar and Hawk aircraft, the former in extensive use in the Royal Air Force and L'Armee de l'Air, and is proving a highly successful project. Already, more than 1,300 Adours have been produced, Jaguars and Hawks are being sold widely, the Japanese produced the engine under licence, and the Indians are about to do so in considerable quantities. Much of the success for all these programmes should rightly be credited to M. Szydlowski who, in addition to his commercial acumen has, throughout the post-World War II years, emphasised his wish and firm intention to collaborate fully with Rolls-Royce and the United Kingdom; there has been ample demonstration that he fulfils this intention.

The Royal Air Force bought 200 aircraft as did the French Air Force, over 150 were sold to India, eighteen to Oman, eighteen to Nigeria and twelve to Ecuador.

As well as the Jaguar which it powered both with reheat (the Sepecat version) and without (the Jaguar International), the Adour has powered the British Aerospace trainer and light attack aircraft the Hawk Mk 1/50, Hawk 60, Hawk 100, Hawk 200, and the McDonnell Douglas naval trainer, the T45A Goshawk, all without reheat and in its licensed form built by Ishi Kawajima Heavy Industries in Japan, the Mitsubishi F1 fighter and T2 trainer. And, apart from Japan, it has also been produced under licence in Finland, India and Switzerland.

While the Adour was being developed and built, the Spey was being adapted for military use. There were early problems as Rolls-Royce struggled with its first attempt to reheat a turbofan engine. Pratt & Whitney encountered similar difficulties as it tried the same with its TF30 engine, which was fitted to the F-111 in the late 1960s. The Spey was fitted to the BAC Buccaneer bomber, to the Nimrod reconnaissance aircraft and, in its reheated form, to the McDonnell Douglas Phantoms, which were purchased by the RAF from the mid-1960s onwards. It was also licensed and developed with Allison in Indianapolis for production as the TF41, used in the Ling Temco Vought A-7 Corsair. A huge amount of effort went into the development of this engine and many Rolls-Royce engineers spent considerable time in Indianapolis.

The A-7 Corsair saw active service in the Vietnam War. Although (apart from this connection) Rolls-Royce was not directly involved in this war, it was such a traumatic event in the recent history of the United States, its effects being felt throughout the world, that we must cover it briefly.

The Vietnam War was unlike the Korean War in the sense that the USA was not able to persuade other nations to support it by sending troops to defend the 'freedom' of an oppressed nation. However, in theory at least, the principle was the same – the defence of a democratically elected form of government against threat from a Communist one.

In 1954, the former French Indo-China broke up into the Communist North and the non-Communist South Vietnam. For the remainder of the 1950s and into the 1960s, South Vietnam came under threat from the North, and the USA, feeling the need to defend the non-Communist states, gradually became more and more embroiled. President John Kennedy authorised the sending of military 'advisers' and, on 24 November 1963, the day before Kennedy's funeral, President Lyndon B. Johnson summoned a group of advisers to talk to Henry Cabot Lodge, the US Ambassador in Vietnam. Johnson told them he was determined not to 'lose Vietnam'. He told the National Security Council that the aim of the USA was to help the South Vietnamese to 'win their contest against externally directed and supported Communist conspiracy'.

As his presidency progressed, Johnson found himself being sucked deeper and deeper into the mire of the Vietnamese conflict. By 1969 the war had become the longest ever fought by the USA and, by this time, protests against the US participation were widespread not only in the USA itself, but also throughout the world – especially on university campuses. By the end of the year, it was announced that 40,000 US personnel had already died in the war. The number for Vietnamese casualties was not announced but must, by then, have been approaching a million.

Eventually, in 1973, President Richard Nixon called a halt and withdrew all troops from South Vietnam, ending their effective state of war (there had never been a declaration of war). The United States Defense Department published the vital statistics since the USA became involved on 8 March 1965. The death toll of the North Vietnamese and Vietcong was 922,290. The South Vietnamese armed forces lost 181,483 men and another 50,000 civilians were killed. The Department stated that the United States war dead numbered 55,337. The war left a scar on the United States and, of course, on Vietnam that would take more than a generation to heal.

In terms of numbers, the military versions of the Spey reached about the same level as the Adour. Domestic manufacture ceased in the early 1990s though the Spey RB 168-807 continued to be built under licence in Italy and Brazil for the Italian–Brazilian AMX close air-support aircraft. We have already seen how Kenneth Keith and Stanley Hooker negotiated a Spey licence in the People's Republic of China, an agreement that led to a small number of Speys being manufactured there.

THE TORNADO

We have seen above how the AFVG project foundered. The UK Defence Minister, Denis Healey, responded by gatecrashing discussions between West Germany, Italy, the Netherlands, Belgium and Canada on the design and production of a variable geometry aircraft. Gradually, the Netherlands, Belgium and Canada withdrew and, in December 1968, the Air Staffs of the UK, West Germany and Italy agreed on the new aircraft's configuration and an Interim Management Organisation was set up. The three main participating companies were BAC, Messerschmitt-Bölkow-Blohm and Fiat, with the RAF saying it would buy 385, the West Germans 600 and the Italians 200.

After announcing the project in the House of Commons on 14 May 1969, Healey told journalists afterwards that there would be a competition for the engine between Rolls-Royce and the USA (in this case, Pratt & Whitney) and that he believed Rolls-Royce could 'win on merit'. Rolls-Royce had two candidates for the engine – a two-spool engine designed at Bristol derived from the work on M-45 with SNECMA for the AFVG, and a three-spool reheated turbofan with similarity in configuration to the RB 211 designed at Derby, the RB 199. There would be formidable competition from Pratt & Whitney and it was decided that all Rolls-Royce's efforts should go into one engine. After much discussion (Gordon Lewis, the chief engineer on the M-45, described it as 'a nasty bit of internal competition'), the RB 199 was chosen, but Bristol was made responsible for the programme and for continuing the collaboration already begun with the German consortium, MTU (MAN Turbo and Daimler-Benz) and with Italy's Fiat Aviazione. No one was under any illusion about the competition – especially as there had been considerable criticism of the problems of the Spey in the McDonnell Phantom.

NAMMA (NATO MRCA Development and Production Management Organisation) only gave the engine companies sixty days from issuing the Request for Proposal, but after a period of intense activity the RB 199 was chosen for the MRCA. It was a big decision for Bristol as 2,000 engines would be required. The company set up to administer the whole project, Panavia, operated very successfully. As Charles Gardner put it:

There is no doubt that Panavia has, organisationally, been the most successful of European collaborative ventures.

In spite of this bold assertion made by Gardner in 1981, there were some worrying moments in the early years of the collaboration. This is clear from the letter, seen earlier, sent to Sir Kenneth Keith by the Controller of Aircraft at the Procurement Executive at the Ministry of Defence, Air Chief Marshal Sir Peter Fletcher, on 5 April 1973, in which he wrote:

The MRCA will provide the RAF with its strike, reconnaissance and air defence capabilities until the 1990s or beyond. If things went wrong there is no aircraft that could be purchased to replace it. There is no engine that could replace the RB199. Although the Governments and the air forces of the three countries are fully behind the programme there is a fragility about it because of the rather unpredictable situation in Germany and, to lesser extent, in Italy. There would not need to be a catastrophe on the engine; if a major development problem, and there will be some, were badly handled at a difficult time the MRCA programme could be in jeopardy.

Fletcher went on to comment in some detail about Rolls-Royce's handling of its collaboration with its partners on the RB 199.

During 1972 General Oblesner, Head of the German Delegation to the NAMMA Board of directors, and his staff became concerned about the fitness of the engine because of evidence from Turbo Union Ltd about slippages, cost increases, technical problems and failures. MOD (PE) were also concerned but, being more familiar than the Germans with the atmosphere that attends complex technical development programmes, were perhaps less apprehensive and less emotional...
    It should be noted that at RB199 project progress meetings held throughout the development period Turbo Union and in particular Rolls-Royce have spoken with frankness about the problems they have had and no particular friction seems to have arisen between the technical agencies involved. In particular Rolls-Royce seemed to have, and still seem to have, good relations with their partners in Turbo Union.

However, Fletcher went on to say that Turbo Union seemed slightly insensitive to the need to give the German authorities reassurance 'because of the situation surrounding the German elections and the uncertain status of the MRCA project at the time'. He was also critical of Rolls-Royce, citing a meeting in December 1972 when the 'Rolls-Royce (71) team spoke in a regrettably arrogant manner, did not address themselves to the main problems, said that their time would have been better spent at Bristol and clearly exacerbated the situation: in fact making the Germans very angry and destroying much confidence in the company and the engine.'

This caused the Germans and Italians to demand a live demonstration of the engine, which took place on 21 February 1973. Fortunately, the test was highly successful, convincing them that the RB 199 was technically sound.

Once the MRCA Tornado was launched, the aircraft plus its RB 199 engine demonstrated how the rapidly advancing technology of aerial combat had dictated a constant reappraisal of military engine technology. The Tornado could loiter at low speed and climb at high speed to altitude. It had to be able to attack enemy aircraft at great heights and also attack at low level, penetrating enemy defences behind the battle lines. It was designed to fly at supersonic speed at 200 feet, following the contours of Europe and thereby keeping under Soviet radar. The engine designers had to consider fuel economy, the ability to accelerate from low to high speed in a few seconds and also the sustained running power to enable the aircraft to remain in the air for long periods. The designers achieved all of these targets in the RB 199.

However, it has to be said that the RB 199 was being asked to do too much: the Tornado was really a compromise, perhaps a politician's compromise aircraft, in that, as a Multi-Role Combat Aircraft, it was trying to be both a ground-attack aircraft and a high-altitude fighter. As Colin Green, Operations Director at Rolls-Royce in 2000, pointed out:

The engine was best for a ground attack role. For combat situations it needed more reheat. The three-shaft configuration in a high performance military engine with significant power offtake limited its ability at very high altitude.

As we shall see in the third volume of this Rolls-Royce history, in the Gulf War in 1991 it was accepted that the Tornado performed well in its ground attack role, though, as the engine was pioneering new technology, the complicated turbine blades experienced problems with the desert dust. As Group Captain Jock Heron, a long-serving RAF pilot and subsequent executive at Rolls-Royce in Bristol, pointed out:

The conditions necessitated a long programme of work on single crystal technology to prolong the life of the blades.

Prospects for the RB 199 in the USA were not good. Both General Electric and Pratt & Whitney had a number of competitive engines, in particular the GE F404.

If the Adour, the RB 199, the military Spey and Pegasus provided most of the turnover for Rolls-Royce's presence in the military engine market, we should not forget the Viper.

Originally designed and developed by Armstrong Siddeley as a short-life power unit for the Australian Jindivik target drone (i.e. an unmanned aircraft), it was developed into a conventional turbojet and enjoyed a long and successful life, powering such aircraft as the Italian Macchi MB-339 trainer, the Yugoslav–Romanian ORAO twin-engined strike aircraft, the Yugoslav Soko Galeb and Jastreb aircraft, and the Indian Air Force's HJT-16 Kiran. Some twenty-nine air forces throughout the world bought the Viper.

HELICOPTER ENGINES

Buying Bristol brought the original de Havilland factory of Leavesden into the fold with its production of small gas turbine engines aimed at helicopters. The leading engine was the Gnome. Designed around a high-pressure-ratio axial compressor of exceptionally small dimension, the Gnome turboshaft engine combined the advantages of high power output, compactness, light weight and low specific fuel consumption. Such characteristics made it an ideal engine for helicopters. Early users were the RAF Whirlwinds, Westland Wessex Mk 2 and Mk 5, Italian Agusta Bell 204B, Agusta 101G and Boeing/Vertol 107. By the 1980s, twenty-one civil and military operators worldwide were using the Gnome.

Following the Gnome came the Gem, which was developed in response to an Anglo–French agreement signed in May 1967 to develop three helicopters. The first two were of French design, the Aerospatiale Puma assault helicopter and Gazelle light observation helicopter. The third would be the Westland WG13 Lynx utility helicopter with its BS360 engine renamed the Gem. The Lynx first flew in March 1971 and the two Gem engines powering it performed almost perfectly. In February 1972 the Gem passed its 150-hour Type Approval Test and in December 1975 production began. It is fair to say that there were some reliability problems. Gordon Page, now Chief Executive of Cobham plc, worked for Rolls-Royce from 1962 until 1989, and in 1984 was made Director, Helicopter Engines. He saw his main task as 'sorting out the Gem' which was a three-shaft engine, in his view 'ridiculous', and which was causing the Army to 'complain like hell'. Many would say that the Gem, largely designed by Fred Morley, was a little too sophisticated. In all fairness to Morley he was recalled to Derby to help on the RB 211 during the crisis of 1970 and 1971, and could not therefore supervise the sorting out of the problems that almost every new engine experiences. The Gem 40 was replaced by the Gem 42 and the Army Air Corps gradually re-equipped their Lynx helicopters with this engine. It became the automatic choice as the European NATO standard equipment on small ships.

From the early 1980s the European Helicopter, EH101, was in its planning stage and the capacity of the aircraft suggested that three engines with power around 1,700 hp would be needed. Certainly the Royal Navy needed a replacement for its Sea Kings. Rolls-Royce talked to Turbomeca and a joint design was launched in April 1981. By the time the EH101 was positively launched, the specification – to meet the demands of the Royal Navy and the Italian navy – called for engines of 2,100 hp, and the earlier design, designated RTM 321, was now revised as Rolls Turbo Meca 322. The engine, similar in concept to the Gem, was much simpler with a single gas generator rotor and only two bearing chambers.

As well as aiming at the EH101, Rolls-Royce also marketed the RTM 322 to power the Sikorsky Black Hawk which was using the GE T700. Gordon Page went to see Bill Paul at Sikorsky, who said to him:

However good your engine is, don't underestimate the task of trying to replace General Electric. You'll have to try it in a Black Hawk yourself.

As a result, Rolls-Royce bought a Black Hawk, which needed a British Government authority. Ironically, the authority was signed by the Conservative Minister of Defence, Michael Heseltine, who was soon to resign over the Westland helicopter affair as he felt that Prime Minister Thatcher was not giving sufficient consideration to a European, rather than a US, solution to future helicopter production. Rolls-Royce demonstrated the Black Hawk with the RTM 322 at both the Paris and Farnborough air shows. Although, as Bill Paul had predicted, GE was not dislodged, there remains the possibility of the RTM 322 powering the new Sikorsky S-92.

In the meantime, the engine was chosen by the Ministry of Defence to power the EH101 Merlin Mk 1 helicopters for the Royal Navy. We shall read more about that in Part Three of this history.

CHAPTER TEN

PRIDE RESTORED

PLANS FOR THE 1980s
'THIS ENGINE MUST WORK BY DECEMBER!'
THE BIG ENGINE MARKET
PRIVATISATION
AIMS AS A PRIVATE COMPANY

PLANS FOR THE 1980s

IN DRAWING UP ITS CORPORATE PLAN for the decade of the 1980s, Rolls-Royce set as its prime objective the consolidation of the position already established with existing civil and military engines, upgrading them where necessary to preserve an adequate competitive advantage.

The RB 211-524, at 50,000 lb thrust, had by this time established a strong position with the Boeing 747. The initial RB 211-524B2, which had entered service in 1976, had demonstrated a fuel consumption level about 7 per cent better than the original Pratt & Whitney-powered version and between 1 and 2 per cent better than the latest Pratt & Whitney- and General Electric-powered versions that had since entered service. An improved 53,000 lb RB 211-524D4, which would be certificated in 1981, would provide a further 5 per cent improvement in fuel consumption and should still provide a 2 or 3 per cent lead over Pratt & Whitney's substantially redesigned JT-9 and General Electric's CF6, which were expected to be available from 1983 onwards. Nevertheless, to consolidate its position Rolls-Royce planned to introduce a 56,000 lb thrust engine in 1985 accompanied by a further 5 per cent fuel consumption improvement. On this basis, Rolls-Royce expected to improve its 747 market share (at 22 per cent in 1979) without recourse to lower prices than the competition.

294

For the Lockheed TriStar and its uprated versions, Rolls-Royce expected to remain the sole supplier of engines and did not anticipate having to commit to specific engine developments.

With regard to Boeing's 767 and Airbus Industrie's A300, Rolls-Royce aimed to be in a position where, from a technical point of view, an RB 211-powered version of either could be delivered within three years of a firm airline commitment. And just as the RB 211-524 was well matched to the 747, so the RB 211-535 was well matched to the Boeing 757 and its likely derivatives. Boeing was envisaging a substantial market for the 757 and Rolls-Royce felt itself in a privileged position but accepted that the market would probably have to be shared with the CF6-32, General Electric's corresponding CF6-6 derivative. As we have seen, Pratt & Whitney launched the PW2037 and GE never produced the CF6-32. The -535 did not have a fuel consumption advantage in theory, although it was expected in practice – thanks to the superior performance retention characteristics of the RB 211 design.

The position on military engines in 1979 was that Rolls-Royce felt that its range, while not having the advantage of the long runs of its US competitors, was nevertheless very appropriate to the needs of the market. The RB 199, in addition to its Tornado application, would meet the requirements of other tactical combat aircraft. The Pegasus, with relatively small developments, would meet the requirements for a dispersible ground-support aircraft aimed at ASR4009 – such as either the Harrier Mark V or the McDonnell Douglas AV-8B.

After considerable teething problems, the Gem helicopter engine was capable of development to competitive performance and reliability. One of its problems was its relatively high manufacturing cost.

The Adour, flown in the Jaguar, Hawk and Mitsubishi F1 and T2, was basically a low cost design though the 50 per cent French manufacturing share increased the cost of production. The Spey, though long in the tooth, was still competitive and was finding new applications where low cost was important, such as the Italian AMX aircraft. A version was still being manufactured in conjunction with Allison for the USA LTV A-7 Corsair ground-support aircraft. As well as this US collaboration, the RB 199 and Adour were collaborative European projects. Although Rolls-Royce would inevitably have to look mainly to the USA, the European links were a valuable balance, particularly in terms of the military side of the business.

Prospects for the RTM 322 had improved since it had become clear that the previously chosen US engine, the T700, was not powerful enough for several other applications for which the new helicopter was intended.

Military sales currently represented over 55 per cent of the company's turnover, amounting to £920 million, of which about a third was exported.

Of the turnover from sales to the military, some 74 per cent was from aircraft applications, 6 per cent on helicopters, 7 per cent on naval gas turbine applications, 9 per cent on nuclear and the remaining 4 per cent on other military sales. There was no doubting the importance of the military sector to Rolls-Royce.

Rolls-Royce had been right to turn to the civil market after the Second World War. In the war-torn world of 1945, it was difficult, except for a visionary few, to imagine a situation where large sections of the more prosperous nations would fly all over the world in search of ever more exotic holidays. This civil market, which also included freight as well as passenger aircraft, had been the big growth market of the 1950s, climaxing in the 60s and 70s, and continuing into the 1980s. Nevertheless, the military market was still extremely valuable.

Rolls-Royce's Corporate Plan covering the years 1982–91 envisaged continued healthy sales to military customers not only in the UK but in many overseas markets throughout the world. This was, after all, the era when the Cold War was still part of every government's thinking and, indeed, confrontation with the Soviet Union was to reach more menacing heights in the 1980s.

Specifically, for the RB 199 on the Tornado IDS and Tornado ADV, Rolls-Royce intended to support British Aerospace and its European partners to secure the launch of the Agile Combat Aircraft (ACA), which had been specified with the RB 199 Mk 104, which was already committed for the Tornado ADV. At the same time, export applications for the RB 199 would be sought in the various single-engined light combat aircraft projects under study requiring engines of RB 199 size and performance. For the Pegasus on the Harrier and AV-8B, Rolls-Royce saw limited further uprating to meet the medium-term requirement for increased thrust in the two aircraft. However, Rolls-Royce should be ready to provide a major performance improvement in the longer term.

The Harrier's outstanding achievements in the Falklands War in the spring of 1982 re-emphasised the dependence of the success of these aircraft on the performance of the Pegasus with its unique vectored thrust concept.

With the Adour in the Jaguar and the Hawk, Rolls-Royce should continue to support British Aerospace's efforts to increase the export sales of the Hawk and continue to pursue opportunities for new aircraft applications. It should also be assumed that an Adour replacement would be needed by the mid-1990s and that its development would need collaboration. The growing success of the Hawk and particularly its selection as the US Navy Trainer was giving the Adour a new lease of life. British Aerospace believed that the longer-term export prospects of the Hawk could be further improved if the hot-day thrust of the Adour 56 could be increased by about another 20 per cent by 1988. There were no commitments beyond the Viper 680, which was used in eight trainer and ground-attack aircraft. The only new application for the military Spey was for the AMX due to start production with the Spey Mk 807 in 1986, but spares would still be provided for the Buccaneer, Phantom and Nimrod. Similarly, there were no plans for further development of the Tyne and no new applications foreseen beyond the G222, which was due to enter service in 1985.

On the civil side, Rolls-Royce was suffering badly. In the twenty-four months to December 1984, only 126 new RB 211s would be delivered, compared with the 350 anticipated even in the low scenario of the previous year's plan. Although air traffic was continuing to grow (4 per cent in 1982), deregulation of the airline industry in the USA had led to severe competition depriving the airlines of the funds to replace their older, noisy and fuel-thirsty aircraft with new aircraft. Even the few sales being made were achieved only with imaginative financing and substantial concessions from the airframe manufacturers.

Specifically, Rolls-Royce planned to launch an RB 211-524D4A for the European Airbus in order to overcome the existing weakness of not having its engines on offer in the most important of Airbus Industrie's products. Such a development would also help the -524's prospects on the 747. Additionally, the company would pursue launch opportunities in the proposed stretched version of the Boeing 757. With RB 211-535, which was already on the Boeing 757, Rolls-Royce would pursue other opportunities – in particular, the McDonnell Douglas MD-100 – and would proceed with an advanced engineering programme to achieve technical readiness for a further 5 per cent specific-fuel-consumption improvement by 1987.

With the RJ500, on which Rolls-Royce was collaborating with the Japanese, the plan was to continue the technology work necessary to produce not only a competitive engine suitably sized for the A320-200 but also capable of being adapted to the smaller fan size required for installation in the 737-300 and -400 or for a redesigned DC-9-80. The Civil Spey, installed in the F28 and Gulfstream, would be upgraded to a high flow derivative, the RB 183-3, to be suitable for the next Gulfstream and the stretched F28. Commitment from Gulfstream Aviation was essential to provide a satisfactory return on the required investment.

The optimism of the Rolls-Royce corporate plan for the 1980s was soon tested in the difficult conditions created by the recession of 1980 and 1981, so that, when the ten-year plan from 1982 to 1991 was drawn up, it had to take into account a deteriorating current situation. Lockheed had decided to terminate production of the L.1011. Current sales in the rest of the civil market (especially the sectors served by the RB 211) were lower than fore-cast, as were sales of military engines and spares, mainly because of delayed requirements for the RB 199 for the Tornado. Overall, turnover was 5 per cent lower in 1982 than 1981 in real terms and was expected to be about 10 per cent lower in both 1983 and 1984, compared with the previously expected rise of 3 per cent per annum.

The termination of the L.1011 left the RB 211 with only two civil applications – the -524 version in the Boeing 747 and the -535 version in the 757. It put extra pressure on securing orders for the -524 in the European Airbus. (The Fokker F28 kept the Spey in civil, as well as military, applications. It kept Fokker going, allowing it to develop the Fokker 100 and Fokker 70.)

As Rolls-Royce looked ahead into the 1980s, there was no doubting the continued fierceness of the competition from General Electric and Pratt & Whitney. At the end of 1980, Pratt & Whitney picked up two large orders from Delta and American Airlines to power its 757-200s, and Lord McFadzean felt it necessary to reassure the Rolls-Royce workforce, saying in the company newspaper that the improved version of the RB 211-535

. . . will achieve a performance in airline service to match that promised by our competitors, and we aim to capture a substantial part of what will be a very large market over the next two decades . . . Rolls-Royce fought hard to win those [Delta and American] orders. We brought forward our planned future improvements in the 535. We made good commercial offers but clearly the competition must have offered a better package; this must have been costly.

As it turned out, American Airlines did not take the 757 at this time and, as we shall see, in the later competition selected Rolls-Royce engines for their aircraft.

Francis Scott McFadzean, created Baron McFadzean of Kelvinside in 1980, succeeded Sir Kenneth Keith, who was also created a baron in 1980 (becoming Lord Keith of Castleacre), as Chairman of Rolls-Royce in that year, 1980. Born in 1915 in Glasgow, McFadzean was a graduate of Glasgow University and the London School of Economics. He had made his career in Shell and was a director of the Shell Petroleum Company from 1964 to 1986. He was also a director of the Beecham Group from 1974 until 1986, and of Coats Patons from 1979 until 1986.

Anthony Sampson wrote of McFadzean in *The Changing Anatomy of Britain*:

Keith was on the board of British Airways, and was later followed by his friend Frank McFadzean, the joint head of Shell who became chairman of British Airways, which he ran for two days a week from the Shell building – to the fury of the airline's professionals . . . Keith in the meantime had taken on another nationalised industry,

Rolls-Royce, which he ran for eight years until 1980 when he was succeeded by (surprise!) Frank McFadzean, now Lord McFadzean.

Ashley Raeburn, a director of Rolls-Royce in the late 1970s and early 1980s, was a great admirer of McFadzean's work at Shell. Of McFadzean's time at Rolls-Royce, Raeburn said:

He was perhaps past his best.

In stark contrast to Pearson, who as Chairman perhaps did not spend enough time in London, McFadzean only visited Derby, the centre of the Rolls-Royce empire, on four occasions during his chairmanship from 1980 until 1983.

It is fair to say that Rolls-Royce still faced many difficult problems when McFadzean took over as Chairman. Keith had boosted morale in the company after the devastating blow of the receivership in 1971 and he had been responsible for improving the marketing efforts of the company. One of his first actions was to beef up the sales, as is clear from this extract of a letter he wrote to Lord Carrington at the Ministry of Defence in July 1973:

The military side of our business is very important, not only as part of the international defence potential, but also as a substantial commercial undertaking. At present we supply ninety-two different Forces in over 60 countries, and our engines power forty-eight different types of military aircraft. The military content of our business is 47 per cent of the total . . . I am convinced that, as in the civil field, our military marketing effort is quite inadequate . . . It is therefore of paramount importance that we have a top executive . . . who not only has a thorough knowledge of military aviation . . . but is himself well known and respected by customer governments.

However, Keith had not tackled some fundamental defects – most notably, chronic overmanning. In 1980, Rolls-Royce still employed 62,000 people, almost as many as the 64,000 of 1971. Sales had grown in the 1970s but inflation, an adverse sterling–dollar exchange rate and a prolonged strike in 1979 had meant low profits in most years and losses of £22 million (about £220 million in today's terms) in 1976 and £58 million (about £348 million today) in 1979. Nevertheless, Rolls-Royce had established itself in the big engine market and in 1980, its 1,000th RB 211 went into production.

McFadzean told the House of Commons Trade and Industry Committee that the decision by Lockheed to terminate TriStar production would mean that Rolls-Royce:

. . . would probably lose a turnover of something approximating to £250 to £300 million in the late 1980s and that would manifest itself in a loss of profit of something between £50 million and £70 million.

McFadzean's reaction was to cut costs, and Rolls-Royce's UK workforce was cut by 6,000 in 1981 and by a similar number in 1982, reducing it to just over 46,000 by the end of that year. Few people have given praise to McFadzean's chairmanship but it is acknowledged that he tackled manpower costs, an area largely ignored while Keith was in office.

To help reduce costs, Rolls-Royce streamlined part of its operations by off-loading production of a wide range of small parts to external companies. This alone brought inevitable redundancies within Rolls-Royce.

And, in 1982, the battle became even fiercer. The estimated market for the Boeing 757 was more than 1,000 aircraft, which meant a market for over 3,000 engines plus spares. Pratt & Whitney's advertising pulled no punches saying:

If you could buy fuel at yesterday's prices the competition's engine might be adequate.

Pratt & Whitney described the RB 211-535C as

. . . a derivative of yesterday's models' and lambasted it as a 1980s engine with 1960s configuration. Less fuel-efficient than Pratt & Whitney's all-new engine. And a derivative doesn't give you much room for improvements.

In what the *Financial Times* described as 'a pained but dignified response', Ralph Robins sent a personal message to 2,000 airline and business executives which said:

I want you to know that we regret that Pratt & Whitney feel it necessary to adopt this approach . . . Rolls-Royce has taken the proven design concept of the RB.211 and incorporated state-of-the-art advanced technology . . .. [Those still in doubt] may wish to seek Boeing's independent assessment of the competing claims for the two engines.

Part of the reason for the fierceness of this competition was the scarcity of orders because of the world recession of the early 1980s. The world had reeled in the mid-1970s when OPEC quintupled the price of oil. It was just recovering when further turmoil in the Middle East, prompted by the overthrow of the Shah of Iran, led to another sharp rise in the price of oil. Once more the whole world had to find the money to pay for a commodity that had become vital in every country's economy. Inflation and recession were the inevitable consequences and once again Britain seemed to suffer more than its competitors, though at least this time the opportunity was taken to push through some much needed changes.

The 1970s had not been a good decade for Britain. Structural economic weaknesses, exacerbated by continuous clashes between trades unions wedded to restrictive practices and poor management hindered by unhelpful legislation and unsupportive Government, had led to a situation where the country was not only declining in relative terms compared with its main industrial competitors, but even in absolute terms. Fortunately, a new Government leader was elected in 1979, and she showed a determination to tackle some of Britain's deep-seated problems.

From February 1975, when she was elected leader of the Conservative Party, until she came to power in May 1979, Margaret Thatcher and her supporters prepared the ground for a revolution in Britain's economic attitudes. Hugo Young, in his biography *One of Us*, summed up her position when she assumed power:

She brought to that post no great technical expertise, but a handful of unshakeable economic principles.

They were not particularly original purposes. But they were commitments made with the fire of the zealot who could not imagine she would ever become a bore. Tax cutting was one of them. In four years as Opposition leader, she had hardly made a single economic speech without alluding to the punitive rates of tax. Usually this was in reference to the upper rates. 'A country's top rate of tax is a symbol,' she said in February 1978, 'very little revenue is collected from people in this country who pay tax at the highest rates. A top rate of 83 per cent is not much of a revenue raiser. It is a symbol of British socialism – the symbol of envy.' Restoring the morale of management, she said around the same time, was the prime requirement. 'No group is more important, and yet none has been so put through the mangle and flattened between the rollers of progressively penal taxation and discriminatory incomes policy.'

So tax cuts were the first objective. They were the traditional Tory nostrum, to which the leader brought her special proselytising zeal. The second was good housekeeping.

The third fundamental therefore defined itself: the control of public spending. And again a kind of pietistic morality went hand in hand with a supposed economic law.

This was music to the ears of businessmen, but to a certain extent – even if not spelt out with quite the same rigour – they had heard it before. It will be remembered that former Conservative Prime Minister Edward Heath had come to power in 1970 making not dissimilar noises, only to buckle under to corporatist solutions as soon as the going became rough. Would this woman be different?

We know now, not least from the memoirs that have been published by several of her colleagues, that Margaret Thatcher did not have the full-blooded backing of the whole of the Cabinet, nor indeed from many leading businessmen, for some of her stronger measures.

The general financial climate from 1980 to 1982 had been the worst since World War II, certainly in terms of corporate failures and rising un-employment. The result of tight monetary policies allied to another sharp hike in the price of oil following the overthrow of the Shah in Iran, which in turn brought a sharp rise in the value of the pound – viewed by this time as a petro-currency thanks to Britain's near self-sufficiency in oil from the North sea – was to create the most difficult trading conditions British companies had experienced since the 1930s. ICI, generally viewed as the bell-wether of British manufacturing industry, cut its dividend and Sir Terence Beckett, the Director-General of the CBI (Confederation of British Industry), promised the Government a 'bare-knuckle fight', saying at the CBI conference:

We have got to have a lower pound – we've got to have lower interest rates.

The Prime Minister remained firm, and declared on the day unemployment passed the two-million mark (Heath had lost his nerve when it passed the one-million mark):

I've been trying to say to people for a very long time; if you pay yourself more for producing less, you'll be in trouble.

Heavy wage inflation, allied to the near doubling of VAT in the first Budget of the new administration introduced within weeks of its winning power, meant that inflation soared again to an average 13.3 per cent in 1979, 18.1 per cent in 1980 and 11.9 per cent in 1981. Thereafter it fell to under 5 per cent in time for the general election of 1983, aided by the overvalued pound and moderate wage settlements induced by an unemployment level not seen since the 1930s.

The result of all this, allied to a very tough attitude towards unions and strikes epitomised by the Government's approach to a national strike in the steel industry (a strike it was determined to win, whatever the cost), meant that the Government's and Thatcher's popularity fell to new depths by the end of 1981. But Thatcher pressed on regardless, and although she had invited people from the whole spectrum of the Party into her Cabinet, she took little notice of what many of them said. Indeed she made it quite clear that:

If you're going to do the things you want to do – and I'm only in politics to do things – you've got to have a togetherness and a unity in your cabinet. There are two ways of making a cabinet. One way is to have in it people who represent all the different viewpoints within the party, within the broad philosophy. The other way is to have in it only people who want to go in the direction in which every instinct tells me we have to go. Clearly, steadily, firmly, with resolution. We've got to go in an agreed and clear direction.

She was not going to 'waste time having any internal arguments'.

Thus, at the depths of the recession and with unemployment still climbing sharply, she strongly supported Sir Geoffrey Howe's Budget in the spring of 1981, which further intensified the squeeze on both companies and the consumer. Initially, Howe had been against any such budget and even the Treasury felt some relaxation of monetary policy was possible. However, Thatcher had now taken on two extra-parliamentary advisers – Sir John Hoskyns (a former Army officer who had built up his own very successful computer software business) and Alan Walters (a former professor at Birmingham University and the London School of Economics, who had established his monetarist principles long before the phrase was heard in political circles; indeed he had predicted the inflation of 1974 as early as 1972, analysing the explosion of credit in that year of the Barber Boom).

Hoskyns and Walters stiffened Thatcher's resolve in the early months of 1981. To them it was a matter of credibility. The Government had to show everyone once and for all that it was not going to move forward 'with enormous and insupportable borrowings'. It had to convince the financial markets that it would get inflation down. And it worked. The financial markets saw that here was a Prime Minister who stuck to her guns, and applauded her toughness and resolution. As many predicted the U-turn that tough-talking Governments of both parties had indulged in since World War II, Thatcher went to the Party conference and told them:

You turn if you want to; this lady's not for turning.

Her determination – some would call it bloody-minded obstinacy – was rewarded. Inflation fell sharply, productivity increased and industry recovered, helped by a falling pound after 1982. In the United States, a leader of similar anti-corporatist solutions also came to power, and the world economy began a long upswing that lasted for the rest of the 1980s.

But it was not only Britain that suffered. In the United States there were also some structural defects in the economy and some painful changes would be necessary to compete in a world where smoke-stack industries would no longer be the growth engines they once were. The electorate also voted in a President committed to a more *laissez-faire* approach, and one of his first acts was deregulation of the airline industry, ending the previously cosy pricing arrangements and enabling entrepreneurial airlines to compete on price. Just as the miners' strike was seen as a turning point in industrial relations in the UK, the US Administration's standing-up to the air traffic controllers when they went on strike was seen as the dawn of a new era in the USA.

This was all good news in the long run. In the short term, it meant economic turbulence and precious few orders for aircraft or aero engines. Orders for new jet airliners in the late 1970s had reached 700 to 800 a year. In 1981, only 333 were ordered and in 1982 only 223. The effect on the aero engine manufacturers was severe. Rolls-Royce had geared itself up to make 300 RB 211s in the early 1980s but was forced to slash its forecasts and, in the event, produced less than a quarter of that number in 1983. Orders in 1982 reached a mere thirty engines for commercial airlines.

Faced with this fall in turnover, while at the same time having to keep investing in the development of new engines, it is not surprising that the three leading engine makers – Rolls-Royce, General Electric and Pratt & Whitney – were keen on collaboration.

'THIS ENGINE MUST WORK BY DECEMBER!'

In the mid-1970s, Rolls-Royce had already collaborated with Pratt & Whitney, MTU of Germany and Fiat on a 25,000 lb thrust engine, the JT-10D. The engine did not represent a technical breakthrough, but was planned as a new design of similar complexity and fuel consumption to the existing larger engines of that time. It was partly aimed at meeting the competition from the General Electric/SNECMA CFM56 and was mainly aimed at Boeing's three-engined medium-range 200-passenger wide-body tri-jet, initially called the 7X7. Both Rolls-Royce and Pratt & Whitney recognised that if such an aircraft materialised, neither could justify launching the required intermediate-size turbofan on their own.

Sir Denis Spotswood, Vice-Chairman of Rolls-Royce and former Chief of the Air Staff, told the board on 24 February 1976 that a meeting had been held the previous week at Norfolk House, Rolls-Royce's London headquarters, to review the technical status of the JT-10D and bring together conclusions of separate meetings held earlier – between Rolls-Royce and Pratt & Whitney and Rolls-Royce and MTU – 'to examine the technical competitiveness of the JT-10D design in relation to other possible configurations, growth potential and market applicability.'

The conclusion of the final meeting was that the approval of the British and German Governments should be obtained as soon as possible for the collaborative agreement to proceed. He continued in his report:

This conclusion was based on the considered judgement that the overall balance of technical advantage lay with the JT10D against the CFM56 and that it was necessary to proceed with the development programme at full speed in order to obtain the maximum market advantage. If we waited until the market was certain we would ensure that the competition obtained a major part of it. It was equally emphasised that the number of variables was such that it remained essential to watch closely for developments of rival 'big twins' over the coming months and to respond rapidly to market intelligence as it emerged.

As it happened, Boeing decided to meet the requirement for this type of aircraft with the twin-engined 767, capable of using improved versions of the existing large turbofans, together with the 757 twin, capable of using their smaller fan derivatives. This decision reduced the potential for the more complex JT-10D to a level at which the investment could not be justified.

Nevertheless, the collaboration with Pratt & Whitney taught Rolls-Royce two lessons. Their respective engineers could work well together, but satisfactory commercial arrangements were difficult to achieve with two companies in competition on other related projects. For example, Pratt & Whitney had the final say on sales concessions and Rolls-Royce would have been compelled to share the costs. This would have been particularly galling in a situation where the JT-10D had been in competition with the RB 211.

For the 1980s there seemed little prospect of further collaboration with Pratt & Whitney as there was no gap in the market. The market from the 757 upwards was covered by the RB 211 in its various versions and, for smaller aircraft, Pratt & Whitney had recently launched two refanned versions of the JT-8D whereas General Electric/SNECMA had the CFM56. Before the JT-10D was abandoned, Rolls-Royce did try to persuade Pratt & Whitney that, instead of launching the refanned JT-8D, it should collaborate on a 20,000 lb thrust derivative of the JT-10D, which would have had much better fuel consumption. However, Pratt & Whitney declined the offer.

The aero engine industry in the early 1980s entered a phase of consolidation following the rapid technical advances from the 1950s to the early 1970s. In the civil field, the top priority was improving performance, reliability and component life on existing large turbofans and developing derivatives for additional aircraft applications. Fuel economy, following the oil price hikes of the mid- and late-1970s, was now paramount and further technology improvements had to be sought. At a fuel price of $1 per US gallon, a 5 per cent advantage in a Boeing 747, even if accompanied by a 15 per cent increase in the price of the engines and spare parts, would still leave the airline with an annual saving of $500,000. Rolls-Royce was in a strong position as the modular design of the RB 211 lent itself to such improvements.

The fear of the early 1970s had been that the market was not large enough to support three engine manufacturers all making turbofans in the 40,000 lb thrust class. However, the gradual expansion of their ranges – to cover 35,000–55,000 lb thrust – created a market sufficiently large for all three to prosper. Rolls-Royce felt confident of its future, especially as 50 per cent of new engine business was expected to come from outside the USA.

In the military field the emphasis was expected to be on reliability, life-cycle costs and ruggedness, rather than only on thrust-to-weight ratio and related performance parameters.

As we have seen, at the end of the 1970s Rolls-Royce was involved in several collaborative programmes on military engines – the RB 199 with MTU and Fiat, the Adour with Turbomeca, and the Spey TF41 with Allison. These three ventures had all achieved considerable success. The RB 199 powered the Tornado and the existing Three-Nations programme would require over 2,000 engines, of which 600 had already been delivered. Nearly 2,000 Adour engines had been built for the Jaguar and made under licence in Japan for the F1 fighter and T2 trainer. The Adour had also been selected for the Hawk trainer. The refanned version of the Spey, the TF41, had replaced the Pratt & Whitney TF30 in the LTV A-7 Corsair ground-attack aircraft, and 1,400 engines had been delivered.

Others had not reached production stage or, if they did, inadequate sales were achieved. During the 1960s, in conjunction with MAN in Germany (by this time MTU), Rolls-Royce had designed and developed several military engine types for vertical take-off applications in which the German Government was interested at the time. Though used in experimental aircraft, as we saw earlier, all were eventually abandoned.

The M-45 civil turbofan was developed jointly with SNECMA for the VFW614 short-haul twin, which failed to sell. SNECMA withdrew halfway through the development programme, leaving Rolls-Royce to complete the outstanding commitments. On the other hand, the development of the Olympus 593 power plant for Concorde with SNECMA was completely successful. Unfortunately, the market envisaged failed to materialise. Finally, in the second half of the 1970s, Rolls-Royce accepted a 35 per cent share in Pratt & Whitney's JT-10D project, a modern fuel-efficient turbofan for the 7X7 tri-jet planned by Boeing. The engine was abandoned when it became clear the 7X7 was not going ahead. The only collaborative venture on civil engines was the recently formed joint company with a Japanese industrial consortium to develop the RJ500 turbofan, of 20,000 lb thrust, aimed at the Spey/JT-8D replacement market. However, Rolls-Royce was considering other collaborative ventures.

At this stage, General Electric looked an unlikely partner. All its versions of its large turbofan, the CF6, were in direct competition with the RB 211. Furthermore, the company had formed strong links with SNECMA on the CFM56 (a new 25,000 lb thrust engine based on a US Government-funded core for the B1 bomber) and the CF6-32 (a CF6 derivative aimed at Boeing's 757). General Electric's most important new military engine, the F404, was in direct competition with the RB 199. Its corporate policy was perceived to be an aggressive attempt to wrest the leadership in the civil aero engine market away from Pratt & Whitney. It had achieved sufficient success to make it likely that any major association with Rolls-Royce would fall foul of the US Justice Department. The only option with General Electric would appear to be the possibility of subcontract manufacture.

Nor did collaboration with Pratt & Whitney seem likely in the short term. The only way Rolls-Royce could see a joint venture happening would be an unexpected technological breakthrough rendering existing large turbofan engines obsolete. In that case, if none of the engine makers could contemplate the necessary investment on their own, collaboration with Pratt & Whitney would be possible and the US Justice Department would probably permit it.

Collaboration on civil engines with a European partner did not seem very likely either. SNECMA was already linked with General Electric on the CFM56 and CF6-32 and on large turbofans for the Airbus A300.

However, economic reality meant that collaboration was essential if new, more powerful, more fuel-efficient engines were to be developed.

In 1981, Rolls-Royce began talks with Pratt & Whitney on a collaborative effort to develop a new engine for the projected 150-seat airliner, widely regarded as the next major civil aviation development. The market for such an aircraft was forecast at between 1,200 and 2,000 over the following twenty years. Pratt & Whitney's PW2025 was somewhat behind Rolls-Royce's RJ500, although this was a smaller engine than would be necessary, but Pratt & Whitney did not intend to be left out of the competition. Nor did General Electric, of course, and its contender was the CFM56-2000 – a joint effort with SNECMA.

Robert Carlson, President of Pratt & Whitney, had said at the Paris Air Show in June 1981 that for the three engine manufacturers to fight it out was to 'court disaster'. In spite of Rolls-Royce's collaboration with Pratt & Whitney in the 1970s ending unfruitfully, Carlson was keen to try again and made an approach to Lord McFadzean early in 1981. However, there were considerable difficulties to be overcome. First, the two companies would have to agree on which parts of the new technologies, embodied in the RJ500 and the PW2025, should be brought together and which discarded. Rolls-Royce was already on the verge of running a demonstrator engine and was not anxious to throw away much of what it had already achieved. Second, there had to be a detailed work and cost sharing arrangement to ensure each partner obtained the maximum return from the total investment of over £500 million.

Third, and most important, existing partners would need to be consulted – the Japanese by Rolls-Royce, and MTU and Fiat by Pratt & Whitney. Discussions began with Pratt & Whitney, Motoren & Turbinen Union in West Germany, Fiat Aviazione in Italy (later renamed Fiat Avio) and three Japanese companies – Ishikawajima Harima Heavy Industries, Kawasaki Heavy Industries and Mitsubishi Heavy Industries. After much discussion in the mid- to late-1970s, Rolls-Royce had joined forces with these three companies to develop a new high-technology engine, the RJ500. This engine was aimed at the 150-seat short-to-medium range airliner. A joint engineering team was set up and two demonstrator engines ran within ten days of each other in 1982 – one in Derby on 23 February and the other in Japan on 4 March. Then, on 11 March 1983, a new company called International Aero Engines was set up. Rolls-Royce and Pratt & Whitney took a 30 per cent share, MTU 11 per cent, Fiat 6 per cent and the three Japanese companies collectively took 23 per cent through a joint organisation called the Japanese Aero Engines Corporation.

Thus, the RJ500 evolved into the IAE V-2500. The Roman 'V' represented the five countries and the '2500' the 25,000 lb initial thrust of the engine. (V-25000 was considered too much of a mouthful.) Commenting on the five countries involved – Britain, the USA, Germany, Italy and Japan, some wag said, 'The engine from the guys who brought you World War Two.'

The cost of the V-2500 was to be spread over seven years. Rolls-Royce's share of the development cost would be £226 million. An ambitious development programme was put in place; as it turned out, as we shall see, perhaps a little too ambitious. The aim was to certify the engine in 1987 and make the first deliveries in 1988. With a thrust initially of 23,000–25,000 lb, the V-2500 would be offered for the A320 and for other 150-seat airliners. It could also be used in new derivatives of the smaller Boeing 737, such as the 737-300, and in some versions of the McDonnell Douglas DC-9.

The new consortium won its first order for the V-2500 in January 1985 when Pan Am ordered it for its sixteen Airbus A320s due to go into service in the late 1980s. The order was worth $125 million and Pan Am took an option on a further thirty-four A320s, which would mean a further $265 million of orders for the consortium.

Flight trials began in May 1988, by which time more than $1 billion of orders had been won, all for the Airbus A320 twin-engined airliner, and the first V-2500 entered service in the spring of 1989.

This all sounds very smooth and simple. The reality was that a crisis developed that threatened the whole programme. Stewart Miller, eventually Director of Engineering and Technology on the Rolls-Royce board (he died tragically young in 1999, within two years of his retirement), was the Rolls-Royce engineer responsible for the project and had to explain to his partners why Rolls-Royce was not able to make the compressor on the engine work properly. During 1986 and 1987, such problems were causing the programme to fall behind. Ralph Robins of Rolls-Royce and Art Wegner of Pratt & Whitney acted as alternate chairmen year by year, and the project probably only survived because these two developed a close rapport. Furthermore, when details of the problems appeared in *Flight* magazine, even the flotation of the company, scheduled for the spring of 1987, was seen to be under threat.

Apart from specific problems the whole collaboration was threatened by the caginess of the US Government over technology transfer. And it was certainly threatened by Rolls-Royce's failure to get the compressor design right. Pratt & Whitney called in its lawyers, who talked of suing Rolls-Royce not only for its investment but also for future lost revenues. Under the circumstances, Stewart Miller had no alternative but to hand over the engineering leadership on the project to Pratt & Whitney.

Pratt & Whitney appointed Tom Harper, described by Miller as 'a cowboy-boot wearing sensible engineer from Florida with no political axes to grind.' Miller went on to say that Harper, and Rolls-Royce's Mike Williams, sorted out the problems. However, this was not before a crisis board meeting in Tokyo in September 1987, when the normally 'quiet and gentlemanly' President of Pratt & Whitney, Art Wegner, became very angry. The whole programme was threatened, and Miller reacted by calling all the engineers on the project into the board room in Derby and saying:

We can no longer afford sequential development. We don't have time. We must attack all the problems on a broad front whatever the cost. We'll worry about the money later.

It's now September. This engine must work by December!

Several teams were assigned to the project and the problems *were* solved. In Robins's view, the breakthrough came one Sunday morning when all the most experienced engineers Rolls-Royce had available, including some who had retired, were called in to study the problem. There was a full-size GA drawing on the wall, and the recently retired Johnny Bush said of the rear stages of the HP compressor: 'There's too much air and not enough metal.' This prompted a radical redesign.

What came to be called 'Miller's War' lasted seventy-one days, less than the ninety days he had given them. When Rolls-Royce finally solved the problems, Pratt & Whitney was sceptical and had to be shown – with the compressor back on a test rig – that the numbers were right. The airworthiness certificates were awarded just in time, and ultimately the compressor became a leader in the industry.

THE BIG ENGINE MARKET

The competitive situation that Rolls-Royce faced in the early 1980s was almost frightening. Although the RB 211-524 had proved itself – especially its fuel advantage – on the Lockheed TriStar and 747, the worldwide recession of the early 1980s and consequential lack of orders meant that its early success could not be fully exploited. Lockheed's decision to cease production of the TriStar left the -524 very exposed, its only aircraft the 747 where it was in competition with Pratt & Whitney's JT-9 and General Electric's CF6-50, both of which were being upgraded to challenge the -524's fuel advantage. The heavier -524 was at a disadvantage in competing on Boeing's 767 and the Airbus A300/310.

Rolls-Royce was hopeful that the -524D4A would achieve more 747 sales and could achieve sales on the A300/310 for customers such as Libya where a 'Europeanised' aircraft was politically necessary.

But the engine's thrust from its 86-inch diameter fan could prove inadequate against the 93-inch fans of Pratt & Whitney and General Electric. Further developments of the -524 looked like being prohibitively expensive.

The -535 was at least only competing with Pratt & Whitney's PW2037, but significant sales were unlikely in the existing climate of low and infrequent airline orders.

As if to emphasise the difficulties, Sam Higginbottom, President of Rolls-Royce Inc. and a former President of Eastern Airlines, told the board on 12 April 1983 that although there were signs of an upturn in the US economy and airline traffic, because of surplus aircraft and severe financial pressures on airlines, he did not expect any significant ordering until well into 1984. He added that ordering was likely to continue to be on disadvantageous terms for manufacturers and if the -535 needed two new major customers it would be necessary for Rolls-Royce to be very competitive to win them.

Furthermore, Wall Street analysts Donaldson, Lufkin and Jenrette said that Rolls-Royce's share of the overall commercial engine market could fall from the 11 per cent it had enjoyed between 1972 and 1982 to only 5 per cent in the period up to 1988. In contrast, General Electric's was expected to rise to 36 per cent, while Pratt & Whitney remained comfortably ahead at 59 per cent. The broker concluded:

Rolls-Royce and the British Government will look at the size of the required investment, consider how heavily the odds are stacked against that investment, and bow out of the high-thrust market.

The reality was not as bad as these dire forecasts predicted. Nevertheless, sales of the RB 211-524 fell from 157 in 1981, to 127 in 1982, to 35 in 1983,

The Armstrong Siddeley Viper was originally designed as a short-life power unit for the Australian Jindivik target drone but was developed into a conventional turbojet and enjoyed a long and successful life, powering aircraft in twenty-nine air forces throughout the world. The Jindivik was used by several air forces including the RAF and the Swedish air force (shown here).

The Gnome turboshaft engine (ABOVE) combined the advantages
of high power output, compactness, light weight and low
specific fuel consumption. These characteristics made it an ideal
helicopter engine and early users were the RAF Whirlwinds,
Westland Wessex Mk 2 (OPPOSITE), Italian Agusta Bell 204B,
Agusta 101G and Boeing/Vertol 107.

The Gem 40 (ABOVE), a three-shaft engine, which, in the view
of some, was 'ridiculous' for such a small engine and whose
initial problems caused the Army Air Corps to 'complain like hell'.
The Gem 40 was replaced by the Gem 42 and the Army gradually
re-equipped their Lynx helicopters (OPPOSITE) with this engine.

The Gulfstream IV.

At Rolls-Royce's Christmas Lunch at the Waldorf Astoria in
December 1982, Ralph Robins (OPPOSITE), Rolls-Royce's
Commercial Director, discussed engines, payment terms and prices
on the Spey with Allen Paulson, Chairman of the Gulfstream
Corporation. Neither had a piece of paper and they were trying to
write it all down on a paper napkin. A guest from Chemical Bank
leaned across the table and said, 'If you guys are trying to do a
deal you better have a piece of paper', and passed across a small
card. The deal, which lasted for many years, was written on that
card which has been in Robins's office in a silver frame ever since.

Lord McFadzean succeeded Lord Keith as Chairman
and reacted to the recession of the early 1980s by cutting
manpower costs.

Sir William Duncan succeeded Lord McFadzean as Chairman of
Rolls-Royce in 1983. Faced with dire economic conditions and
charged by the Thatcher Government with preparing the
company for privatisation, he saw little alternative to
collaboration with other aero engine makers.

The V-2500 engine. The '2500' stood for 25,000 lb thrust,
while the 'V' stood for the five nations collaborating –
the UK, USA, Germany, Italy and Japan. This led some
wag to quip – 'The engine from the guys who
brought you World War Two.'

Stewart Miller. After giving Rolls-Royce the enormously
successful engine, the RB 211-535, he was given the
responsibility for perfecting and delivering the V-2500.
After almost interminable problems with the compressor,
the deadline was achieved.

An Airbus 320 powered by two V-2500 engines.

Prime Minister Margaret Thatcher's vision was to free British
industry from what she considered to be the dead hand of
the State. She insisted that Rolls-Royce be privatised. Here she
is in 1986 at Rolls-Royce (appropriately near the new Control
Room!) with, from left to right, Trevor Salt, Ralph Robins,
Sir Francis Tombs and Alan Jackson.

Stuart John, chief engineer of Cathay Pacific, chose Rolls-Royce engines for the rapidly expanding Cathay Pacific Fleet.

A Cathay Pacific 747, powered by Rolls-Royce
RB 211-524 B2s, landing at Hong Kong airport.

Sir Francis Tombs succeeded Duncan when the latter died
in office in November 1984. With his background in
the electrical power generating business, Tombs was used
to taking the long-term view and took an altogether
more robust stance on Rolls-Royce's future prospects as
an independent company.

to 25 in 1984, before recovering slightly to 41 in 1985. Sales of the RB 211-535 were erratic but still not enough to support an aero engine manufacturer with large overheads. Only five were sold in 1981, 40 in 1982, 37 in 1983, 23 in 1984 and 75 in 1985.

It was against this background that Rolls-Royce felt obliged to consider collaboration with one or other of its two major competitors, on both the big engine category (that is, -524 and above) and on the -535. It seemed impossible with Pratt & Whitney because of the -535/PW2037 competition and the likelihood of specifically restricted collaboration with the lower thrust engine now designated V-2500.

Discussions on possible collaboration with General Electric had taken place at various levels over a number of years and had covered a wide variety of projects. In the early 1980s, serious discussion took place on possible collaboration on an engine for the 150-seat aircraft, in parallel with discussions with Pratt & Whitney. In the event, General Electric decided that it only wanted to develop variants of the CFM56 for this market and was not prepared to join in producing a new engine. This led Rolls-Royce to join with its other partners on the V-2500 project.

Rolls-Royce developed the RB 211-535 after assessing the potential market in the 1980s and concluding there would be demand for an engine in the 32,000–35,000 lb thrust range. From 1966–76, air traffic growth had been 9.8 per cent per annum. From 1977–91, it was expected to slow to 7 per cent. Rolls-Royce did not expect a marked increase in the sale of engines from the depressed levels of the 1970s until the 1980s. The older aircraft would not be seriously obsolete until then as the airlines did not have the financial surpluses to enable them to replace their fleets until it was essential to do so.

However, a new generation of aircraft would be required and most of them would be for the short- and medium-haul market – an area covered in the 1970s by the Boeing 727 and 737. The engine size that would most effectively suit the new aircraft in this segment would be in the 32,000–35,000 lb thrust range and Rolls-Royce intended to develop the RB 211-535 to meet the requirement. In the 1970s, Rolls-Royce had only a very small share of this market.

Rolls-Royce continued its development of the -535 and it entered service as the launch engine in the Boeing 757 twin-engined, short-to-medium range airliner with Eastern Airlines and British Airways in January 1983. This engine was the RB 211-535C and it was further developed so that its E4 variant, of 40,000–43,000 lb thrust, entered service in October 1984.

In spite of these developments, Sir William Duncan, the Chairman who succeeded McFadzean, was concerned about Rolls-Royce's ability to survive without further collaborative agreements and in March 1983, just before he took over as Chairman, he met Brian Rowe, the head of General Electric's aero engine activities, and identified prospects for further discussions on how they might work together. In May, the two companies began talking about collaboration in the big engine market and on the -535; then, in June, Alan Newton, Rolls-Royce's Director of Engineering, met Brian Rowe to discuss all the possibilities.

It was clear to Newton that General Electric was concerned that if Rolls-Royce launched the further development of the -524, designated the RB 211-600, which was being publicised at the time, it could have a significant effect on the big engine market as there would be three engines – the Pratt & Whitney PW4000, General Electric's CF6-80C2 and the RB 211-600 – all roughly the same technical standard and all competing for the same business. Rowe maintained that Rolls-Royce should join General Electric on the CF6-80C, which was already committed and launched, rather than make the large investment in the RB 211-600. The CF6-80C had a bigger potential market, while Rolls-Royce technology would improve the standard of the engine and give it greater long-term potential.

A further meeting took place at Buckingham Gate between Newton, Rowe and Duncan, and Rolls-Royce expressed interest in joining the CF6-80C programme if a satisfactory deal could be worked out. Rolls-Royce made these points of principle:

* Rolls-Royce was in the big engine market and must continue to support and develop the engines committed to its existing customers.
* Any deal on the CF6-80C must be combined with General Electric taking a significant share of the RB 211-535 since Rolls-Royce judged that, just as it would bring technology and market to the -80C, General Electric should also bring technology and market to the -535. Furthermore, it was vital that customers saw the deal as reciprocal and not Rolls-Royce opting out of a market sector.
* Rolls-Royce's objective was equal participation in both programmes, and any follow-up programmes. Rolls-Royce would bring far more to the deal than General Electric's existing partners, in terms of market and technology, and therefore could not accept the sort of deal and conditions accepted by SNECMA, Volvo and MTU as risk-sharing subcontractors.

These basic principles were accepted by Brian Rowe, and a number of further meetings and discussions took place. The main issues to be resolved were the size of the entry fee, the magnitude of the further costs contribution – covering marketing, product support, programme management, warranty, sustained engineering and overheads – as well as the definition of the list of parts that each company could manufacture from its partner's engine.

Agreement was reached whereby Rolls-Royce would initially take a 15 per cent share of the -80C programme, while General Electric took 15 per cent of the -535E4 programme, with both companies expecting to increase their programme shares to 25 per cent in 1988. The two sides assumed that, with collaboration, the -80C would achieve 50 per cent of the big engine market and that Rolls-Royce's 40 per cent share for the -535 of the 757 market would be increased to 50 per cent. In the ten years to 1993, Rolls-Royce expected that the collaboration would add £860 million to its turnover and £171 million to its profits and, over a twenty-year period, Rolls-Royce expected the collaboration to add a sum of no less than £2.7 billion to turnover and £890 million to profits, after allowance for all costs – for example, research and development for the initial and prospective versions of the collaborative engines.

In putting the case for the collaboration to the Rolls-Royce board, the architects of the plan said:

We assess the big engine market as the largest sector in the future. It is also the one on which much has been invested in the past decade. We consider the CF6-80C programme is technically sound and realistic. We judge that we can manufacture -80C parts with an acceptable margin. In the longer term both companies have complementary technology which should eventually lead to better designs for both the 80C and the E4. In the shorter term, collaboration will provide much needed support for the 535, particularly with customers who have traditionally favoured GE. This will at least consolidate our market and, more likely, increase it in this sector.

Reaction from Pratt & Whitney to the deal was 'statesmanlike' if not ecstatic. The Justice Department in the USA told Phil Gilbert (Rolls-Royce's long-time legal adviser in the USA) and General Electric's lawyers that the Department would not object to the proposal, and Ralph Robins reported to his fellow directors that customers initially saw the deal in a positive light but 'more considered reactions would follow when, in due course, airlines re-assessed their fleet plans.'

The deal was announced on Friday 3 February 1984 and Sir William Duncan said, at a press conference, that he saw the partnership as a 'watershed' for Rolls-Royce in the development of big engines. He continued:

I see it as a very important policy decision.

Rolls-Royce's results for 1983 showed the pressure under which the company was operating and the reason why the board went along with the proposed partnership with General Electric. Research and development costs ate up all the group's operating profits, leaving a loss of £57 million even before interest payable. The attributable loss was £193 million. By contrast, the Aero Engine Division of General Electric increased its profits in 1983 by 22 per cent to $196 million. Rolls-Royce's problem was that it needed to maintain a full 'R&D' commitment on sales half those of General Electric.

No sooner was the ink dry on the agreement with General Electric than the Chairman, Sir William Duncan, died, on 5 November 1984. While a new chairman was sought and considered by the Government, Sir Arnold Hall took over and Ralph Robins was appointed Managing Director. Sir Francis Tombs, already a non-executive director, took over from Hall early in 1985.

Born and brought up in the 'Black Country' town of Walsall in the West Midlands [like the author], Tombs had moved between the public and private sectors. After leaving school at the age of sixteen, he had begun his working life at GEC (now Marconi), gaining engineering qualifications and a degree in economics in his own time. After several moves between the public and private sectors, he became Chairman of the South of Scotland Electricity Board in 1974 and Chairman of the Electricity Council in 1977. After resigning in 1980, he helped rescue both the engineering companies – Weir Group, and Turner and Newall – from near bankruptcy. Such varied experience prepared him admirably for the chairmanship of Rolls-Royce – especially in its run-up to privatisation.

In the view of both Tombs and Robins, the agreement with General Electric was potentially disastrous for Rolls-Royce if it prevented the company from offering improved versions of the RB 211-524, which were now being urgently requested by the company's existing airline customers. The customers felt that they had selected the engine on the basis that it would continue to be developed in the usual way.

During 1984 and 1985, it became clear that upgraded versions of the -524, in particular the RB 211-524 D4D, could be delivered to meet the customer requirements for the 747-400 and 767 without increasing the fan diameter (this was very important in view of the agreement). Rolls-Royce therefore began to talk to its customers about this improved version of the -524 and, of course, word reached GE.

On 16 December 1985, the influential business magazine *Forbes* had written an article under the heading 'Ambivalence', which said *inter alia*:

Chairman Sir Francis Tombs may be endangering this advantage [of being the third world-class aero engine maker, with an 11 per cent share of commercial airliner engine sales, against 51 per cent for Pratt & Whitney and 32 per cent for GE] by an ambivalence toward sharing the wealth and the glory.

Right now, Rolls is infuriating the top people at GE's jet engine division by threatening a partnership agreed to in 1984. How is Rolls threatening this partnership? By talking to airlines and airframe makers about developing a larger version of its RB211 that would directly compete with the engine it is making jointly with GE. Not surprisingly, the GE engine bosses are angrily demanding to know whether Rolls is serious about partnerships.

The article went on to outline the fearsome cost of developing a brand new engine – $1 billion plus another to get into full production – and reminded its readers what happened when Rolls-Royce got it wrong with developing the RB 211.

The fearsome economics of the business has led increasingly to partnership deals similar to the one between GE and Rolls. There is a reason, after all, that the US-based market leaders are both divisions of large, diversified companies. Pratt & Whitney was floundering for years until Harry Gray came in and created UTC [United Technologies Corporation] around it, with air conditioners and elevators to carry the jet engine business through the bleak periods. [This was not a correct analysis of the situation. Gray used the strong cash-flow of Pratt and Whitney at that time to purchase other businesses such as Otis for the UTC group.] Or take GE's jet business. Without the support of light bulb sales it's safe to say it could have been closed more than once when engine sales declined to disaster level.

Rolls is going to have to make up its mind if it's in business to put its insignia on everything or to make money.

It did at least quote Tombs's riposte:

It may be comforting to be part of a large corporation but in my view it's not necessary. And [going it alone] allows us to be single-minded about our business.

Sam Higginbottom, who had strongly favoured the General Electric deal, was soon sending distressed messages to London:

We have received further complaints from Brian Rowe and Dick Smith. Douglas had advised of Rolls-Royce pending visit to discuss installing D4D in MD-11 and had queried GE as to whether that meant Rolls-Royce was no longer involved with 80C2. British Caledonian had told Brian of Rolls-Royce's efforts to interest BC in D4D and queried Brian re status of collaboration. Brian queried – has Rolls-Royce made a decision to go forward with D4D and not told GE, or is Rolls-Royce continuing to misbehave while it considers what to do? In either event, Brian is distressed by Rolls-Royce's behaviour!
    Conclusion and recommendation:
    Rolls-Royce should decide on its course ASAP and announce it!

Robins had, in fact, been to visit Brian Rowe at General Electric Engine Group's headquarters in Cincinnati in November 1985. Clearly, after a frank exchange of views, there were still some points to be resolved. Rowe wrote to Robins on 23 December. He said that General Electric was very disappointed to see Rolls-Royce actively promoting the -524 D4D against the CF6-80C2.

I feel these activities have violated the spirit of our entire relationship.

Robins had pointed out that Rolls-Royce had to try to satisfy its existing RB 211 customers; he also pointed to the clause in the contract that allowed both parties to 'independently continue to diligently pursue' the CF6-50 and the -524. However, Rowe maintained that continuing with RB 211-524 derivatives was not in the spirit of the agreement.

Rowe used both carrot – 'If there is anything General Electric could do to help Rolls-Royce in improving its position for going private, I would be happy to explore this with you.' – and stick – 'Ralph, if by your current actions you hurt the -80C2 program, this could have serious implications for us. Please give this serious consideration.'

By January 1986, Robins was suggesting that Rolls-Royce might consider a revised agreement that took account of the new, more powerful -524, and indications from General Electric were that it might accept this too. However, lawyer Phil Gilbert expressed caution on the basis that the US Department of Justice might not accept it. He looked at all possibilities and concluded in a memorandum to Robins on 13 January 1986:

If Rolls-Royce decides to pursue the expanded 524 program, the present collaboration cannot be continued and must be abandoned or revised. The parties are, on the one hand, collaborating and at the same time are competing through the 524 and the CF6-50. What is new in the situation is that Rolls-Royce is now plainly capable of doing more with the 524 engine than anyone anticipated a year ago, and the 524 can now compete also with the CF6-80C. It seems clear that the competition preserved under clauses 2.6 and 6.5 was, at the time, considered to be between the then existing 524 and the CF6-50 only, since no one expected an extended thrust 524 to compete with the CF6-80C.

Discussions between General Electric and Rolls-Royce continued through-out the first half of 1986 as both companies attempted to find an acceptable formula. Looming in the second half of 1986 was the decision by British Airways on which engine to choose to power its new Boeing 747-400s. On 18 July, General Electric issued a press release which said, *inter alia*:

EVENDALE, Ohio, July 18, 1986 – Rolls-Royce has completed its first production General Electric CF6-80C2 high bypass turbofan engine at its factory in Derby, England.

The engine has been produced under a joint agreement of the two companies signed in 1984. The CF6-80C2 has been in production by General Electric for more than one year.

The Rolls-Royce CF6-80C2 is being shipped to Rohr Industries in Toulouse, France, for installation of the nacelle and thrust reverser. Airbus Industrie, also in Toulouse, will then install the engine on an A310-300 aircraft.

Based on current predictions, the engine just shipped is the first of 400 CF6-80C2s to be produced by Rolls-Royce through 1994. Two additional CF6-80C2 engines are currently undergoing testing in Derby, England. Rolls-Royce has a 25 per cent share in the -80C2 program . . .

'The CF6-80C2 has the built-in thrust growth needed to satisfy the increased capability of the 747. Higher thrusts of the CF6-80C2 also are timed for the new versions of the 767, A300/A310, and the MD-11,' Rowe said. 'Rolls will be a major participant in these programs as the market grows.'

Rowe further stated, 'The Rolls-Royce shipment of its first production CF6-80C2 is timely in that British Airways is about to decide on the engine for its new 747s. The choice will be either the GE–Rolls CF6-80C2, the PW 4056, or the Rolls-Royce RB211-524 D4D. A selection of the GE–Rolls engine by British Airways is important to the success of the CF6-80C2 and the profitability of the Rolls-Royce/General Electric team.

So far, the CF6-80C2 has been ordered by eleven customers for all eight current widebody applications: the Boeing 747-400, 747-300, 747-200 (Air Force One), 767-300, 767-200ER, and the Airbus Industrie A300-600, A310-300 and A310-200 advanced aircraft. The customer list includes Air India, Alia – The Royal Jordanian Airline, All Nippon Airways, CAAC-Shanghai, Kenya Airways, Lufthansa German Airlines, Thai International Airways, US Government, VARIG and one unannounced airline.

On the face of it, all was progressing well. However, as Roger Eglin, the business news editor of the *Sunday Times,* pointed out on 27 July:

This partnership appears to be under severe strain. To its evident dismay, GE finds itself in ferocious competition, not just with its American rival, Pratt & Whitney, but with its collaborator, Rolls-Royce, for a $480 million order to supply British Airways with engines for its fleet of new 747 jets . . .. It is a measure of how seriously this battle is being fought that GE and Rolls are taking the lead in arranging finance for the whole deal, aircraft as well as engine.

Eglin went on to point out that General Electric had told British Airways that if BA bought the CF6, General Electric would make sure Rolls-Royce's share would be increased to 33 per cent and that all the engines would be made in Derby. Rolls-Royce was unhappy that it only found out about this offer indirectly. Rolls-Royce told Eglin that the RB 211 was what mattered. A British Airways choice of the RB 211 would be worth six times anything coming from a General Electric victory. He concluded:

Quite clearly sparks are about to fly between two (former?) partners.

Eglin had been given information by Rolls-Royce, which told him that a British Airways choice of the RB 211-524 D4D would provide 20,000 man-years of work to the UK. About 87 per cent of the material and accessories for the -524 were sourced in the UK, whereas 50 per cent of the material for the Rolls-Royce share of the CF6-80C2 came from the USA. Over the previous five years, the -524 had won more than 25 per cent of 747 sales and Rolls-Royce expected to maintain at least that share.

In the event, Rolls-Royce won the order for the engines for the sixteen Boeing 747-400s ordered by British Airways and, when questioned by the *Financial Times,* Tombs said:

We made up our minds to win on straight technical and commercial grounds. Our collective agreement did not in any way preclude competition between the General Electric CF6-80C and the RB211-524 D4D engines. Indeed any such undertaking would have fallen foul of US anti-trust requirements and therefore it is hardly surprising that the agreement recognised that we would continue to develop the 524 engine. We have not broken the letter or the spirit of our collaborative agreement with General Electric.

Tombs pointed out that the original agreement had been amended in February 1986 after the -524 D4D was launched, and continued:

Among the amendments agreed was a reduction in our revenue of sales of CF6-80 engines for Boeing 747s. We admire General Electric's capability and have enjoyed a mutually beneficial relationship. But we both live in a highly competitive industry and we for our part make no complaint about the fact that the pioneering work being carried out by General Electric on such new engines as the prop-fan will, if successful, compete directly with RB211-535 which forms part of our joint collaboration.

He concluded by saying:

We are a successful member of the trio of engine makers in the western world with such an overall capability and intend to remain so.

From this point, relations between Rolls-Royce and General Electric deteriorated further and, on 13 November, Robins reported to his board that the agreement with General Electric was now in serious trouble and that subcontract work on the CF6-80C2 might be all that was possible in the future.

Sir Francis (now Lord) Tombs told the board that it was difficult to achieve a better understanding with General Electric but that he and Robins would try when they met Rowe and Ed Hood, who was Rowe's boss at General Electric, on 17 November.

On Wednesday 19 November it was officially announced that the engine pact between the two companies was at an end. Throughout the period of collaboration, Rolls-Royce had made a significant contribution to the large CF6-80 programme, assembling twelve engines and manufacturing 350 LP turbine modules and HP compressor casings. On the other hand, GE made very little of the 535, mainly subcontracting their share back to Rolls-Royce, though they did assemble and test six engines. This was probably due to their heavy commitment to their own engines at that time.

Sam Higginbottom expressed his disappointment. He saw no alternative to the ending of the agreement and suggested that a senior Rolls-Royce executive call on the US Justice Department to ensure that the company's record of events was clearly understood.

Higginbottom was not the only one to be disappointed. As we have seen, Engineering Director Alan Newton had been closely involved in the negotiations and he was dismayed at the breakdown. Mike Neale, Director-General of Engines at the Ministry of Defence, and the focal point for Government relations with Rolls-Royce, also thought at the time that the collapse of the agreement was a great pity. He did not think, given Rolls-Royce's cost base, that the company could compete on its own against General Electric and Pratt & Whitney. However, when he saw how Rolls-Royce set about reducing its costs in the 1990s, he changed his mind. He told the author in 1999:

I didn't think a British company would be able to do it, but Rolls-Royce did.

And, finally, General Electric itself was very upset. This was not surprising for they must have expected the agreement to strengthen substantially their position in the civil engine market. However, both Tombs and Robins had become increasingly aware of the importance of meeting their airline customers' requirements across the whole civil aircraft range, if they were to stay competitive in the long term.

In spite of this disagreement with General Electric and the continuing highly competitive market conditions, Rolls-Royce was not completely friendless. It still had many customers around the world, even if most of them had not been throwing orders around in the depressed years of the early 1980s. One good friend was Cathay Pacific, whose chief engineer was Stuart John. He had joined Cathay in April 1977 when it was operating twelve second-hand Boeing 707s and two Lockheed TriStars on which, dare it be said, the Rolls-Royce RB 211s had been having some problems. In February 1978, Cathay decided to buy a brand new Boeing 747 and the Chairman, Duncan Bluck, told John to recommend an engine. John visited Rolls-Royce, having made his own way to Derby, and he visited Pratt & Whitney and General Electric, both companies sending an executive jet to meet him when he landed in the USA. In spite of this disparity of treatment, John wrote a report for Bluck, recommending the Rolls-Royce RB 211-524B2. The present Chairman of Rolls-Royce, Sir Ralph Robins, then the Commercial Director, took something of a gamble by offering all the concessions on this one aircraft engine order that would normally only have been offered on an order for tens of engines. It was a gamble that was to pay handsome dividends. Cathay expanded aggressively through the 1980s buying two 747s a year, always with Rolls-Royce engines. It also took its TriStar fleet from two to seventeen – all, of course, powered by Rolls-Royce RB 211 engines.

How did Rolls-Royce fare? After the company's results for 1983 – which showed an attributable loss of £193 million – were announced in early April 1984, 'Lex' in the *Financial Times* wrote:

Its balance sheet, meanwhile, makes any thought of privatisation a distant pipe-dream. Reserves are in deficit to the tune of over £200 million and net published debt totals over £350 million.

However, as we have seen on many occasions in the story of Rolls-Royce, two years is a long time in the aerospace industry. As we have seen, the early 1980s was an extremely grim time in the aerospace industry coping with a worldwide recession. However, by the mid-1980s the world economy was growing again, aided by a collapse in the price of oil and the consequent decline in inflation. On 23 April 1986, 'Lex' was writing:

As a dummy run for the prospectus, the 1985 annual report of Rolls-Royce Ltd is all that any share salesman could wish . . . More than most other privatisation candidates, Rolls has already set about drawing its accounts into a form that will prove digestible to stock market analysts. There are no arcane policies on inflation accounting such as the market will have to assimilate with British Gas, nor any inflation accounting come to that. Unlike British Aerospace, Rolls sets its development expenditure straight against current profits; there is not the smallest trace of capitalised development costs.

Rolls is also fortunate in its timing. After a spluttering flight through the early eighties there is the beginning of a stable profit record on which to float.

Whenever a company floats on the London stock market, auditors produce what is called a Long Form Report which – after a questionnaire running into hundreds of questions, and detailed examination of the company's financial affairs – forms the basis for the prospectus put before potential shareholders. Many a yacht in St Tropez and a mistress's flat in Mayfair have been uncovered by this probe, though not, of course, in the case of Rolls-Royce. Nevertheless, the Long Form Report, stretching to nearly 500 pages, drawn up by Coopers & Lybrand and sent to Mr M.K. O'Shea of the Air Division of the Department of Trade and Industry, the directors of Rolls-Royce and the directors of the merchant bank Samuel Montagu, did not pull any punches when describing the competitive problems Rolls-Royce had faced and still would face as a private company.

The last five years, from 1980 to 1985, have been a period of significant change for the Group. The recession in the airline industry was both deeper and longer than expected. There was a substantial fall in the workload. In 1980 a major manpower reduction programme was commenced by the then Chairman, Lord McFadzean, and a fall in employee numbers from 62,000 in 1980 to 41,000 at the end of 1984 was achieved with no serious industrial relations disruption. The costs of this restructuring were high in human terms and in the immediate financial cost, but it was a programme without which the Company could not have survived as a viable enterprise. Nonetheless, the period 1981 to 1983 saw combined pre-tax losses of £191 million (before charging restructuring costs of £129 million). Much of this was due to the escalating spend borne by Rolls-Royce in research and development, which rose from £62 million in 1981 to 131 million in 1982 and 131 million again in 1983. The main reason for this was the cost of the –535C. Development of the –535E4 was initiated in 1981 and the engine entered service in 1984. Savings in research and development spend have been made in 1984 and 1985, although the total net cost continues at some £100 million per annum.

The report went on to emphasise the absolute importance of the US market [reminding us that Huddie and Pearson were correct in their realisation of this fact twenty-five years earlier]. It also pointed out that the profitability of this market had been eroded, not only by the recession of the early 1980s but also by the deregulation of 1978 which led to the setting up of a large number of new airlines using second-hand aircraft and enjoying low labour costs and offering low fares. The established airlines hit back but profitability suffered. As a result, aircraft purchases were deferred and greater emphasis was laid on fuel efficiency. At the same time, airlines sought lower maintenance costs by reducing spare parts inventories and by repairing, rather than purchasing, components. Intense competition for new orders meant aggressive pricing concessions and engine manufacturers, including Rolls-Royce, had little choice but to participate.

In 1945, the whole of Rolls-Royce's aero engine output went to the military market. As we have seen, the company soon realised how important the civil market would be and had made strenuous efforts and invested heavily to establish itself in that market, especially in the USA. Nevertheless, at the time of privatisation, the military market was still of vital importance to the company. After the takeover of Bristol, Rolls-Royce became the sole UK supplier of aero engines to the military market and it tackled this in several ways, making sales to the UK armed forces and overseas countries' armed forces and collaborating on joint ventures in Europe and with Allison in Indianapolis in the USA.

The home market was obviously important to Rolls-Royce but the UK military forces had made it clear, time and again, that they were prepared to go to overseas suppliers if British companies could not match their requirements in terms of specification, quality, reliability or price. Lockheed Hercules C130 transport aircraft powered by Allison T56 engines were used by the UK military forces and some Phantoms were powered by GE J79 engines. As well as its home market, Rolls-Royce would have to look to military sales around the world.

Rolls-Royce's success in the US military market was limited but not insignificant. Under an inter-governmental Memorandum of Understanding in 1975, bids from the UK were not given a 20–25 per cent loading as would be required by the 1933 Buy American Act. As we have seen, Rolls-Royce has been successful in selling the Pegasus for the Harrier AV-8A and later the AV-8B to the US Marine Corps. A respectable 9 per cent of engines in combat aircraft in the USA in 1986 were Rolls-Royce designed – either the Pegasus or the TF41 which was based on the Spey and produced in conjunction with Allison for the A-7 Corsair.

Tuesday 28 April 1987 – 'Impact Day' as it is called in the City, with the announcement of the price of shares that were to be offered to the public again after sixteen years of Government ownership – was a proud day; especially for those in the company who could still remember the awful day, 4 February 1971, and the dark, uncertain period that followed.

At Stansted Airport there was a spectacular display of aircraft and engines spanning seventy years, from the famous Bristol Fighter powered by the Falcon in World War I, to the Concorde supersonic airliner with its Olympus 593, as well as an RB 211-powered Qantas 747 which flew up to Stansted from Heathrow just after it had flown into the UK from Australia. Rolls-Royce took the opportunity to entertain important guests and customers as well as many employees.

The Prospectus was also published on that day, and in it the Chairman, Sir

Francis Tombs, was able to say that the group as a whole had earned a record pre-tax profit in the year to end 1986 of £120 million and could announce:

With outstanding orders worth £3.1 billion, the prospects for 1987 are encouraging. We have a strong order book, there is a large civil and military engine market to be satisfied, and we have a broad product range with which to compete. Add to this the steady growth in productivity over the past five years, improved financial controls, an experienced management team and a skilled workforce, and I am confident that we have the ingredients for a successful future.

Tombs, along with the rest of the board and senior management, could hardly wait to move from the state to the private sector. As he told the author in 2000:

Rolls-Royce virtually went to sleep for sixteen years. It wasn't their fault, they just couldn't get decisions quickly enough.

On privatisation – 'Impact Day' – Rolls-Royce aero engines were in service with more than 270 airlines, 700 executive and corporate operators and 110 armed services worldwide. The company also had over 175 industrial customers operating gas turbines for power generation, gas and oil pipeline pumping and other industrial uses. Its gas turbines powered the naval vessels of twenty-five countries.

Most importantly, in view of the recent decision concerning the General Electric agreement, the Prospectus pointed out that, in the civil aero engine market:

Rolls-Royce pursues a strategy whereby it can offer an engine to compete within each principal sector of the market, either alone (as in the case of the RB-211 family of engines and the Tay) or in collaboration with other manufacturers (such as the International Aero Engines V2500 now under development).

And, of course, the company was truly international, or *global* as it would be termed in 2000. In the previous five years, 70 per cent of sales had been to airframe manufacturers or other customers outside the UK. Rolls-Royce had long committed itself to a world, not just a UK, market. Rolls-Royce personnel criss-crossed the globe constantly seeking new opportunities and selling the company's expertise. Every new development, increasingly expensive, had to be considered in terms of its potential in as many markets as possible. Conditions in the 1990s and beyond would demand such an approach. Fortunately, Rolls-Royce was prepared for them.

With an eye to the disaster that had befallen the company in 1971, the

Prospectus said of the substantial resources devoted to investment into research and development of advanced technology:

Since the mid-1970s, it has aimed to reduce the technological risk and high cost of engine design and development, where practicable, by proving technologies in advance of the high expenditure phase of an engine project, and by making technologies transferable between engines.

Over the immediately preceding years, Rolls-Royce had been spending well in excess of £100 million annually on research and development – necessary to enable the company to meet the increasingly complex technological advances in aero engine development. The Prospectus stressed that the company was engaged in an exceptionally long-term business, in which improvements in engines were dependent on advances in technology. The period between the initial concept of any aero engine through to its entry into service could be as much as ten years, of which about four years would cover the likely emergence of a potential market and undertaking the preliminary design studies on the best type of product to meet it, and another four to six years on the overall development programme once the company had committed itself to the concept. This was the average time for a civil aero engine, but a military engine, because of its frequently greater complexity stemming from its different performance goals, to say nothing of the political niceties involved, could take even longer.

However, once into production, the programme might extend for thirty years or longer – including the sale of spare parts, which normally continued for many years after production of a particular engine had ceased. For example, the Rolls-Royce Dart, which first ran in 1946 and which we read about in Chapter Two, was manufactured for 40 years until 1986, and there are still 2,000 engines flying 55 years later in 2001. The industrial version of the Avon, whose life began during the Second World War, is still being made and sold. Thus, decisions taken at a particular time could have an important effect on the company's business for many years into the future.

This also meant a significant impact on costs, explaining why an entirely new engine could cost as much as £1 billion, while a major derivative could cost several hundred million pounds. Much of this money, of necessity, had to be spent in the preliminary research, design, development, testing and initial production stages, before completed engines could be delivered to customers, and significant financial returns emerge from the programme. It was often impossible, therefore, for any aero engine manufacturer to guarantee the extent of the ultimate profitability of any specific engine venture, because it was looking so far into the future with every engine it undertook.

This, in turn, implied the fundamental need for a company structure that was sufficiently flexible to take account of all conceivable possibilities throughout the life cycle of a given engine. One of the reasons for the downfall of Rolls-Royce in the early 1970s was that the management structure itself proved insufficiently flexible to take account of changing circumstances, including the increasing technological difficulties encountered in the development of the original RB 211 engine, and with it the rising costs of that programme.

The Prospectus also pointed out that during the period of recovery of the company from the bankruptcy of 1971, it became essential for a strong central management to be established, keeping close control of costs and activities throughout all parts of the company. This led to a tendency for the company to be run autocratically, but although this may have proved unpalatable to some, it was essential for several reasons.

First, it met the direct requirements of the Government, as the sole shareholder, for tough, strong, central management, keeping tight control of costs.

Second, it ensured a strong central strategic product thrust, with no possibilities of individual divisions launching into expensive ventures without the strictest surveillance of market opportunities and potentialities.

Third, it ensured that everyone in the company knew precisely what the overall targets were, and how they could and would be achieved, with the entire company working along clearly defined product lines, without necessarily stifling initiatives.

The net effect was to ensure that even while still in the throes of recovery, the company could survive the severe business recession in worldwide commercial aviation of the early 1980s. During that period, the company's strategy was to broaden its civil aero engine product range and to prepare to take advantage of the anticipated eventual upturn in demand for the air transport business when airline profitability returned.

Significant improvements in productivity had been achieved by the company through a manpower reduction programme, coupled with continued investment in manufacturing facilities and automated equipment. As we have seen, between 1980 and 1984, employee numbers had been reduced from about 62,000 to about 41,000 through a programme of voluntary severance. This had been accomplished without industrial relations disruption, while steps had also been taken to improve the control and cost effectiveness of research and development activities. By 1984, when the anticipated resurgence of demand began to occur, the company was in a strong position to take advantage of the new opportunities.

The practicalities of the flotation were that 801,470,588 ordinary shares were offered at a price of 170p a share, valuing the company at a price of just over £1.36 billion.

AIMS AS A PRIVATE COMPANY

Rolls-Royce's long-term objective in the civil aero engine market was to win at least a 30 per cent share of the large turbofan market, which it estimated at over £1 billion in the years up to the end of the century. Rolls-Royce was determined to compete with an engine in each market sector so that it could take advantage of increases in demand in whichever sector they occurred, while its risks would be limited by not depending on any one part of the market.

This determination to produce a 'family' of engines was a key strategic decision. The investment in an original engine concept would be heavy but, by having commonality of design, maintenance techniques and logistical support in the family of engines produced, that heavy investment would be protected.

And what would the new civil aircraft of the 1990s be? After lengthy discussions with the commercial airlines in the late 1980s, the aircraft manufacturers began to develop a new breed of large jet airliners. One of them was the medium-haul, high-density, four-engined A340 of Airbus Industrie, which would enter service in early 1993. Another was its stable companion, the twin-engined A330, which would become operational in 1994. In the USA, Boeing, which had dominated the long-range market with its -400 version of the 747, talked to the airlines about a new, medium- to long-range aircraft, which it began building in 1991 and which entered service in 1995 as the twin-engined 777. Smaller than the 747, the 777 was wanted by the airlines for long routes with a slightly lower traffic density. From the Douglas division of McDonnell Douglas came the MD-11, a derivative of the three-engined DC-10.

Rolls-Royce had an engine to meet all the demands of the market in the 1990s. The RB 211-524 high-thrust engine satisfied the requirements of the Boeing 747-400 series. The lower-thrust RB 211-535 was perfect for the twin-engined, medium-range Boeing 757s, and the short- to medium-range Airbus A320 was covered by the V-2500 engine built by the five-nation International Aero Engines consortium in which Rolls-Royce had taken a 30 per cent share. The Tay could meet the short-range requirements of the Fokker 100.

We shall read about Rolls-Royce's success with its family of Trent engines in Part Three of this history and we shall realise the significance of Rolls-Royce's breakthrough into the US market again with the RB 211-535 as the launch engine for the 757 after the disaster of February 1971. We shall see how the company broadened its base by acquiring the British engineering company NEI, the US aero engine manufacturer Allison, and its old friend, Vickers. And we shall also see how Vickers developed Rolls-Royce and Bentley motor cars and how it sold the division to VW.

In looking back at the forty-two years covered by this book we would have to conclude that it had been a tough period for Rolls-Royce. If we turn back to 1945, Rolls-Royce, like Britain, could justifiably feel proud of itself. A huge effort had been made in winning the war and respect was high for both the company and the country. The more astute realised that past achievements were useful but that the world moved on, and that only by producing what the new generation wanted would that respect be maintained.

Perhaps, again just as its country, Rolls-Royce was a little complacent in the 1950s, but certainly by the 1960s its two leaders, Sir Denning Pearson and Sir David Huddie, realised where the future lay, and that was throughout the world. Their attempts to establish the company there were bold but, in the short term, disastrous. Nevertheless, by the end of this book, there were distinct signs that Rolls-Royce would become a force to be reckoned with in that demanding market.

In Part Three we shall see whether the company fulfilled that ambition.

BIBLIOGRAPHY OF
BOOKS CONSULTED

Adams, Jad, *Tony Benn*, Macmillan, 1992

Banks, F. Rodwell, *I Kept no Diary*, Airlife, 1978

Barnett, Corelli, *The Audit of War*, Pan, 1996

Baxter, Alan, *Olympus – the first forty years*, Rolls-Royce Heritage Trust (Historical Series Vol. 15*), 1990

Bennett, Martin, *Rolls-Royce and Bentley, The Crewe Years*, Foulis, 1994

Bird, Anthony, and Hallows, Ian, *The Rolls-Royce Motor Car*, B.T. Batsford Ltd., 1984

Birtles, Philip, *De Havilland*, Jane's Publishing, 1984

Birtles, Philip, *Lockheed L1011 TriStar*, Ian Allan, 1989

Birtles, Philip, *Boeing 757/767/777*, Ian Allan, 1992

Blakey, George, *The Post-War History of the London Stock Market*, Mercury Business Books, 1993

Bobbitt, Malcolm, *Rolls-Royce Silver Shadow Bentley T-Series*, Veloce Publishing, 1996

Bobbitt, Malcolm, *Rolls-Royce and Bentley*, Sutton Publishing Ltd., 1997

Bowyer, Chaz, *History of the RAF*, Hamlyn, 1985

Brabazon, Lord, of Tara, *The Brabazon Story*, private publication

Braybrook, Roy, *Harrier: The Vertical Reality*, The Royal Air Force Benevolent Fund Enterprises, 1996

Brendon, Piers, *The Motoring Century: The Story of the Royal Automobile Club*, Bloomsbury, 1997

Brooks, David S., *Vikings at Waterloo*, Rolls-Royce Heritage Trust (Historical Series Vol. 22*), 1997

Bruce-Gardyne, Jock, and Lawson, Nigel, *The Power Game*, Macmillan, 1976

Brummer, Alex, and Cowe, Roger, *Hanson*, Fourth Estate, 1994

Brummer, Alex, and Cowe, Roger, *Weinstock*, HarperCollins, 1998

Campbell, John, *Margaret Thatcher, Vol. One*, Jonathan Cape, 2000

Churchill, Winston, *The Second World War, Vols. I–VI*, Cassell, 1951

Clarke, Peter, *Hope, Glory, Britain 1900–1990*, Penguin, 1996

Cook, Ray, *Armstrong Siddeley, The Parkside Story 1896–1939*, Rolls-Royce Heritage Trust (Historical Series Vol. 2*), 1988

Davies, Peter, and Thornborough, Anthony, *The Harrier Story*, Arms & Armour Press, 1996

De Havilland, Sir Geoffrey, *Sky Fever*, Hamish Hamilton, 1961

Donne, Michael, *Rolls-Royce, Leader of the Skies*, Frederick Muller, 1980

Douglas, Sholto, *Years of Command*, Collins, 1966

Eden, Sir Anthony, *Full Circle*, Cassell, 1960

Evans, Harold, *Vickers against the Odds, 1956–77*, Hodder and Stoughton, 1978

Fedden, Sir Roy, *Britain's Air Survival – an appraisement and strategy for success*, Cassell, 1957

Furse, Anthony, *Wilfrid Freeman*, Spellmount, 1999

Gardner, Charles, *British Aircraft Corporation*, B.T. Batsford, 1981

Gaskell, Keith, *British Airways*, Airlife, 2000

Gilbert, Martin, *A History of the Twentieth Century, Volume Two: 1933–1951*, HarperCollins, 1998

Gilbert, Martin, *A History of the Twentieth Century, Volume Three: 1952–1999*, HarperCollins, 1999

Gilchrist, Peter, *Boeing 747-400*, Airlife, 1998

Golley, John, *Whittle, the true story*, Airlife, 1987

Gough, Keith, *The Vital Spark – the development of aero-engine sparking plugs*, Rolls-Royce Heritage Trust (Technical Series Vol. 2*), 1991

Gray, Robert, *Rolls on the Rocks*, Panther, 1971

Green, Geoff, *Bristol Aerospace since 1910*, Geoff Green, 1985

Gunston, Bill, *By Jupiter! – the life of Sir Roy Fedden*, Royal Aeronautical Society, 1978

Gunston, Bill, *Fedden – the life of Sir Roy Fedden*, Rolls-Royce Heritage Trust (Historical Series Vol. 26*), 1998

Gunston, Bill, *Rolls-Royce, Aero Engines*, Patrick Stephens, 1989

Gunston, Bill, *The Development of Jet and Turbine Aero Engines*, Patrick Stephens, 1997

Hague, Douglas, and Wilkinson, Geoffrey, *The IRC – An Experiment in Industrial Intervention*, George Allen & Unwin, 1983

Harker, R.W., *Rolls-Royce from the Wings*, Oxford Illustrated Press, 1976

Harker, R.W., *The Engines were Rolls-Royce*, Collier Macmillan, 1980

Harvey-Bailey, Alec, and Evans, Michael, *Rolls-Royce, the pursuit of excellence*, Sir Henry Royce Memorial Foundation (Historical Series Vol 3*), 1984

Harvey-Bailey, Alec, *Rolls-Royce – The Derby Bentleys,* Sir Henry Royce Memorial Foundation (Historical Series Vol. 5*), 1985

Harvey-Bailey, Alec, *Rolls-Royce, Hives, The Quiet Tiger,* Sir Henry Royce Memorial Foundation (Historical Series Vol. 7*), 1986

Harvey-Bailey, Alec, *Rolls-Royce, Twenty to Wraith,* Sir Henry Royce Memorial Foundation (Historical Series Vol. 8*), 1986

Harvey-Bailey, Alec, *Rolls-Royce, the sons of Martha,* Sir Henry Royce Memorial Foundation (Historical Series Vol. 14*), 1989

Harvey-Bailey, Alec, *Rolls-Royce, Hives' Turbulent Barons,* Sir Henry Royce Memorial Foundation (Historical Series Vol. 20*), 1993

Hastings, Max, *The Korean War*, Pan Macmillan, 2000

Hattersley, Roy, *Fifty Years On,* Little, Brown & Company, 1997

Hayward, Keith, *Government and British Civil Aerospace,* Manchester University Press, 1983

Healey, Denis, *The Time of my Life,* Michael Joseph, 1989

Heathcote, Roy, *The Rolls-Royce Dart,* Rolls-Royce Heritage Trust (Historical Series Vol. 18*), 1992

Hennessy, Peter, *Never Again, Britain 1945–1951,* Jonathan Cape, 1992

Hennessy, Peter, *The Prime Minister, The Office and its Holders Since 1945,* Allen Lane, Penguin, 2000

Hogg, Sarah, and Hill, Jonathan, *Too Close to Call,* Little, Brown & Company, 1995

Hooker, Sir Stanley, *Not much of an Engineer,* Airlife, 1984

Howarth, T.E.B., *Prospect and Reality, Great Britain 1945–1955,* Collins, 1985

Howe, Geoffrey, *Conflict of Loyalty,* Macmillan, 1994

Huenecke, Klaus, *Jet Engines,* Airlife, 1997

Hutton, Will, *The State We're In,* Jonathan Cape, 1995

Ingells, Douglas J., *L-1011 TriStar and the Lockheed Story,* Aero Publishing Inc., 1973

Ingham, Bernard, *Kill the Messenger,* HarperCollins, 1991

Irving, Clive, *Wide-Body, The Triumph of the 747,* William Morrow & Company, 1991

Isaacs, Jeremy, and Downing, Taylor, *Cold War,* Bantam Press, 1998

Jenkins, Roy, *A Life at the Centre,* Macmillan, 1991

Jenkins, Roy, *The Chancellors,* Macmillan, 1998

Johnson, Christopher, *The Economy under Mrs Thatcher,* Penguin, 1991

Jones, Glyn, *The Jet Pioneers, the birth of jet-powered flight,* Metheun, 1989

Keith, Sir Kenneth, and Hooker, Sir Stanley, *The Achievement of Excellence,* Newcomen Society of North America, 1976

Laqueur, Walter, *Europe since Hitler,* Holt, Rinehart and Winston, 1970

Lea, Ken, *Rolls-Royce – the first cars from Crewe,* Rolls-Royce Heritage Trust, (Historical Series Vol. 23*), 1997

Lewis, Peter, *British Bomber since 1914,* Bodley Head, 1980

Lewis, Peter, *British Fighter since 1912 – Sixty years of design and development,* Putnam & Co., 1979

Lloyd, Ian, *Rolls-Royce – The Years of Endeavour,* Macmillan, 1978

Longyard, William H., *Who's Who in Aviation History*, Presidio Press, 1994

McCarthy, Roy, *Rolls-Royce Limited*, Odhams (for Rolls-Royce), 1957

MacCrindle, R.A., QC, and Godfrey, P., FCA, *Rolls-Royce Limited – Investigation under Section 165(a)(i) of the Companies Act 1948*, Her Majesty's Stationery Office, 1973

MacGregor, Ian, *The Enemies Within*, William Collins, 1986

Marshall, Sir Arthur, *The Marshall Story*, Patrick Stephens, 1994

McCullough, David, *Truman*, Simon & Schuster, 1992

Miller, Donald, and Sawers, David, *The Technical Development of Modern Aviation*, Ron Hedge & Kegan Paul, 1968

Miller, William J., *Memoirs*, private publication, 1980s

Mönnich, Horst, *The BMW Story*, Sidgwick & Jackson, 1991

Morgan, Kenneth O., *Callaghan – a Life*, Oxford University Press, 1997

Morgan, Kenneth O., *The People's Peace, British History 1945–89*, Oxford University Press, 1990

Nahum, Andrew, Foster-Pegg, R.W., and Birch, David, *The Rolls-Royce Crecy*, Rolls-Royce Heritage Trust, (Historical Series Vol. 21*),1994

Needham, Richard, *Battling for Peace*, Blackstaff Press, 1998

Neumann, Gerhard, *Herman the German*, William Morrow and Company, 1984

Newhouse, John, *The Sporty Game*, Alfred A. Knopf, 1982

Nicholson, Rupert, *Rolls-Royce Ltd., Recollections of the receiver and manager*, KPMG, 1993

Nockolds, Harold, *Lucas – The First 100 Years*, David & Charles, 1977

Nockolds, Harold, *The Magic of a Name*, G.T. Foulis, 1959

Oldfield, David, *The European Fighter Aircraft*, PhD thesis at the University of Bath, 2000

Owen, Geoffrey, *From Empire to Europe*, HarperCollins, 1999

Paxman, Jeremy, *Friends in High Places*, Michael Joseph, 1990

Pearson, Sir Denning, *The Development and Organisation of Rolls-Royce*, Rolls-Royce, 1964

Penrose, Harald, *Adventure with Fate*, Airlife, 1984

Plowden, Lord, *Report of the Committee of Inquiry into the Aircraft Industry*, Her Majesty's Stationery Office, 1965

Postann, M.M., *British War Production*, HMSO, 1952

Pratt & Whitney, *The Pratt & Whitney Aircraft Story*, Pratt & Whitney, 1950

Price, A.B., *Rolls-Royce: The Cars and their Competitors 1906–65*, B.T. Batsford, 1988

Reed, Arthur, *Britain's Aircraft Industry. What went right? What went wrong?* J.M. Dent & Sons, 1973

Rimmer, Ian, *Rolls-Royce and Bentley – Experimental Cars*, Rolls-Royce Enthusiasts Club, 1986

Robotham, W.A., *Silver Ghosts and Silver Dawn*, Constable, 1970

Rolls-Royce plc, *The Jet Engine*, Rolls-Royce, 1986

Rubbra, A.A., *Rolls-Royce Piston Aero Engines – a designer remembers*, Rolls-Royce Heritage Trust (Historical Series Vol. 16*), 1990

Sampson, Anthony, *The Changing Anatomy of Britain*, Hodder and Stoughton, 1983

Schlaiffer, Robert, and Heron, S.D., *Development of Aircraft Engines and Development of Aviation Fuels,* Harvard Business School, 1950

Scott, J.D., *Vickers,* Weidenfeld and Nicolson, 1962

Serling, Robert J., *Legend and Legacy,* St Martin's Press, 1992

Sharp, Cecil M., *DH – A History of De Havilland,* Airlife, 1982

Shaw, Robbie, *Airbus A300 & 310,* Airlife, 1991

Shaw, Robbie, *Tri-Jets,* Osprey Aerospace, 1996

Shaw, Robbie, *Boeing 737-300 to -800,* Airlife, 1999

Shepherd, Robert, *Iain Macleod,* Hutchinson, 1994

Shirer, William, *The Rise and Fall of the Third Reich,* Secker and Warburg, 1963

Sims, Charles, *Royal Air Force, The First 50 Years,* Adam & Charles Black Ltd., 1968

Sked, Alan, and Cook, Chris, *Post-War Britain,* Penguin, 1993

Smith, Sir Alex, *Lock up the swings on Sundays,* The Memoir Club, 1998

Smith, David, *From Boom to Bust,* Penguin, 1993

Sonnenburg, Paul, and Schoneberger, William A., *Allison, Power of Excellence 1915–1990,* Coastline Publishers, 1990

Stokes, Peter, *From Gipsy to Gem – with diversions 1926–86,* Rolls-Royce Heritage Trust (Historical Series Vol. 10*), 1987

Taylor, Douglas R., *Boxkite to Jet – the remarkable career of Frank B Halford,* Rolls-Royce Heritage Trust (Historical Series Vol. 28*), 1999

Tebbitt, Norman, *Upwardly Mobile,* Weidenfeld & Nicolson, 1988

Thatcher, Margaret, *The Downing Street Years,* HarperCollins, 1993

Thatcher, Margaret, *The Path to Power,* HarperCollins, 1995

Tiratsoo, Nick (Ed.), *From Blitz to Blair,* Weidenfeld & Nicolson, 1997

Tombs, Sir Francis, *A History of Rolls-Royce,* GKN Lecture, 1991

Venables, David, *Napier, the first to wear green,* G.T. Foulis/Haynes Publishing, 1998

Ward, Commander 'Sharkey', *Sea Harrier over the Falklands,* Leo Cooper, 1992

Wedgewood-Benn, Tony, *Diaries – Years of Hope,* Hutchinson, 1994

Weinberg, Gerhard L., *A World at Arms,* Cambridge University Press, 1994

Whitney, Daniel D., *Vees for Victory! The story of Allison V-1710, 1929–1948,* Schiffer Military History, 1998

Whittle, Sir Frank, *Jet, the story of a pioneer,* Frederick Muller, 1953

Wilde, Geoffrey, *Flow matching of the stages of axial compressors,* Rolls-Royce Heritage Trust (Technical Series Vol. 4*), 1999

Williams, David E., *A View of Ansty, 1935–1982,* Rolls-Royce Heritage Trust, 1998

Wilson, Stewart, *Viscount, Comet and Concorde,* Aerospace Publications, 1996

Worswick and Ady, *The British Economy in the 1950s,* Oxford University Press, 1952

Young, Hugo, *One of Us,* Macmillan, 1990

Ziegler, Henri, *La Grande Aventure de Concorde,* Bernard Grasset, 1976

Zigmunt, Joan, *Allison – The People and the Power,* Turner Publishing Company, 1997

* The Historical Series is a joint venture between the Rolls-Royce Heritage Trust and initially, the Sir Henry Royce Memorial Foundation but latterly, The Rolls-Royce Enthusiasts Club.

INDEX

INDEX

Head, Dennis 177, 255, 261–2, 266
Healey, Denis 252, 289
Hearle, Frank 47
Heath, Edward 144, 146, 181, 238,
    246–7, 251, 252, 301
Heathcote, Roy 36–7, 40
helicopters 68, 244, 267, 292–3, 295
Hennessey, Peter 214
HMS *Hermes* 282
Heron, Jock 291
Heseltine, Michael 293
Hibbard, Hall 107
Hickman, Bob 200
Higginbottom, Sam 310, 315, 319
high bypass ratio 105, 316–17
Hill, Sir Robert 214
Thomas Hill (Rotherham) Ltd. 200–1
Hinckley, Fred 101, 104
Hispano-Suiza 90
Hitler, Adolf 16
Hives, Ernest ii, 34–5; Aero/Chassis
    divisions 157; Aircraft Production
    Ministry 7; Atomic Energy Authority
    211–12; Boeing 46; Bureau of
    Aeronautics 33; diesel engines 195–6;
    Edwards 139; gas turbines viii, 3,
    9–10, 12–13, 14–16; Gwinn 26–7, 28;
    Hooker 82–3, 84; peerage 35; Pratt &
    Whitney 23–4, 27–8; pump gland
    failures 16–17; Rickover 213–14; Silver
    Dawn 162; Taylor 25–6; Tinling 9;
    Trans-Canada Airlines 15; Vickers
    39–40; Westinghouse 31–3; Whittle
    6–7, 8; Wilks, Spencer 10
Hobbs, L.S. 24, 26, 27
Holder, Douglas 234–5
Hollings, John 186, 212
Hood, Ed 319
Hooker, Stanley 81–5, 101; Avon 44–5;
    Barnoldswick 11; Bristol Aeroplane
    Company 81–3; China 266; Clyde 36;
    Concorde 91–2; Harker on 13; Hives
    82–3, 84; jets 10–11; Olympus 85–7;
    RB 211 230, 236; Rolls-Royce board
    235; Spey 61–2; TSR2 88; V/STOL
    283–4; VTOL 274, 279–80; Whittle 6,
    10; Wibault 278
Hooper, Ralph 279–80
Hooper & Co. 165, 168
Horlick, Admiral 223–4
Horner, H. Mansfield 23, 27–8, 29, 31
Hoskyns, Sir John 303
HMS *Hotham* 221
Howe, Sir Geoffrey 303
Hucknall 9, 37, 136
Huddie, David ii, 327; collapse 136;
    Douglas 123; Gilbert 24, 33, 69;
    Lockheed 115, 151–2, 154; McDonnell

Douglas 232–3; Metcalfe 129, 134–5;
    Nicholson 101–2; Perkins 124–5; price
    cuts 117; RB 178 109; RB 211 122,
    129–30, 133, 134; responsibility 155–6;
    Robins 121; Ryan X-13 274; Trans-
    Canada Airlines 15; tri-jet 119;
    Westinghouse 33, 34, 69
Hyfil fan blades 115, 128, 131, 134, 135,
    136, 239

industrial unrest 248–50, 255, 256–7;
    dockers' strikes 248–9; miners' strikes
    246, 249–50, 251–2, 304
inflation vi, 187, 251–2, 268, 300–1, 302
intercooler pump gland failures 16–17
International Harvester 198, 202
HMS *Invincible* 223, 282
Iran, Shah of 184, 206, 208, 247, 300, 302
Irving, Clive 239, 240

Jackson, Denis 61, 63
Jaguar 95, 285–7, 296
Jamison, R.R. 85
Japan 224, 287, 297, 308
Jay, Douglas 119
Jenkins, Roy 93
jet engines: Allis-Chalmers 20–1;
    components 186; Griffith 3, 6, 275–6;
    Hives viii, 14–16; Hooker 10–11; surge
    problems 8; USA 19–20; Whittle 3–6,
    7; *see also* gas turbine engines
John, Stuart 319–20
Johnson, Kelly 85
Johnson, Lyndon 111, 112, 119–20
Johnson, W.E.P. 9
Jordan, King of 208
Joseph, Sir Keith 269–70
JT-8D 305
JT-10D 304–6

Keir, Jim 155
Keith, Sir Kenneth: Benn 252; big engine
    market 240–1; British Airways 261–2;
    Callaghan 263; Carrington 299; China
    61–3; Conservatives 61, 237–8; contacts
    245–9; Eltis 265; Joseph 269–70;
    Labour 253–4, 255–6, 268; MRCA
    290–1; National Enterprise Board 257,
    258; Pratt & Whitney 248; Richardson
    254–5; Varley 259, 262; Wilson 253–4,
    268
Kennedy, John 288–9
Keynes, J.M. 164
Kilcarr, J. Kenneth 263
Kilner, Hew 39
Kindersley, Lord 184
KLM 260–1
Kolk, Frank 108, 117–18

337